*Flora M. Li, Arokia Nathan, Yiliang Wu,
and Beng S. Ong*

**Organic Thin Film Transistor
Integration**

Flora M. Li is currently a Senior Scientist at Polymer Vision, Eindhoven, The Netherlands. Prior to this, she was a Research Associate/NSERC Postdoctoral Fellow at the Centre of Advanced Photonics and Electronics (CAPE) in the Electrical Engineering Division of the University of Cambridge, UK. She received her Ph.D. degree in Electrical and Computer Engineering from the University of Waterloo, Canada, in 2008. She was a Visiting Scientist at Xerox Research Centre of Canada (XRCC) from 2005–2008. Her research interests are in the field of nano- and thin-film technology for applications in large area and flexible electronics, including displays, sensors, photovoltaics, circuits and systems. Dr. Li has co-authored a book entitled CCD Image Sensors in Deep-Ultraviolet (2005), and has published articles in various scientific journals.

Arokia Nathan holds the Sumitomo/STS Chair of Nanotechnology at the London Centre for Nanotechnology, University College, London, and is a recipient of the Royal Society Wolfson Research Merit Award. He is also the CTO of Ignis Innovation Inc., Waterloo, Canada, a company he founded to commercialize technology on thin film silicon backplanes on rigid and flexible substrates for large area electronics. He has held Visiting Professor appointments at the Physical Electronics Laboratory, ETH Zürich and at the Engineering Department, University of Cambridge, UK. He received his Ph.D. in Electrical Engineering from the University of Alberta. He held the DALSA/NSERC Industrial Research Chair in sensor technology, was a recipient of the 2001 Natural Sciences and Engineering Research Council E.W.R. Steacie Fellowship, and was awarded the Canada Research Chair in nano-scale flexible circuits. Professor Nathan has published extensively in the field of sensor technology and CAD, and thin film transistor electronics, and has over 40 patents filed/awarded. He is the co-author of two books, Microtransducer CAD and CCD Image Sensors in Deep-Ultraviolet, and serves on technical committees and editorial boards in various capacities.

Yiliang Wu received his Ph.D. degree in polymer science from the Tokyo Institute of Technology, Japan, in 1999. After two years of postdoctoral studies at Queen's University, Kingston, Canada, he joined the Xerox Research Centre of Canada in 2001 as a research scientist working on organic transistor materials design and process. Currently, Dr. Wu is a Principal Scientist at the centre, leading Xerox's Printable Electronic Materials project. He is the holder of over 85 US patents and has authored/co-authored 78 peer-reviewed papers and two book chapters.

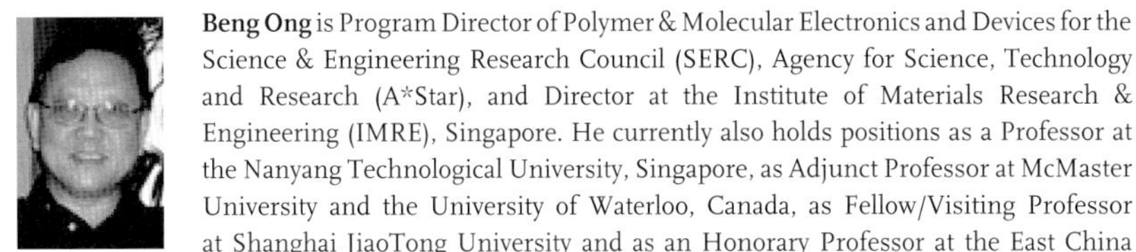

Beng Ong is Program Director of Polymer & Molecular Electronics and Devices for the Science & Engineering Research Council (SERC), Agency for Science, Technology and Research (A*Star), and Director at the Institute of Materials Research & Engineering (IMRE), Singapore. He currently also holds positions as a Professor at the Nanyang Technological University, Singapore, as Adjunct Professor at McMaster University and the University of Waterloo, Canada, as Fellow/Visiting Professor at Shanghai JiaoTong University and as an Honorary Professor at the East China University of Science and Technology in Shanghai. Prior to his relocation to Singapore in 2007, Professor Ong was a Senior Fellow of the Xerox Corporation, and a manager of Advanced Materials and Printed Electronics at the Xerox Research Centre of Canada. Professor Ong currently has a patent portfolio of 187 US patents and about 100 refereed papers on enabling materials, processes, and integration technologies. Many of the technology programs he managed at Xerox have won awards including the Nano-50 Award for Materials Innovation (2007) and for Nanotechnology Commercialization (2007), the ACS Innovation Award (2006), the Connecticut Quality Improvement Gold Award (2006), and the Wall Street Journal Technology Innovation Runner-up Award (2005).

Flora M. Li, Arokia Nathan, Yiliang Wu,
and Beng S. Ong

Organic Thin Film Transistor Integration

A Hybrid Approach

WILEY-VCH Verlag GmbH & Co. KGaA

The Authors

Dr. Flora M. Li
University of Cambridge
Electrical Engineering
9 JJ Thomson Avenue
Cambridge CB3 0FA
United Kingdom

Currently at
Polymer Vision –Wistron
Kastanjelaan 1000, building SFH
5616 LZ, Eindhoven
The Netherlands

Prof. Arokia Nathan
University College London
London Centre of Nanotechnology
17-19 Gordon Street
London WC1H 0AH
United Kingdom

Dr. Yiliang Wu
Xerox Research Centre Canada
2660 Speakman Drive
Mississauga, Ontario L5K 2L1
Canada

Prof. Beng S. Ong
Institute of Materials Research and
Engineering (IMRE)
3 Research Link
Singapore 117602
Singapore

Library of Congress Card No.: applied for

British Library Cataloguing-in-Publication Data
A catalogue record for this book is available from the British Library.

Bibliographic information published by the Deutsche Nationalbibliothek
The Deutsche Nationalbibliothek lists this publication in the Deutsche Nationalbibliografie; detailed bibliographic data are available on the Internet at <http://dnb.d-nb.de>.

Cover Design Adam Design, Weinheim
Typesetting Laserwords Private Limited, Chennai, India
Printing and Binding Strauss GmbH, Mörlenbach

Printed in the Federal Republic of Germany
Printed on acid-free paper

ISBN: 978-3-527-40959-4

Flora Li dedicates this book to her extraordinarily amazing family, for their unconditional love and unwavering support: David, Adda, Christina, Ben, and 婆婆

Contents

Preface

Organic semiconductors offer great promise for large area, low-end, lightweight, and flexible electronics applications. Their technological edge lies not only in their ease of processability but in their ability to flex mechanically. This makes them highly favorable for implementation on robust substrates with non-conventional form factor. Since its proof of concept in the early 1980s, progress in organic electronics has been impressive with performance attributes that are competitive with the inorganic counterparts. In particular, organic electronics is attractive from the standpoint of complementing conventional silicon technology, thriving in a different market domain that targets lower resolution, cost-effective mass production items such as identification tags, smart cards, smart labels, and pixel drivers for display and sensor technology.

While the material properties and processing technology for organic semiconductors continue to advance and mature, progress in organic thin film transistor (OTFT) integration and its scalability to large areas has not enjoyed the same pace. A major driving force behind this technology lies in the ability to manufacture low-end, and disposable electronic devices. This in turn demands a fabrication process that allows high volume production at low cost. The process should be able to produce stand-alone devices, device arrays, and integrated circuits of acceptable operating speed, functionality, reliability, and lifetime. However, this comes with its fair share of challenges, which we have attempted to address in this book. It is intended as a text and/or reference for graduate students in Electrical Engineering, Materials Science, Chemistry, and Physics, and engineers in the electronics industry.

Most of the results presented here stem from research conducted at the Giga-to-Nano Labs, University of Waterloo, and the Xerox Research Centre of Canada (XRCC), which granted access to its high quality, high performance, stable organic semiconductor materials. We acknowledge the contributions of several colleagues in these laboratories whose expertise ranged from materials processing and TFT integration to circuit and system design. We especially thank Prof. A. Sazonov (University of Waterloo), Dr Yuri Vygranenko (Instituto Superior de Engenharia de Lisboa), Dr D. Striakhilev (Ignis Innovation Inc.), Prof. P. Servati (University of British Columbia), Dr S. Koul (General Electric), Dr M.R.E. Rad (T-Ray Science), Dr C.-H. Lee (Samsung Electronics), Dr G. Chaji (Ignis Innovation Inc.),

Dr K. Sakariya (Apple Computers), Dr S. Sambandan (PARC), Dr H.-J. Lee (DALSA Inc.), Dr K. Wong (University of Waterloo), R. Barber (University of Waterloo), Dr G.-Y. Moon (LG Chemicals), Dr I.W. Chan (ETRI).

We would also like to acknowledge the support of other colleagues: Prof. W.I. Milne, Dr. P. Beecher, and Dr C.-W. Hsieh of University of Cambridge, A. Ahnood and J. Stott of University College London, and Prof. G. Jabbour and Dr H. Haverinen of Arizona State University and Oulu University.

The text has evolved from a series of courses offered to graduate students in Electrical Engineering as well as doctoral dissertations covering different aspects of large area electronics. The scope of this book is to advance OTFT integration from an engineering perspective, and not material development, which is the strength of chemical physicists. By assimilating existing materials, techniques and resources, the book explores a number of approaches to deliver higher performance devices and demonstrate the feasibility of organic circuits for practical applications. Much of the material in the book can be presented in about 30 hours of lecture time. The text begins with an assessment of organic electronics and market opportunities for OTFT technology. The latter is further described in Chapter 2, examining device architectures and material selection. Strategies to enable circuit integration are presented in Chapter 3, while Chapter 4 explores optimization of gate dielectric composition and structure. Interface engineering methodologies for OTFTs to enhance the dielectric/semiconductor and contact/semiconductor interfaces are described in Chapters 5 and 6. Chapter 7 presents examples of functional circuits for active-matrix display and other applications. Chapter 8 concludes with a glimpse of future challenges related to OTFT integration.

This book would not have been possible without the support of various institutions and funding agencies: University of Waterloo, Xerox Research Centre of Canada, University College London, University of Cambridge, Nanyang Technological University, Natural Sciences and Engineering Research Council of Canada, Ontario Centres of Excellence, and The Royal Society.

Cambridge, London, Toronto,
Singapore 2010

Flora M. Li, Arokia Nathan,
Yiliang Wu, and Beng S. Ong

Glossary

Abbreviations

AC	alternating current
AFM	atomic force microscopy
Ag	silver
Al	aluminum
Al_2O_3 or AlO_x	aluminum oxide
ALD	atomic layer deposition
AMLCD	active-matrix liquid crystal display
AMOLED	active-matrix organic light emitting diode
a-Si:H or a-Si	amorphous silicon
Au	gold
BCB	benzocyclobutene
C60	fullerene
CMOS	complementary metal oxide semiconductor
CNT	carbon nanotube
CT	charge transfer
CTC	charge transfer complex
Cu	copper
$C-V$	capacitance–voltage characteristics
CVD	chemical vapor deposition
D6HT	dihexyl-sexithiophene
DC	direct current
DFH-4T	diperflurorohexylquarter-thiophene
DIP	dual in-line package
DOS	density of states
Dpi	dots per inch
EDM	electro-discharge machining
E-Paper	electronic paper
ERDA	elastic recoil detection analyses
$F_{16}CuPc$	hexadecafluoro-phthalocyanine
F8T2	poly(9,9′-dioctyl-fluorene-co-bithiophene)
FTIR	fourier transform infrared spectroscopy
GIXRD	grazing-incidence X-ray diffraction

HF	hydrofluoric acid
HMDS	hexamethyldisilazane
HOMO	highest occupied molecular orbital
IC	integrated circuit
ICP	inductively coupled plasma
IEEE	Institute of Electrical and Electronics Engineers
IJP	inkjet printing
IP	ionization potential
$I-V$	current–voltage characteristics
LCD	liquid crystal display
LUMO	lowest unoccupied molecular orbital
MIS	metal-insulator-semiconductor
MOS	metal-oxide-semiconductor
MNB	2-mercapto-5-nitro-benzimidazole
Mo	molybdenum
MOSFET	metal oxide semiconductor field effect transistor
MTR	multiple trapping and release model
N_2	nitrogen
NH_3	ammonia
NMOS	n-channel or n-type metal oxide semiconductor
NW	nanowire
O_2 plasma	oxygen plasma
ODTS	octadecyltrichlorosilane
OFET	organic field effect transistor
OLED	organic light emitting diode
OTFT	organic thin film transistor
OTS or OTS-8	octyltrichlorosilane
P3HT	poly(3-hexylthiophene)
PA	polyacetylene
PANI	polyaniline
PBTTT	poly(2,5-bis(3-alkylthiophen-2-yl)thieno[3,2-*b*]thiophene)
PCBM	phenyl-C61-butyric acid methyl ester
PECVD	plasma enhanced chemical vapor deposition
PEDOT:PSS	poly(3,4-ethylene dioxythiophene) doped with polystyrene sulfonic acid
PEN	poly(ethylene naphthalate)
PET	poly(ethylene terephthalate)
Ph.D.	doctor of philosophy
PI	polyimide
PMMA	poly(methyl methacrylate)
PPV	poly(*p*-phenylene vinylene) or polyphenylene vinylene
PQT	poly(3,3‴-dialkylquaterthiophene)
Pt	platinum
PT	polythiophene
PTV	poly(thienylene vinylene)

PVA	polyvinyl acetate or polyvinyl alcohol
R&D	research and development
RCA clean	a standard set of wafer cleaning steps; RCA = Radio Corporation of America
RF	radio frequency
RFID	radio frequency identification
RIE	reactive ion etching
SAM	self-assembled monolayer
SiH_4	silane
SiN_x	silicon nitride
SiO_2	silicon dioxide
SiO_x	silicon oxide
SnO_2	tin oxide
TFT	thin film transistor
TiO_2	titanium oxide
UV	ultraviolet
UW	University of Waterloo
XPS	X-ray photoelectron spectroscopy
XRCC	Xerox Research Centre of Canada
ZnO	zinc oxide

Mathematic Symbols

φ_B	injection barrier
Φ_M	work function of the electrode (metal)
[N]/[Si]	nitrogen to silicon ratio, to describe stoichiometry or composition of SiN_x
μ_{FET}	field effect mobility
C_i	gate capacitance per unit area
C_S	storage capacitor
E_G	band-gap energy
f_{max}	maximum switching frequency
g_m	transconductance
I_D	drain current
I_G	gate current
I_{leak}	leakage current
I_{OFF}	off current
I_{ON}	on current
I_{ON}/I_{OFF}	on/off current ratio
I_S	source current
IP_S	ionization potential of the semiconductor
L	channel length
$R_{CONTACT}$	contact resistance
S	inverse subthreshold slope ($V\,dec^{-1}$)

τ	transit time
V_{BG}	bottom-gate voltage
V_{DD}	positive supply voltage
V_{DS}	drain-source voltage
V_{GS}	gate-source voltage
V_{ON}, V_{SO}	onset voltage or switch-on voltage
V_{SS}	negative supply voltage
V_{T}	threshold voltage
V_{TG}	top-gate voltage
W	channel width

Definitions

Definitions of selected terms cited from Wikipedia webpage. *http://en.wikipedia.org/wiki/Main_Page.*

Alkanes (also *Alkyl*)	Chemical compounds that consist only of the elements carbon (C) and hydrogen (H) (i.e., hydrocarbons), wherein these atoms are linked together exclusively by single bonds (i.e., they are saturated compounds) without any cyclic structure (i.e., loops). An alkyl group is a functional group or side-chain that, like an alkane, consists solely of singly-bonded carbon and hydrogen atoms.
Charge transfer complex (CT complex)	An electron donor–electron acceptor complex, characterized by electronic transition(s) to an excited state. In this excited state, there is a partial transfer of elementary charge from the donor to the acceptor. A CT complex composed of the tetrathiafulvalene (TTF, a donor) and tetracyanoquinodimethane (TCNQ, an acceptor) was discovered in 1973. This was the first organic conductor to show almost metallic conductance.
Conductive polymer (also *conducting polymer*)	Polymer that is made conducting, or "doped," by reacting the conjugated semiconducting polymer with an oxidizing agent, a reducing agent, or a protonic acid, resulting in highly delocalized polycations or polyanions. The conductivity of these materials can be tuned by chemical manipulation of the polymer backbone, by the nature of the dopant, by the degree of doping, and by blending with other polymers. Conductive polymer is an organic polymer semiconductor, or an organic semiconductor.

Conjugated polymer A system of atoms covalently bonded with alternating single and double carbon–carbon (sometimes carbon–nitrogen) bonds in a molecule of an organic compound. This system results in a general delocalization of the electrons across all of the adjacent parallel aligned p-orbitals of the atoms, which increases stability and thereby lowers the overall energy of the molecule.

Dielectric (also *insulator*) A non-conducting substance, that is, an insulator. Although "dielectric" and "insulator" are generally considered synonymous, the term "dielectric" is more often used when considering the effect of alternating electric fields on the substance while "insulator" is more often used when the material is being used to withstand a high electric field. Dielectric encompasses the broad expanse of nonmetals (including gases, liquids, and solids) considered from the standpoint of their interaction with electric, magnetic, of electromagnetic fields. In this book, the terms "dielectric" and "insulator" are used interchangeably.

Electrode (also *contact*) An electrical conductor (e.g., metallization) used to make contact with a nonmetallic part of a circuit (e.g., a semiconductor). The gate/source/drain metal *layer* of the TFT is referred to as an electrode. The *connection* between the source/drain metal layer and the semiconductor layer (i.e., when we speak of the interface) is referred to as the "contact." In this book, the terms "electrode" and "contact" are used almost interchangeably.

Insulator (also *dielectric*) A material that resists the flow of electric current. It is an object intended to support or separate electrical conductors without passing current through itself. An insulation material has atoms with tightly bonded valence electrons. The term electrical insulation often has the same meaning as the term dielectric.

Mobility (also *carrier mobility*, *field-effect mobility*, *effective mobility*) The state of being in motion. *Carrier mobility* is a quantity relating the drift velocity of electrons or holes to the applied electric field across a material; this is a material property. *Field-effect mobility* or *effective mobility* describes the mobility of carriers under the influence of the device structure in field-effect transistors. Field-effect mobility is device-specific, not material-specific, and includes effects such as contact resistances, surface effects, and so on.

Organic compounds	Chemical compounds containing carbon-hydrogen (C–H) bonds of covalent character.
Organic electronics (also *plastic electronics*)	A branch of electronics that deals with conductive polymers, plastics, or small molecules. It is called "organic" electronics because the polymers and small molecules are carbon-based, like the molecules of living things. This is as opposed to traditional electronics which relies on inorganic conductors such as copper or silicon.
Organic semiconductor (also *polymer semiconductor*)	Any organic material that has semiconductor properties. Both short chain (oligomers) and long chain (polymers) organic semiconductors are known. There are two major classes of organic semiconductors, which overlap significantly: organic charge-transfer complexes, and various "linear backbone" polymers derived from polyacetylene. This book focuses on the investigation of polymer organic semiconductors; thus, in most cases, the term "organic semiconductor" and "polymer semiconductor" are used interchangeably.
OTFT (also *OFET*)	An organic thin film transistor (OTFT) or organic field effect transistor (OFET) is a field effect transistor using an organic semiconductor in its channel.
Plastic	A general term for a wide range of synthetic or semi-synthetic polymerization products. Plastics are polymers, that is, long chains of atoms bonded to one another.
Polymer	A substance composed of molecules with large molecular mass composed of repeating structural units, or monomers, connected by covalent chemical bonds.

1
Introduction

Organic semiconductor technology has attracted considerable research interest in view of its great promise for large area, low-end, lightweight, and flexible electronics applications [1]. Owing to their processability advantages and unique physical (i.e., electrical, optical, thermal, and magnetic) properties, organic semiconductors can bring exciting new opportunities for broad-impact applications requiring large-area coverage, mechanical flexibility, low-temperature processing, and low cost. Thus, organic semiconductors have appeal for a broad range of devices including transistors, diodes, sensors, solar cells, and light-emitting devices. Figure 1.1 depicts a number of application domains that can benefit from the versatility of organic electronics technology [2]. Since their proof of concept in the 1980s, the impressive development in organic semiconductor materials has led to performance properties that are competitive with amorphous silicon (a-Si), increasing their suitability for commercial applications [3].

The transistor is a fundamental building block for all modern electronics; transistors based on organic semiconductors as the active layer are referred to as organic thin film transistors (OTFTs). A number of commercial opportunities have been identified for OTFTs, including flat panel active-matrix liquid crystal displays (LCDs) or active matrix organic light-emitting diode displays (AMOLEDs), electronic paper (e-paper), low-end data storage such as smart cards, radio-frequency identification (RFID) and tracking devices, low-cost disposable electronic products, and sensor arrays; more applications continue to evolve as the technology matures [4]. Figure 1.2 illustrates a few commercial opportunities envisioned for OTFTs.

The unique features which give organic electronics a technological edge are simpler fabrication methods and the ability to mechanically flex. Fabrication of organic electronics can be done using relatively simple processes such as evaporation, spin-coating, and printing, which do not require high-end clean room laboratories. For example, solution-processable organic thin films can be deposited by spin coating, enabling fast and inexpensive coverage over large areas. Inkjet printing techniques can be used to deposit soluble organic inks. In addition, low-temperature processing and the mechanical flexibility of organic materials make them highly favorable for implementation on robust substrates

Organic Thin Film Transistor Integration: A Hybrid Approach, First Edition. Flora M. Li, Arokia Nathan, Yiliang Wu, and Beng S. Ong.

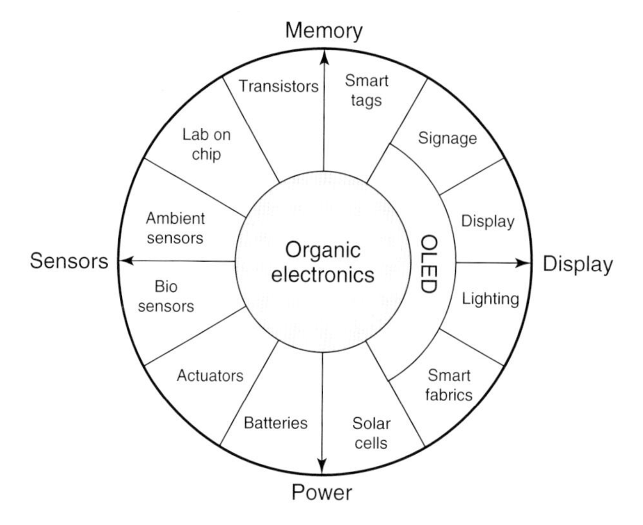

Figure 1.1 A broad range of products and technologies inspired by organic electronics [2].

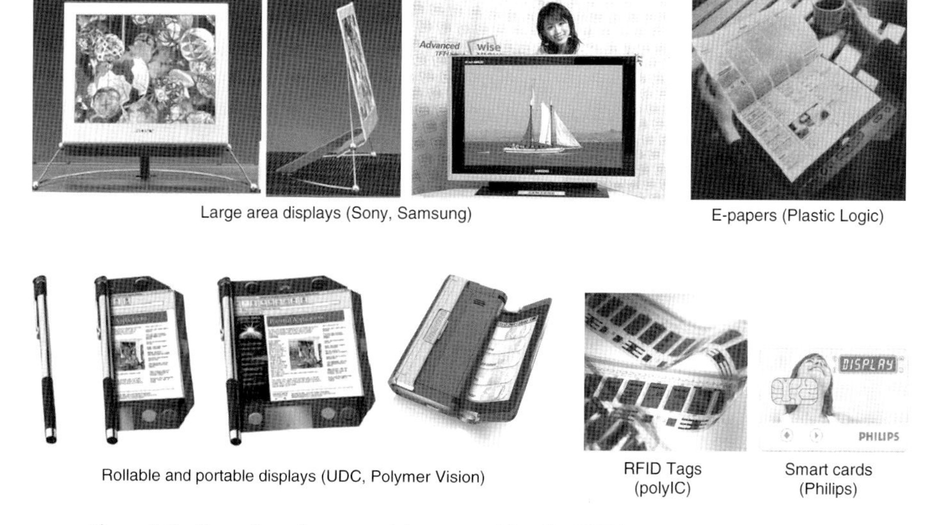

Figure 1.2 Examples of commercial opportunities for OTFTs.

with non-conventional form factors. In general, organic electronic devices are not expected to compete with silicon devices in high-end products, because of their lower speed as compared to silicon. Thus organic electronics is intended to complement conventional silicon technology. It is expected to thrive in a different market domain targeting lower resolution, cost-effective mass production items such as identification tags, smart cards, and pixel drivers for display and sensor technology.

1.1 Organic Electronics: History and Market Opportunities

Historically, organic materials (or plastics) were viewed as insulators, with applications commonly seen in inactive packaging, coating, containers, moldings, and so on. Research on the electrical behavior of organic materials commenced in the 1960s [5]. Photoconductive organic materials were discovered in the 1970s and were used in xerographic sensors. The announcement of conductive polymers in the late 1970s [6], and of conjugated semiconductors and photoemission polymers in the 1980s [7], gave new impulse to the activity in the field of organic electronics. Polyacetylene was one of the first polymers reported to be capable of conducting electricity [8], and it was discovered that oxidative doping with iodine causes the conductivity to increase by 12 orders of magnitude [9]. This discovery and the development of highly-conductive organic polymers was credited to Alan J. Heeger, Alan G. MacDiarmid, and Hideki Shirakawa, who were jointly awarded the Nobel Prize in Chemistry in 2000 for their 1977 discovery and development of oxidized, iodine-doped polyacetylene.

The continued evolution of organic semiconductor materials from the standpoint of electrical stability, processability, functionality, and performance is enabling realization of high-performance devices in laboratory environments [10–14]. The advancement in organic semiconductor materials is starting to prompt the transition of the technology from an academic research environment to industrial research and development (R&D). The shift toward industrial R&D is aided by the establishment of several government-sponsored research initiatives [15, 16], the founding of various organic electronics driven associations and companies [17–19], and the development of IEEE standards for the testing of organic electronics devices [20]. The increased cooperative efforts between academia, industry, and government are vital to the development of a strong materials and manufacturing infrastructure [21–26].

The outlook for low-cost production of organic electronics is a key driver for market opportunities in this area. To achieve these cost targets, low-cost materials, cost-effective processes, and high-volume manufacturing infrastructure are required. The development of high-volume roll-to-roll manufacturing platforms for fabrication of organic circuits on continuous, flexible, low-cost substrates, has been reported. These platforms are based on the integration of lithography, vacuum deposition, and printing technologies. It has been forecast that an organic semiconductor fabrication facility can be built for far less than the cost of a silicon semiconductor fabrication facility [3]. The high cost of silicon-based foundries can be attributed to the sophisticated wafer processing and handling equipment, high-resolution lithography tools, wafer testing equipment, clean-room environment, and costly chemical distribution and disposal facilities. In contrast, the cost reduction forecast for an organic electronic manufacturing facility is expected to be derived from lower materials cost, less sophisticated equipment, simpler manufacturing technologies, less stringent demands on clean-room settings, and reduced

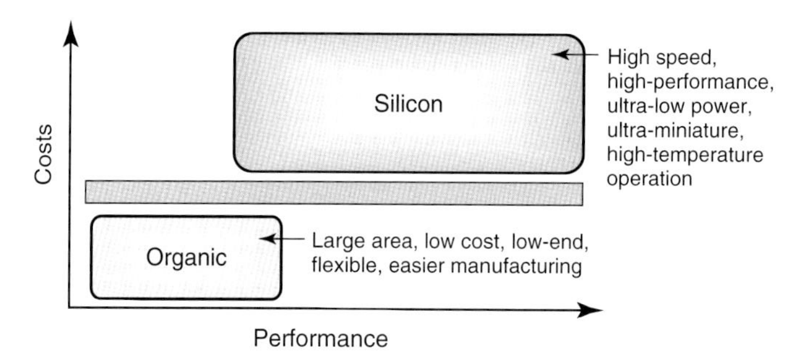

Figure 1.3 Illustration of cost versus performance comparison of silicon technology and organic semiconductor technology.

waste output. However, the potential savings in the manufacturing cost of organic electronics come with the trade-off of lower performance.

Figure 1.3 provides a conceptual view of the cost-and-performance sectors served by silicon technology and organic semiconductor technology. It must be noted that organic semiconductor devices do not offer the same electrical performance as silicon devices. While silicon technology is aimed for high-end, high performance, and high processing power electronic products, organic semiconductor technology appeals for lower-end, cost-effective disposable electronics products.

One of the most frequently discussed opportunities for organic electronics is their integration as the driver backplane of flexible displays. Specifically, printed organic semiconductor materials are strong candidates for novel electrically active display media. The same applies to radio frequency interrogation devices. An overview of these OTFT-based applications and their current market status is presented next. Note that, at present, a-Si thin film transistors (TFTs) and polycrystalline silicon (poly-Si) TFTs are the key backplane technologies used in flat panel display products. Therefore, OTFTs are not intended to displace a-Si TFTs in large-area high-resolution flat panel displays. Instead, they will have a bigger impact on lower-cost flexible displays and e-paper applications. The key features of OTFT and a-Si TFT are compared in Table 1.1.

1.1.1 Large-Area Displays

The application of OTFTs for large area displays has been demonstrated by a number of companies and research institutions. For example, Plastic Logic Ltd. demonstrated the integration of an OTFT-driven backplane to a Gyricon display in 2003 [27]. The active-matrix display backplane was inkjet printed and drove a 3000-pixel display that was fabricated on glass. In early 2007, the world's first factory was built to produce plastic electronic devices [18].

A number of corporations have also invested in R&D for OTFT-driven large area displays. Examples include Sony, Samsung, Kodak, LG Philips, Motorola, 3M, and

Table 1.1 Comparison of OTFTs and amorphous silicon (a-Si) TFTs.

	OTFT	a-Si TFT
Material	Organic semiconductor as active layer; p-type, n-type, ambipolar	a-Si as active layer n-type
Processing	Spin-coat, print, evaporation. Low temperature (e.g., room temperature)	Plasma enhanced chemical vapor deposition (PECVD) $T < 350\,^{\circ}\mathrm{C}$
Mobility	Can be comparable to a-Si	$\sim 1\,\mathrm{cm}^2\,\mathrm{V}^{-1}\,\mathrm{s}^{-1}$
Substrate and form factor	Variety of substrates and flexible form factor Mechanically flexible	Glass (most common), Plastic (in development)
Mechanical flexibility	Bendable	Fragile and brittle
Electrical stability	Rapid degradation, but degradation stabilizes (may be favorable for devices that turn on for a longer time)	Slower bias-induced degradation, but degradation does not stabilize
Pros	Potentially no clean-room, lower cost	More mature and stable
Cons	Process challenge; device performance, stability, and lifetime	Mechanical flexibility (stress) Higher processing temperature
Key applications	Numerous: displays, RFID tags, sensors, disposable electronics	Circuits for large-area displays and sensors array backplane
Outlook	New opportunities: smaller/flexible displays, disposable electronics, smart textiles	Continue to excel in AMLCD, AMOLED, active-matrix sensor technologies

Hewlett-Packard. In 2007, Sony demonstrated a 2.5 in. AMOLED display driven by OTFTs [28]; LG Philips' LCD Division presented a high resolution active-matrix liquid crystal display (AMLCD) with an OTFT-driven backplane fabricated using solution processing [29]; and Samsung Electronics reported an active-matrix display using printed OTFTs [30].

1.1.2 Rollable Displays

The mechanical flexibility of organic materials makes them particularly attractive for rollable or flexible displays. Polymer Vision, a spin off from Royal Philips Electronics, was a pioneer in demonstrating the capability of rollable displays, which were produced by combining ultrathin flexible OTFT-driven active-matrix backplane technology and flexible electronic ink (E-Ink) display technology. In January 2008, Polymer Vision introduced their first rollable display product, called *Readius®*, a pocket-sized device, combining a 5″ rollable display with high speed connectivity. The Readius® demonstrated a merger of the reading-friendly strengths of electronic-readers with the high mobility features of mobile phones, along with instant access to personalized news and information [31]. Demand for larger mobile displays is accelerating as telecom players push mobile content and mobile advertisements. The solution is to unroll the display when needed and simply store it away when not in use. Therefore, rollable display enabled devices are expected to be an emerging commodity for new generations of portable communication devices, thus presenting exciting commercial opportunities for OTFT-driven display backplane technology.

1.1.3 Radio Frequency Identification (RFID) Tag

One of the frequently promoted applications for organic electronics is the RFID tag. The RFID tag is a wireless form of automated identification technology that allows non-contact reading of data, making it effective for manufacturing, inventory, and transport environments where bar code labels are inadequate. Advantages of organic-based RFID tags over silicon-based tags include mechanical flexibility (e.g., bendable) and direct fabrication onto large area substrates using simple printable methods. The attractiveness of printed organic semiconductor materials and manufacturing platforms has drawn the involvement of several companies (e.g., 3M, Siemens), start-ups (e.g., OrganicID, ORFID Corp.), and research institutions to develop technology for organic-based RFID tags [25, 26]. For example, a 64-bit inductively-coupled passive RFID tag on a plastic substrate was demonstrated, operating at 13.56 MHz and with a read distance of over 10 cm. These specifications are approaching item-level tagging requirements, paving the way for low-cost high-volume production of RFID tags, with the potential to replace barcodes [19, 32, 33].

1.1.4 Technological Challenges

Organic electronics have reached early stages of commercial viability. Personal electronic devices incorporating small displays based on organic light-emitting diodes (OLEDs) are now available. However, many challenges still remain that are currently hindering the wide adoption of OTFTs in electronic devices. The shortcomings of OTFTs include limited charge carrier mobilities, high contact resistance, relatively higher operating voltages, device reliability issues (e.g., stability, shelf-life under operation), and limited availability of robust/mature patterning techniques and fabrication processes that are compatible with organic thin films. These technical challenges can be grouped into two categories: device performance and device manufacture.

1.1.4.1 Device Performance

One of the limitations of organic semiconductor materials, compared to silicon technology, is their intrinsically lower mobility. Most organic semiconductor thin films are composed of a mixture of polycrystalline and amorphous phases. The hopping process between molecules in the disordered regions often limits charge-carrier mobilities in organic semiconductor films [34]. To improve charge transport and device mobility, the disordered phase must be suppressed. The two most common approaches considered are:

- Tuning the molecular structure of the organic semiconductor during material preparation. Examples include modifying molecular parameters (e.g., regioregularity, molecular weight, side-chain length, doping level) during material synthesis, and altering processing conditions during material deposition (e.g., thermal annealing, solvent selection, film thickness, deposition methods, and parameters) [10–12].
- Exploiting interfacial phenomena to improve molecular ordering of the semiconductor layer during device processing. Examples of interfacial phenomena include semiconductor alignment using self-assembled monolayers (SAMs), surface-mediated molecular ordering, surface dipoles, physical alignment, and photoalignment [35]. Surface control using SAMs is a well-known technique for such interface modifications and can provide microscopically good interface regulations.

We will expand on the latter approach, where the interfaces are engineered to enhance device performance. An OTFT has two critical interfaces: the interface between the gate dielectric and the organic semiconductor, and the interface between the source/drain contacts and the organic semiconductor. The two interfaces dictate charge transport and charge injection in OTFTs, respectively, thus having an overriding influence on the device characteristics. We will investigate interface modification techniques for these two device interfaces, with an attempt to enhance device performance.

One of the areas of interest lies in the integration of a plasma-enhanced chemical vapor deposited (PECVD) gate dielectric with a solution-processable organic semiconductor for OTFT fabrication. Here, we study PECVD silicon nitride (SiN_x) as the gate dielectric. Past experiments have reported limited OTFT performance with a PECVD SiN_x gate dielectric, and have attributed this to surface roughness and unfriendly (or non-organic-friendly) interfaces of SiN_x. However, PECVD SiN_x as a gate dielectric has excellent dielectric properties. It is scalable to large areas and can be deposited at low temperatures, making it compatible with plastic substrates. We believe the critical factors to enable integration of SiN_x in organic electronics lie in identifying a suitable SiN_x composition and an agreeable interface modification process. Strategies to address these factors and to enable the use of SiN_x while delivering acceptable device performance remain the ultimate goal.

1.1.4.2 Device Manufacture

Since a major driving force behind OTFT technology is the manufacture of low-end, low cost, and disposable electronic devices, this demands a fabrication process that allows high volume production at low cost. Moreover, the process should be able to produce stand-alone devices, device arrays, and integrated circuits (ICs) of acceptable operating speed, functionality, reliability, and lifetime. However, an integration process that can meet the above requirements for production of high yield, stable OTFTs is, currently, non-existent.

Conventional photolithography processes for manufacturing silicon-based microelectronics are not completely amenable to organic electronics. Although photolithography has the advantage of producing high resolution, complex device structures with excellent precision, the process must be modified to ensure compatibility with organic materials. Advanced printing techniques (e.g., inkjet printing or nanoimprinting) that take advantage of the solution-processability of organic materials are favorable for achieving the goals of low-cost and high-volume production. Inkjet printing technology is particularly attractive because it offers the advantages of fast, direct imaging and single-step print processing, precise deposition of the organic ink only where it is needed (thus reducing waste), compatibility with flexible substrates, large area processing, and high material usage efficiency. However, because the requirements of printing electronic functions are very different from those of printing visual images, the adaptation of inkjet systems for processing organic electronic devices will require extensive optimization of printing parameters and processing conditions; in addition, technological concerns such as layer continuity and multilayer registration must be resolved. Inkjet printed organic devices with good performance have been demonstrated; however, low device yield is an issue. We will address the challenges in OTFT manufacture by exploring a hybrid manufacturing approach that combines a photolithography process with a novel inkjet printing technique. This delivers an integration strategy with workable manufacturing yields while lowering costs compared to conventional processes.

1.1.5 Scope and Organization

As the material properties and processing technology for organic electronics continue to advance and mature, the next phase of development is directed at integrating OTFTs into circuits and systems. It is not within the scope of this book to review organic semiconductor material development (which is the strength of chemical physicists), but rather to advance OTFT research from an engineering and integration perspective. By utilizing and assimilating existing materials, techniques and resources, we explore a number of approaches to deliver higher performance devices and demonstrate the feasibility of organic circuits for practical applications. The key focus areas include:

- Development of OTFT fabrication strategies to enable circuit integration (Chapter 3);

- Optimization of PECVD gate dielectric composition and structure to improve OTFT performance (Chapter 4);
- Investigation of interface engineering methodologies to enhance the dielectric/semiconductor interface and the contact/semiconductor interface (Chapters 5 and 6);
- Finally, the scientific and technical knowledge acquired from these investigations is applied to demonstrate the integration of OTFTs into functional circuits for active-matrix display and RFID applications (Chapter 7).

The results presented here stem primarily from research conducted at the Giga-to-Nano (G2N) Labs, University of Waterloo, in collaboration with the Xerox Research Centre of Canada (XRCC), which granted access to its high quality, high performance, stable organic semiconductor material. In particular, Xerox's solution-processable poly(3,3‴-dialkylquarterthiophene) (PQT-12) polymer semiconductor forms the basis for the majority of the OTFT experimental work [10] discussed in this book.

This book is divided into eight chapters. We begin with an introduction to organic electronics and market opportunities for OTFT technologies in Chapter 1. Chapter 2 examines the OTFT technology in greater depth, with a review of fundamental properties of organic semiconductors and a discussion of OTFT operation, device architectures, and material selection. Chapter 3 presents integration strategies to enable the fabrication of OTFT circuits.

With the aim of improving OTFT performance, optimization of PECVD gate dielectrics is explored in Chapter 4. Interface engineering strategies to improve charge transport by dielectric/semiconductor interface treatment methods and to enhance charge injection by contact/semiconductor interface modification techniques are given in Chapters 5 and 6, respectively. The objectives for these investigations are to enhance OTFT characteristics via functionalization of the gate dielectric material and the device interfaces, and to develop a better understanding of the materials and interfaces for OTFTs.

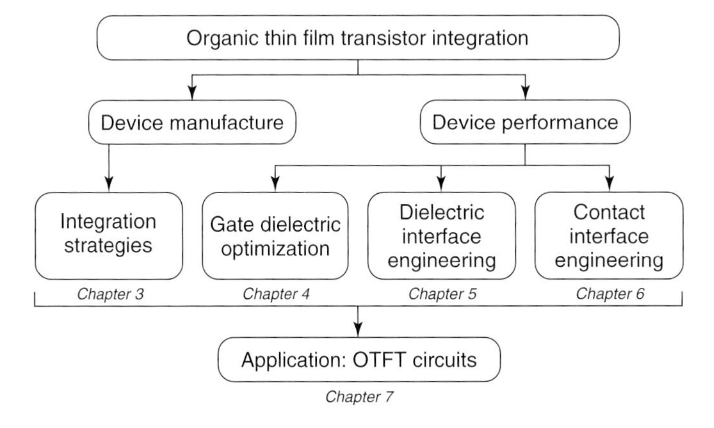

Figure 1.4 Illustration of the organizational structure of the book.

Chapter 7 demonstrates integration of OTFTs into functional circuits. Finally, Chapter 8 presents a summary of the outlook and future challenges related to OTFT integration. The structural design of the book is summarized in Figure 1.4, which illustrates the flow of the various topics related to advancing device manufacture, device performance, and OTFT circuit integration.

References

1. Reese, C., Roberts, M., Ling, M.M., and Bao, Z. (2004) Organic thin film transistors. *Mater. Today*, **7** (9), 20.
2. Orgatronics (2007) Products and technologies inspired by organic electronics. *http://www.orgatronics.com/* (accessed 2008).
3. Chason, M., Brzis, P.W., Zhang, J., Kalyanasundaram, K., and Gamota, D.R. (2006) Printed organic semiconducting devices. *Proc. IEEE*, **93** (7), 1348.
4. Afzali, A., Dimitrakopoulos, C.D., and Breen, T.L. (2002) High-performance, solution-processed organic thin film transistors from a novel pentacene precursor. *J. Am. Chem. Soc.*, **124**, 8812.
5. Shaw, J.M. and Seidler, P.F. (2001) Organic electronics: introduction. *IBM J. Res. Dev.*, **45** (1), 3.
6. Nordén, B. and Krutmeije, E. (2000) The 2000 Nobel Prize in Chemistry. *www.nobel.se/chemistry/laureates/2000/press.html; http://nobelprize.org/nobel_prizes/chemistry/laureates/2000/adv.html* (accessed 13 January 2011).
7. Horowitz, G. (1998) Organic field-effect transistors. *Adv. Mater.*, **10** (5), 365.
8. Ito, T., Shirakawa, H., and Ikeda, S. (1974) Simultaneous polymerization and formation of polyacetylene film on the surface of concentrated soluble Ziegler-type catalyst solution. *J. Polym. Sci. Chem. Ed.*, **12** (1), 11.
9. Chiang, C.K., Fincher, C.R., Park, Y.W., Heeger, A.J., Shirakawa, H., Louis, E.J., Gau, S.C., and MacDiarmid, A.G. (1977) Electrical conductivity in doped polyacetylene. *Phys. Rev. Lett.*, **39**, 1098.
10. Ong, B.S., Wu, Y., and Liu, P. (2005) Design of high-performance regioregular polythiophenes for organic thin-film transistors. *Proc. IEEE*, **93** (8), 1412.
11. McCulloch, I., Bailey, C., Giles, M., Heeney, M., Shkunov, M., Sparrowe, D., Suzuki, M., and Wagner, R. (2005) Stable semiconducting thiophene polymers and their field effect transistor properties. *Materials Research Society Spring Meeting, Symposium I: Organic Thin-Film Electronics*, p. I7.10.
12. McCulloch, I., Heeney, M., Bailey, C., Genevicius, K., MacDonald, I., Shkunov, M., Sparrowe, D., Tierney, S., Wagner, R., Zhang, W., Chabinyc, M.L., Kline, R.J., McGehee, D., and Toney, M.F. (2006) Liquid-crystalline semiconducting polymers with high charge-carrier mobility. *Nat. Mater.*, **5**, 328.
13. Lin, Y., Gundlach, D., Nelson, S., and Jackson, T. (1997) Pentacene-based organic thin-film transistors. *IEEE Trans. Electron. Devices*, **44** (8), 1325.
14. Sirringhaus, H., Kawase, T., Friend, R.H., Shimoda, T., Inbasekaran, M., Wu, W., and Woo, E.P. (2000) High-resolution inkjet printing of all-polymer transistor circuits. *Science*, **290**, 2123.
15. National Institute of Standards and Technology (NIST) (2000) Project Brief: Printed Organic ASICs: A Disruptive Technology. *http://jazz.nist.gov/atpcf/prjbriefs/prjbrief.cfm?ProjectNumber=00-00-4209* (accessed 2010).
16. National Institute of Standards and Technology (NIST) (2000) Project Brief: Printed Organic Transistors on Plastic for Electronic Displays and Circuits. *http://jazz.nist.gov/atpcf/prjbriefs/prjbrief.cfm?ProjectNumber=00-00-4968* (accessed 2010).
17. Organic Electronics Association (2010) Organic Electronics Association. *http://www.oe-a.org* (accessed 2010).
18. BBC News (2007) (Jan 3: 2007) UK in Plastic Electronics Drive. *http://news.bbc.co.uk/2/hi/business/6227575.stm* (accessed 2010).

19. PolyApply (2004) The application of polymer electronics towards ambient intelligence (POLYAPPLY), Framework Programme 6 (FP6), European Commission. *http://www.polyapply.org; http://cordis.europa.eu/fetch?CALLER=FP6_PROJ&ACTION=D&DOC=1&CAT=PROJ&QUERY=011985f2cac0:9e7e:5ed47c46&RCN=71238* (accessed 2010).
20. IEEE Standard (2004) P1620™. *IEEE Standard for Test Methods for the Characterization of Organic Transistors and Materials. http://grouper.ieee.org/groups/1620/.*
21. NAIMO Project (2004) Nanoscale Integrated processing of self-organizing Multifunctional Organic Materials (NAIMO), Framework Programme 6 (FP6), European Commission. *http://www.naimo-project.org; http://cordis.europa.eu/fetch?CALLER=FP6_PROJ&ACTION=D&DOC=1&CAT=PROJ&QUERY=012d802fee5f:0925:066e573d&RCN=74348* (accessed 2010).
22. Jackson, T., Lin, Y., Gundlach, D., and Klauk, H. (1998) Organic thin-film transistors for organic light-emitting flat-panel display backplanes. *IEEE J. Sel. Top. Quantum Electron.*, **4** (1), 100.
23. Huitema, H., Gelinck, G., van der Putten, J., Kuijk, K., Hart, C., Cantatore, E., Herwig, P., van Breemen, A., and de Leeuw, D. (2001) Plastic transistors in active-matrix displays. *Nature*, **414**, 599.
24. Crone, B., Dodabalapur, A., Lin, Y., Filas, R.W., Bao, Z., and La-Duca, A. (2000) Large-scale complementary integrated circuits based on organic transistors. *Nature*, **403**, 521.
25. Rotzoll, R., Mohapatra, S., Olariu, V., Wenz, R., Grigas, M., Shchekin, O., Dimmler, K., and Dodabalapur, A. (2005) 13.56 MHz organic transistor based rectifier circuits for RFID tags. *Mater. Res. Soc. Symp. Proc.*, **871E**, I11.6.1–I11.6.6.
26. Baude, P.F., Ender, D.A., Haase, M.A., Kelley, T.W., Muyres, D.V., and Theiss, S.D. (2003) Pentacene-based radio-frequency identification circuitry. *Appl. Phys. Lett.*, **82**, 3964.
27. Burns, S., Kuhn, C., MacKenzie, J.D., Jacobs, K., Stone, N., Wilson, D., Devine, P., Chalmers, K., Murton, N., Cain, P., Mills, J., Friend, R.H., and Sirringhaus, H. (2003) Active-matrix displays made with ink-jet-printed polymer TFT. Poster presented at the *Society for Information Display International Symposium Exhibition, and Seminar*, Baltimore.
28. OSAD(2007) *Organic Semiconductor Analyst (OSA) Direct–a Weekly Newsletter on the Organic Semiconductor Industry*, Monday, 22 October 2007, Vol.5 (39). Cintelliq Ltd, *www.cintelliq.com/docs/OSAD20071022_sample.htm* (accessed 2008).
29. Kang, H. (2007) High resolution OTFT-AMLCD using solution process. Poster presented at the *Printed Electronics USA 2007*, San Francisco.
30. Lee, S. (2007) Active matrix display using TFT array. Poster presented at the *Printed Electronics USA 2007*, San Francisco.
31. Polymer Vision Ltd. (2007) Polymer vision pioneers world's first production of rollable displays. Press release, December 10, 2007.
32. Heremans, P., Verlaak, S., and McLean, T. (2004) in *Printed Organic and Molecular Electronics* (eds P. Gamota, P. Brazis, K. Kalyanasundaram, and J. Zhang), Kluwer Academic Publishers, Boston, Dordrecht, New York, London.
33. Myny, K., Beenhakkers, M.J., van Aerle, N.A.J.M., Gelinck, G.H., Genoe, J., Dehaene, W., Heremans, P. Robust digital design in organic electronics by dual-gate technology. 2010 IEEE International Solid-State Circuits Conference Digest of Technical Papers (ISSCC), 7-11 February 2010, pp. 140–141.
34. Hiroshiba, N., Kumashiro, R., Komatsu, N., Suto, Y., Ishii, H., Takaishi, S., Yamashita, M., Tsukagoshi, K., and Tanigaki, K. (2007) Surface modifications using thiol self-assembled monolayers on Au electrodes in organic field effect transistors. *Mater. Res. Soc. Symp. Proc.*, **965**, paper 0965-0S08-03.
35. Park, Y.D., Lim, J.A., Lee, H.S., and Cho, K. (2007) Interface engineering in organic transistors. *Mater. Today*, **10**, 46–54.

2
Organic Thin Film Transistor (OTFT) Overview

By definition, organic materials describe a large class of chemical compounds whose molecules contain carbon. The dividing line between organic and inorganic is somewhat controversial and historically arbitrary, but, generally speaking, organic compounds have carbon–hydrogen bonds, and inorganic compounds do not [1]. Until about 40 years ago, all carbon-based organic compounds and polymers[1)] were regarded as insulators. Organic polymer materials (or plastics) were used as inactive packaging and insulating materials. This narrow perspective rapidly changed as a new class of polymer known as a *conductive polymer* was discovered in the 1960–1970s [2–4]. Today, there is a tremendous research effort focused on using conductive polymers for electronic fabrication. Depending on the resistivity levels, conductive polymers can behave as semiconductors (referred to as "*organic semiconductors*"), or they can be highly doped to behave as conductors or metals.

An *organic semiconductor* is loosely defined as any organic material that has semiconductor properties. Both short chain (oligomers) and long chain (polymers) organic semiconductors are known. There are two major classes of organic semiconductors, with considerable overlap: organic charge-transfer complexes, and various "linear backbone" polymers derived from polyacetylene. Charge transfer (CT) complexes are obtained by pairing an electron donor molecule with an electron acceptor molecule, and are characterized by electronic transition(s) to an excited state. One example of a CT complex is a tetrathiafulvalene-tetracyanoquinodimethane (TTF-TCNQ) crystal, where TTF serves as a donor and TCNQ serves as an acceptor. The TTF-TCNQ crystal was the first organic conductor found to show almost metallic conductivity (in 1972). Organic semiconductors based on linear-backbone polymers are obtained from doped conjugated polymers. In this book, discussion of "organic semiconductors" pertains to this class of polymer-based organic semiconductors.

1) Polymer is a long, repeating chain of atoms, formed through the linkage of many molecules called *monomers*. In this report, the discussion of polymers is limited to organic polymers (although inorganic polymers also exist).

Organic Thin Film Transistor Integration: A Hybrid Approach, First Edition. Flora M. Li, Arokia Nathan, Yiliang Wu, and Beng S. Ong.

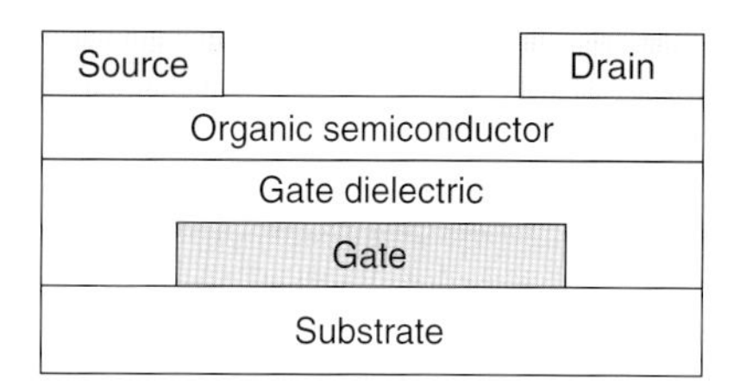

Figure 2.1 Cross-section of a basic OTFT structure, in bottom-gate and top-contact configuration.

Transistors based on organic semiconductors as the active layer to control current flow are commonly referred to as organic thin film transistors (OTFTs), or sometimes organic field effect transistors (OFETs). The simplest OTFT configuration is shown in Figure 2.1. Generally speaking, a thin film transistor (TFT) is composed of three main parts: a thin semiconductor layer, a dielectric (or insulator), and three electrodes (gate, source, and drain). The source (S) and drain (D) electrodes directly contact the semiconductor, whereas the gate (G) electrode is separated from the semiconductor by a dielectric layer. The gate turns the device on and off with an applied voltage, and thus controls the current flow (I_{DS}) in the semiconductor between the source and drain electrodes.

This chapter presents an overview of OTFT technology. The organic semiconductor, which is the heart or foundation of the OTFT, is introduced in Section 2.1. The fundamental properties of organic semiconductors and recent progress in material development are reviewed. The basic operation and characteristics of the OTFT are discussed in Section 2.2, along with the parameter extraction techniques used for characterizing OTFT performance. The device architectures and material systems, specific to the context of this book, are examined in Sections 2.3 and 2.4, respectively. *Please note that the terms "organic" and "polymer" are used interchangeably in this book. Generally speaking, "organic" covers a wider scope of materials, where "polymer" is a subset of organic. The book focuses primarily on polymer semiconductors.*

2.1 Organic Semiconductor Overview

The operation and performance of OTFTs are largely dictated by the characteristics of the active organic semiconductor layer. In particular, the mobility of the device is related to the efficiency of charge transport through the semiconductor channel. This section begins with a review of the unique conjugated chemical structure that gives rise to electrical conduction in organic semiconductor materials. Typical charge transport mechanisms in organic materials are presented. These models pertain primarily to non-crystalline (or disordered) organic semiconductors (e.g., polymers); the transport mechanism in well-organized organic molecular crystals is different, and is outside the scope of this book. The recent progress in material development and classifications of organic semiconductors are considered.

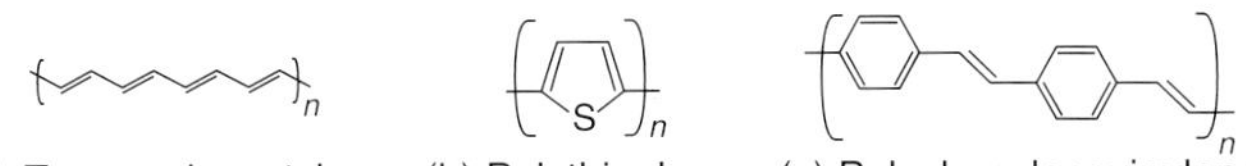

Figure 2.2 Chemical structure of three conjugated polymers: (a) polyacetylene, (b) polythiophene, and (c) polyphenylenevinylene [7].

2.1.1 Basic Properties

Polyacetylene was one of the first polymers reported to be capable of conducting electricity [5], and it was discovered that oxidative doping with iodine causes the conductivity to increase by 12 orders of magnitude [6]. The "doped" form of polyacetylene had a conductivity of 10^5 S m^{-1}. As a comparison, an insulator such as teflon has a conductivity of 10^{-16} S m^{-1}, and a metal such as silver or copper has a conductivity of 10^8 S m^{-1} [4]. However, polyacetylene reacts rapidly and irreversibly with oxygen, is insoluble in organic solvents, and is difficult to process. Progress was made with the discovery of polythiophene (PT) and polyphenylenevinylene (PPV), which exhibited better characteristics than polyacetylene [7]. Figure 2.2 shows the chemical structure of these three conductive polymers.

All electrically conductive polymers share two principal properties. The first is the presence of "conjugated double bonds" along the backbone of the polymer. In conjugation, the polymer consists of alternate single and double bonds between the carbon atoms. This alternating structure can be observed in polyacetylene, displayed in Figure 2.2a. The second property is that the polymer must be "doped", implying that electrons are removed (through oxidation) or introduced (through reduction). These extra holes or electrons can move along the molecule to contribute to electrical conductivity [4].

In the conjugated double bond structure, the single bond is a sigma (σ) bond, and the double bond consists of a σ-bond and a pi (π) bond [8]. σ-bonds, the strongest type of covalent bonds, require that both atoms give an electron from the s orbital. Thus, the electrons that form the σ-bond are attached to the two nuclei and are localized. π-bonds are a direct sharing of electrons between the p orbitals of two atoms. π-bonds are weaker than σ-bonds because their orbitals are further away from the positively charged nucleus. Normally, the electrons that form a π-bond are localized. However, in conductive polymers, π-orbitals of the neighboring double bonds overlap due to the conjugated structure, as illustrated in Figure 2.3a. This overlapping results in weakly localized (or "delocalized") π-electrons that can move from one bond to another or move along the entire molecule. Therefore, delocalization, accomplished by the continuous overlapping π-orbitals of the conjugated backbone, makes the conduction of charge carriers along the polymer chain (i.e., intramolecular transport) possible [8].

The system of alternating double and single bonds in the conjugated backbone gives rise to a separation of bonding and anti-bonding states, resulting in the

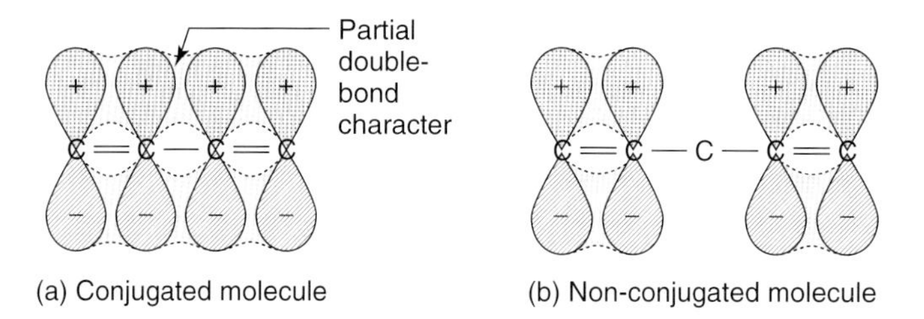

Figure 2.3 (a) Conjugated and (b) non-conjugated structure of an organic molecule [8].

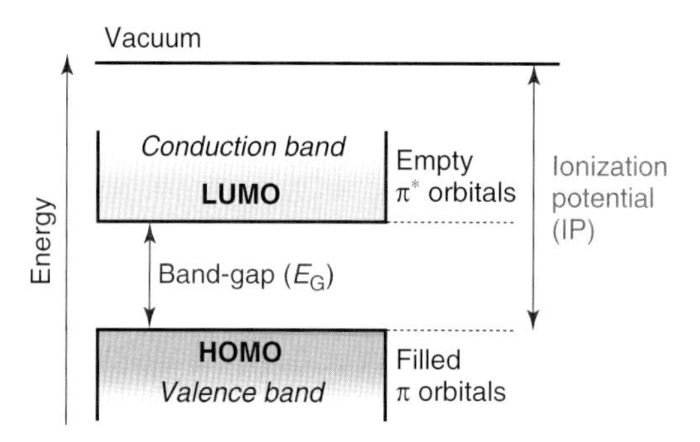

Figure 2.4 Representative energy band diagram of an organic semiconductor.

formation of a forbidden energy gap and a spatially delocalized band-like electronic structure, as illustrated in Figure 2.4. The highest occupied molecular orbital (HOMO) consists of bonding states of the π-orbitals with filled electrons, and is analogous to the valence band in silicon. The lowest unoccupied molecular orbital (LUMO) consists of empty higher energy anti-bonding (π^*) orbitals, and is analogous to the conduction band. The energy difference between the HOMO and LUMO defines the band-gap energy (E_G). E_G depends on the chemical structure of the repeating unit, and generally decreases with the number of repeat units in the chain [4]. The E_G of conjugated polymers is typically in the energy range of 1–4 eV. This band-like structure, along with low electronic mobility, is responsible for the semiconducting properties observed in conjugated polymers. Due to the disordered nature of organic materials, conduction mainly takes place via phonon-assisted hopping and polaron-assisted tunneling between localized states; this is in contrast to crystalline semiconductors (e.g., silicon) where conduction occurs in energy bands through delocalized states.

Most conductive polymers in their neutral state are wide-band-gap semiconductors and exhibit very low conductivities [7]. To increase electrical conductivity, doping is required. The conductivity of a conjugated polymer can be modified by chemical doping or electrochemical doping, where oxidation and reduction reactions are used to achieve p-type (electron removal) and n-type (electron addition) doping, respectively [4]. These doping processes generate mobile charge carriers which move in an electric field, giving rise to electrical conductivity. In most cases, conductive polymers are doped by oxidative reactions; thus, p-type conductive polymer materials are more common, with holes as the majority transport carriers.

Highly conjugated organic materials can work as semiconductors because of their strong π-orbital overlap. When an electron is added or a hole is injected, the resultant charge becomes delocalized across the conjugated system. This injected charge acts as a carrier for current conducting through the molecule and through the organic semiconductor thin film. Transistors based on organic semiconductors as the active layer to control current flow are commonly referred to as *organic thin film transistors*.

2.1.2 Charge Transport

In conductive polymers, the charge transport is relatively easy along a conjugated molecular chain (i.e., intramolecular transport) because of the strong π-orbital overlap. However, due to the amorphous structure of polymer materials and the weak intermolecular interaction, charge transport between molecules (i.e., intermolecular transport) is more difficult. A polymer semiconductor often contains polycrystalline phases intermixed in disordered phases. These disorder segments tend to hinder charge transport. Typically, the charge transport between molecules is described as a thermally activated tunneling of charge carriers, commonly referred to as *"hopping"* [9, 10]. Hopping of carriers from chain to chain occurs between localized states; this rather inefficient conduction mechanism is responsible for the limited carrier mobility in amorphous organic semiconductors.

A main difference between the delocalized transport in a crystalline inorganic semiconductor and the localized "hopping" transport in an amorphous organic semiconductor is that, in the former, the transport is limited by phonon scattering, whereas, in the latter, it is phonon assisted [9]. Accordingly, the charge mobility decreases with temperature in conventional crystalline semiconductors, and the reverse relationship applies for organic materials. The temperature dependence of the mobility (μ) of such "hopping" transport generally follows the form:

$$\mu = \mu_0 \exp\left[-(T_0/T)^{1/\alpha}\right] \tag{2.1}$$

where T is temperature, T_0 and μ_0 are material parameters, and α is an integer ranging from 1 to 4 [9]. The boundary between localized and delocalized transport processes is often taken at mobility near $0.1{-}1\,\text{cm}^2\,\text{V}^{-1}\,\text{s}^{-1}$. The mobility in highly ordered molecular crystals is close to this limit, thus there is controversy as to whether the conductivity in these materials should be described by localized or delocalized transport.

In addition to temperature dependence, the mobility of organic semiconductors becomes field dependent at a high electric field (e.g., $>10^5\,\text{V}\,\text{cm}^{-1}$) [9]. This phenomenon occurs through a Poole–Frenkel mechanism, in which the applied field modifies the potential near the localized states in such a way as to increase the tunnel transfer rate of carriers between sites. The general field-dependence of the mobility is described by:

$$\mu(E) = \mu(0)\exp\left[\frac{q}{kT}\beta\sqrt{E}\right] \tag{2.2}$$

where $\mu(0)$ is the mobility at zero field ($E = 0$), $\beta = (\varepsilon/\pi\varepsilon\varepsilon_0)^{1/2}$ is the Poole–Frenkel factor, and E is the magnitude of the electric field.

Gate voltage dependent mobility is often observed in organic semiconductors. This dependence stems from the fact that as the gate voltage increases, injected charge-carriers tend to fill the traps, so trapping becomes less efficient and charge transport improves. This behavior is described by the multiple trapping and release (MTR) model, which assumes delocalized transport is limited by a distribution of traps near the band edge [11]. The model also predicts thermally activated mobility. Interestingly, as the quality of OTFT devices improves, gate bias dependence and thermally activated mobility tend to be less encountered, which suggests these are defect-induced effects [11].

The charge transport and mobility in organic semiconductors are also limited by macroscopic factors such as poor contacts between different crystalline domains in the material or disorder in the material [4]. Therefore, molecular ordering and alignment of the conjugated segments in a thin film plays a critical role in attaining efficient charge transport between molecules, as discussed next.

2.1.3 Microstructure and Molecular Alignment

Most organic semiconductor thin films are composed of a mixture of polycrystalline and amorphous (disordered) phases. Charge transport via hopping of carriers between molecules in the disordered regions often limits charge carrier mobility. Thus, it is important to suppress the disordered phases and grain boundaries to achieve efficient charge transport. By ensuring the polymer chains or the organic molecules lie/stack close together in an orderly fashion, more efficient carrier transport can be achieved [4]. However, organic semiconductors tend to have poor self-organizing properties, due to their weak London or Van der Waals intermolecular bonds. Therefore, clever molecular design and proper interface preparation are needed to achieve a well-stacked and well-ordered intermolecular structure in the polymer film.

The microstructure of an organic thin film can be characterized using grazing incidence X-ray diffraction (GIXRD) and transmission electron diffraction measurements. An example is shown in Figure 2.5 for poly(3,3‴-dialkyl-quarterthiophene) (PQT-12) organic semiconductor thin film, where GIXRD and transmission electron diffraction were used to analyze molecular parameters and the π–π stacking structure. Transmission electron diffraction analysis in Figure 2.5c indicates a π–π stacking distance of 3.7 Å (along the face-to-face direction). The XRD pattern in Figure 2.5b renders information on interchain ordering, where the diffraction peaks correspond to an interlayer distance of ~17.3 Å (i.e., intermolecular spacing along

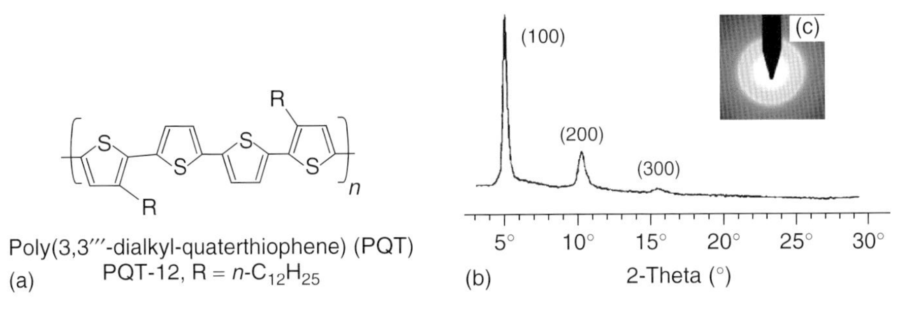

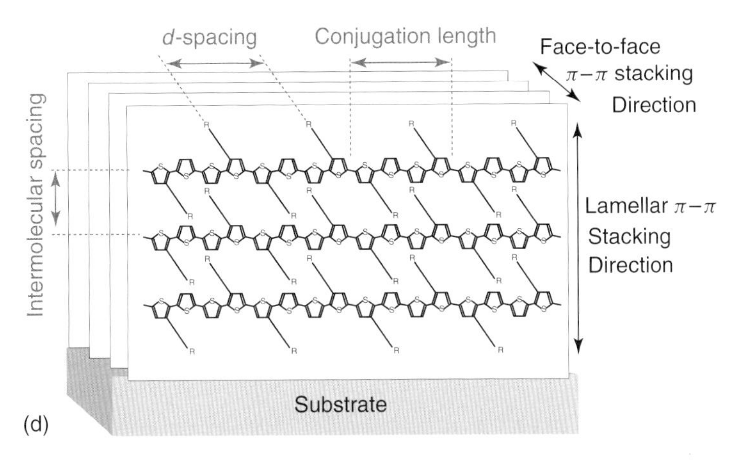

Figure 2.5 (a) Chemical structure of PQT-12 organic semiconductor. (b) GIXRD and (c) transmission electron diffraction pattern of a PQT-12 film. (d) Schematic illustration of lamellar $\pi-\pi$ stacking in a PQT-12 film (adapted from [13]).

the lamellar $\pi-\pi$ stacking direction) [12].[2] These results signify the formation of lamellar $\pi-\pi$ stacking structural order in the thin film as well as a preferential orientation of the lamellar (100) axis normal to the substrate. Accordingly, it can be concluded that PQT-12 possesses an excellent ability to organize into highly ordered lamellar $\pi-\pi$ stacking structures (Figure 2.5d) whose orientation to the substrate could be manipulated through proper alignment layers and techniques [12, 13].

In addition to stacking order, the orientation and alignment of the polymer chains in the device structure have a strong influence on the performance of polymer OTFTs. It was reported that poly(3-hexylthiophene) (P3HT) may adopt

2) According to Bragg's law $2d\ \sin(\theta) = n\lambda$, for the diffraction peak at $2\theta = 5.1^\circ$, the interlayer distance (d) of $\sim$17.3 Å can be calculated. Assuming $n = 1$ (measurement done in air) and $\lambda = 1.54$ Å (wavelength of the X-ray used in XRD measurement).

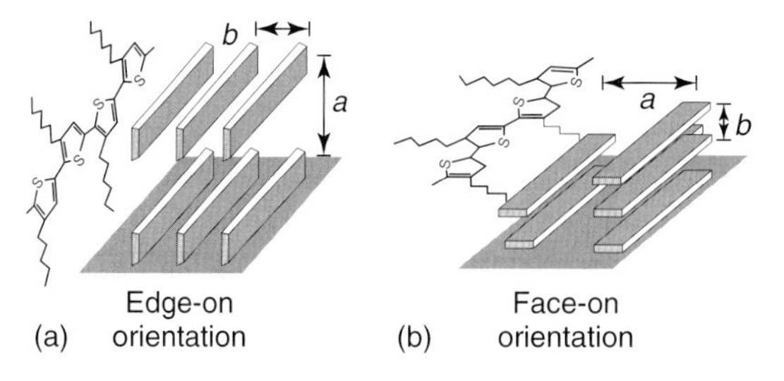

Figure 2.6 (a,b) Two different orientations of ordered P3HT domains with respect to the dielectric substrate surface (adapted from [14]).

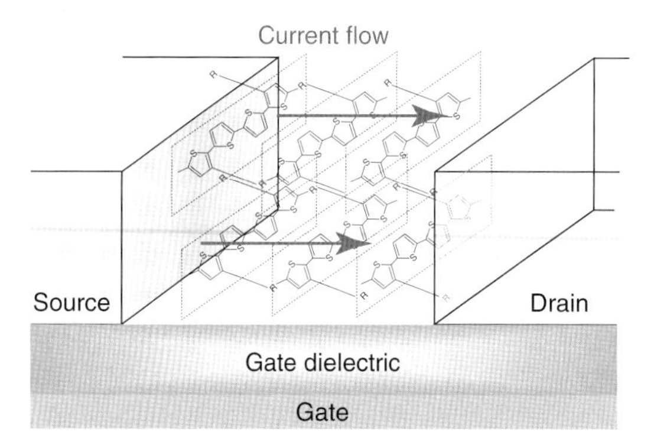

Figure 2.7 Schematics of ideal alignment of organic semiconductor building blocks with strong $\pi-\pi$ stacking in a TFT configuration (adapted from [15]).

two different orientations depending on the processing conditions and dielectric surface properties (Figure 2.6): (i) with thiophene rings oriented edge-on (i.e., perpendicular to) relative to the dielectric surface, or (ii) with the chains oriented face-on (i.e., parallel to) the surface [14]. The mobility differed by more than a factor of 100, and higher device mobility was achieved for P3HT OTFTs with the edge-on arrangement.

The alignment of the polymer chains relative to the electrodes of an OTFT is also important. The ideal alignment of the organic molecules is such that charge transport is parallel to the substrate with in-plane electrodes, where the strong $\pi-\pi$ stacked building blocks are uniaxially aligned in a direction parallel to the current flow in the channel region, as illustrated in Figure 2.7 [15].

In many cases, the molecular structure and morphology of the organic semiconductor are largely determined by the underlying surface properties. In the case of bottom-gate OTFTs, careful preparation and functionalization of the dielectric surface prior to semiconductor deposition is essential to facilitate device enhancements. The impact of dielectric surface conditions on the electrical characteristics of OTFTs is studied in Chapters 4 and 5.

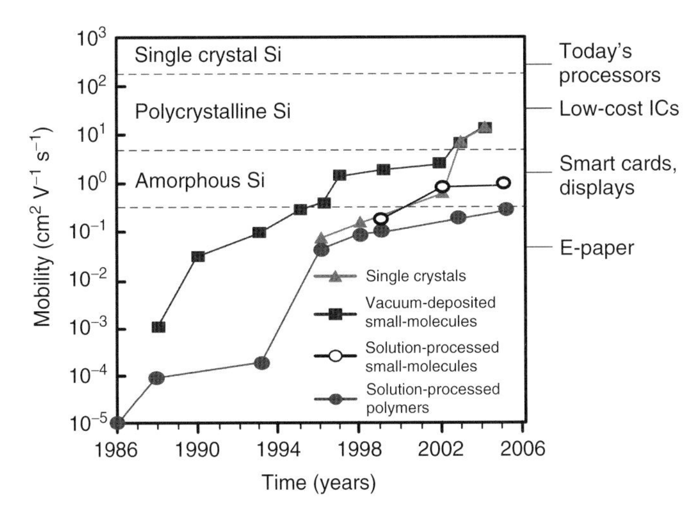

Figure 2.8 Progress in the performance of organic semiconductors (adapted from [3, 18]).

2.1.4
Material Development and Classifications

In the last two decades, the use of organic semiconductors in TFTs has gained considerable interest due to their potential application in low-cost integrated circuits (ICs). Most of the initial efforts were focused on increasing the mobility and on/off ratio of the OTFTs by improving existing materials, synthesizing new materials, experimenting with innovative chemistry and processing methods, and by improving the self-assembly and ordering of materials [3]. Figure 2.8 summarizes the mobility improvements achieved in various organic semiconductors. Since the research on OTFTs commenced in the 1980s, dramatic progress in device performance has been witnessed in the last two decades. In particular, the mobility of PT has improved by five orders of magnitude [3]. OTFTs using pentacene as the organic semiconductor material have achieved mobilities comparable to that of amorphous silicon (a-Si:H) ($\mu \sim 1\,\text{cm}^2\,\text{V}^{-1}\,\text{s}^{-1}$). Amorphous silicon is commonly used for the fabrication of TFTs to drive pixels in flat-panel active-matrix liquid crystal displays (AMLCDs) and active-matrix organic light-emitting diode displays (AMOLED) [16, 17]; some of these applications coincide with the applications envisioned for OTFTs. As the mobility of organic semiconductors improves and approaches that of a-Si:H, OTFTs can offer a promising alternative to a-Si:H TFT technology, especially for applications involving flexible plastic substrates (e.g., flexible displays).

Organic semiconductor materials can generally be categorized into three main classes: small-molecules, oligomers, and polymers [3]. Small-molecules are characterized by smaller molecular mass, shorter molecular chains, and fewer carbon atoms than polymers. On the other hand, polymers possess longer chains and larger molecular mass, contain more carbon atoms, and have good solubility. The properties of oligomers are intermediate between those of small-molecules and polymers. In terms of processability, small-molecule and oligomer compounds usually exhibit limited solubility in organic solvents; as a result, they are normally deposited by vacuum evaporation. In contrast, polymer semiconductors have higher solubility, which makes them particularly well-suited for simpler and lower cost solution-based processing methods such as spin-coating and inkjet printing. Table 2.1 lists the key properties and a representative chemical structure for each type of organic semiconductor. Additional information on small molecule and polymer organic semiconductors, as well as n-type organic materials, is presented next.

2.1.4.1 **Small Molecules**

Evaporated small-molecule organic semiconductors typically exhibit higher mobility and better semiconducting characteristics than polymer semiconductors, owing to higher molecular ordering and the ability to form a well-ordered film. The most frequently studied p-type small-molecule organic semiconductor is pentacene, because of its chemical and thermal stability as well as superior mobility compared to other organic materials. The field-effect mobility of vacuum-deposited pentacene-based TFTs is typically reported to be in the range $0.5{-}1.5\,cm^2\,V^{-1}\,s^{-1}$, and as high as $5\,cm^2\,V^{-1}\,s^{-1}$, which is comparable to devices using a-Si:H as the semiconductor material [19, 20]. The reported on/off current ratio of such devices was $10^5{-}10^8$. The high performance of pentacene is attributed to its ability to form single-crystal-like or polycrystalline films upon vacuum deposition onto gently heated (about 80 °C) substrates.

The insolubility of pentacene may present as a shortcoming in terms of processability. Researchers at IBM have developed solution-processable pentacene based on the precursor route, with an initial mobility ($0.02\,cm^2\,V^{-1}\,s^{-1}$) significantly lower than vacuum-deposited pentacene, due to a lower degree of molecular ordering [21]. More recently, Kawasaki *et al.* reported an OTFT that was made using an inkjet-printed pentacene channel layer having a mobility of $0.15\,cm^2\,V^{-1}\,s^{-1}$ (one of the highest values reported for inkjet-printed OTFTs) and a current on/off ratio of 10^5 [22]. The mobility depends strongly on the film morphology, which can be controlled by the processing conditions, such as solution concentration, substrate temperature, solvent composition, and environmental conditions during film drying.

These results showed that deposition of good quality pentacene film requires vacuum evaporation. This approach is undesirable if low-cost organic devices are eventually to be fabricated. To build cost-effective organic electronics on a mass scale, solution-processing techniques that operate at low temperature and normal room conditions are preferred. This can reduce the cost and complexity of the manufacturing infrastructure, and enhance compatibility with a wider variety of flexible substrates. Polymer-based OTFTs, fabricated by spin-on, drop-casting, or printing techniques, can satisfy these specifications.

Table 2.1 Properties, representative chemical structure, and typical mobility of the three main classes of organic semiconductors. The mobility of silicon is included for comparison [3, 23].

Classes	Properties	Representative chemical structure	Typical mobility ($cm^2 V^{-1} s^{-1}$)
Small-molecules	Low molecular mass (~20–40 carbon atoms) Size: ~1 nm Polycrystalline Vapor-deposited	pentacene	~1 (>5 was reported)
Oligomers	Short molecular chains (made of only a few monomers or structural units)	α,ω-dihexyl-sexithiophene (α,ω-DH6T) C_6H_{13} S S S S S S C_6H_{13}	<1
Polymers	Long molecular chains (made of repeating units of the monomer) High molecular mass (>50 carbon atoms) Size: ~10–1000 nm Semi-crystalline or amorphous Solution processed (high solubility)	Regioregular poly(3-hexylthiophene) (P3HT) R R R R R R S S S S S S $R = C_6H_{13}$	≤0.1
Silicon	Inorganic semiconductor	Crystalline silicon Polycrystalline silicon Amorphous silicon (a-Si:H)	300–1000 50–100 ~1

2.1.4.2 **Polymers**

Polymer semiconductors exhibit structural stability, tunable electrical properties, and solubility, which are customizable by designing and shaping the polymer chain structures. The conjugation length and rotational freedom determine the semiconductor polymer functionality and environmental sensitivity, while the alkyl side chains determine polymer solubility [24]. The mobility of polymer semiconductors stands roughly one order of magnitude below that of vacuum-deposited small molecules, and is attributed to the poorer molecular ordering of solution-processed polymer materials [25]. Devices fabricated using solution processable regioregular poly(3-hexylthiophene) (RR-P3HT) resulted in field-effect mobility typically of the order of 10^{-5} to $10^{-2}\,cm^2\,V^{-1}\,s^{-1}$, depending on the processing conditions (e.g., in air or oxygen-free environments, annealing), dielectric material, and dielectric/semiconductor interface properties. The highest mobility reported was $0.1\,cm^2\,V^{-1}\,s^{-1}$ from a soluble RR-P3HT, with greater than 98.5% head-to-tail (HT) linkages [26, 27]. The chemical structure of a HT RR-P3HT is displayed in Figure 2.9b [26]. HT is one of the two possible arrangements for the 3-alkyl substituents (represented by "R" in Figure 2.9b) in the P3HT structure;[3] the other arrangement is known as head-to-head (HH), as depicted in Figure 2.9a. If P3HT consists of both HH and HT moieties, it is "regiorandom"; if only one type of arrangement is present (either HH or HT), it is "regioregular". High regioregularity is critical for higher molecular ordering, and thus high mobility, in polymer OTFTs. In contrast, regiorandom P3HT films usually yield lower mobility due to a higher degree of disorder in the regiorandom structure [26, 28]. Alternative techniques to improve the mobility of P3HT OTFT have also been reported, including optimizing the choice of solvent, thermal annealing, optimizing the chain length, and using chemical treatment of the substrate surface prior to deposition of the polymer semiconductor film [14, 28, 29]. Poly(9,9′-dioctyl-fluorene-co-bithiophene) (F8T2) is another relatively air-stable polymer, and displayed a mobility of 0.01–$0.02\,cm^2\,V^{-1}\,s^{-1}$ after special interface preparation and high-temperature annealing [30].

A notable advance in the conjugated regioregular PT family is the development of poly(3,3‴-dialkyl-quaterthiophene)s (PQTs) by Ong *et al.* at the Xerox Research Centre of Canada (XRCC), with considerably enhanced oxygen resistance, solution processability, and self-assembly [13]. The typical device mobility is 0.07–$0.12\,cm^2\,V^{-1}\,s^{-1}$ with a current on/off ratio of 10^6 under ambient conditions. Most of the research work presented in this book is conducted using a PQT-12 polymer semiconductor. Through dielectric and interface optimization, a mobility of $0.2\,cm^2\,V^{-1}\,s^{-1}$ and an on/off ratio of 10^8 are obtained for PQT-12 OTFT on a silicon nitride (SiN_x) gate dielectric (see Chapter 5). A more recent

3) An alkyl is a functional group of an organic chemical that contains only carbon and hydrogen atoms, which are arranged in a chain. Its general formula is C_nH_{2n+1}, and it is usually found attached to other hydrocarbons.

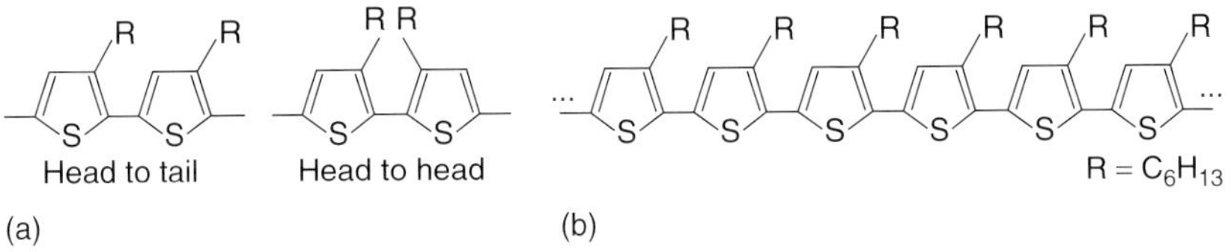

Figure 2.9 (a) Two types of arrangements for P3HT. (b) Head-to-tail regioregular P3HT [23].

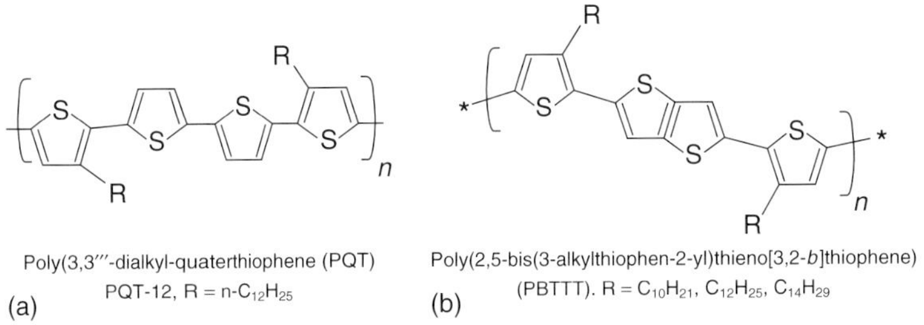

(a) Poly(3,3'''-dialkyl-quaterthiophene (PQT) PQT-12, R = n-$C_{12}H_{25}$

(b) Poly(2,5-bis(3-alkylthiophen-2-yl)thieno[3,2-*b*]thiophene) (PBTTT). R = $C_{10}H_{21}$, $C_{12}H_{25}$, $C_{14}H_{29}$

Figure 2.10 Chemical structure of (a) PQT and (b) PBTTT [13, 31].

advancement in solution-processable semiconductors is a new liquid crystalline poly(2,5-bis(3-alkylthiophen-2-yl)thieno[3,2-*b*]thiophene) (PBTTT) polymer, developed by Merck Chemicals. PBTTT transistors showed good mobilities and stability in the range 0.2–0.6 $cm^2\ V^{-1}\ s^{-1}$ [31]. The chemical structures for PQT and PBTTT are presented in Figure 2.10.

2.1.4.3 **n-Type Semiconductors**

To make p–n junction diodes, bipolar transistors, and complementary circuits, n-type organic semiconductors are required. The majority of organic semiconductors exhibit p-type characteristics, which means that they primarily transport holes (h^+) rather than electrons, and that holes are more easily injected than electrons. Comparatively, there are considerably fewer reports on n-type organic semiconductors [32–34]. In general, most n-type organic semiconductors have relatively lower field-effect mobilities, poorer stability, and limited solubility compared to p-type polymers. Examples of n-type materials include fullerene (C_{60}) and *N,N'*-dialkyl-3,4,9-10-perylene tetracarboxylic diimide derivates, which currently have the highest electron mobility of up to 0.5 $cm^2\ V^{-1}\ s^{-1}$ [11, 35]. A major problem with most n-type organic semiconductors is their high instability in ambient conditions, especially oxygen and moisture [20, 25]. A limited number of n-type compounds with improved air stability was identified (e.g., hexadecafluoro-phthalocyanine ($F_{16}CuPc$), diperflurorohexylquarter-thiophene (DFH-4T)); these materials typically have lower mobilities. n-Type compounds also tend to have lower solubility than p-type polymers; phenyl-C_{61}-butyric acid methyl ester (PCBM) is one of the few soluble n-type polymers [36].

More extensive research efforts are required to improve the properties of n-type organic semiconductors. The recent discovery of ambipolar functionality in some organic materials, where a single semiconductor layer exhibits both n-type and p-type transport capabilities, appears promising for the implementation of complementary logic OTFT circuits and is expected to increase the versatility of OTFTs [17, 37]. Another route to achieve ambipolar conduction is to interpenetrate a network of two compounds, one p-type and one n-type. Meijer *et al.* used a PPV:PCBM blend to fabricate ambipolar OTFTs and complementary metal oxide semiconductor (CMOS)-like inverters [38].

From a manufacturing perspective, to truly realize the advantages of organic materials in device applications, liquid phase processing techniques by spin-coating, casting, or printing are strongly desired. Soluble polymer semiconductors combined with large-area printing techniques are particularly favorable from the manufacturing cost standpoint. This book focuses on solution-processable polymer semiconductors. More details on the selection of device materials considered in this research are given in Section 2.4.

2.1.5
Sensitivity to Environmental Influences

The high sensitivity of most organic semiconductors to ambient conditions or environmental influences, such as oxygen, humidity, light, and temperature, poses a huge limitation on the performance and stability of OTFTs [16,39–41]. For example, atmospheric oxygen and moisture, as well as photo-induced oxidation, have a doping effect on the organic material, causing an increase in the conductivity, which subsequently degrades the on/off current ratio [28]. Moreover, the current–voltage ($I-V$) characteristics of OTFTs tend to shift with bias-stress, thermal-stress, and exposure to atmosphere [16]. This results in changes in important device parameters such as conductivity, mobility, threshold voltage, on/off current ratio, and subthreshold characteristics.

The effect of air exposure on the conductivity of P3HT was studied by Allport *et al.* [42], they measured the $I-V$ characteristics of a P3HT/aluminum diode after fabrication (in vacuum) and after one week of exposure to air in a clean environment. Increases in both the forward and reverse currents were observed after exposure to air, and were attributed to an increase in doping concentration and in conductivity. Many polymer semiconductors easily absorb oxygen and water molecules in air, and often become increasingly p-type doped. As a result of (photo-induced) oxidative doping, the free carrier density increases and the conductivity of the material increases [42].

The uncontrolled increase in conductivity can lead to severe deterioration of the performance of OTFTs, possibly resulting in a higher off-current and lower on/off current ratio. Abdou *et al.* [43] studied the interaction of oxygen with conjugated polymers and observed that oxygen forms a reversible CTC with the polymer. It was suggested that the CTC is largely responsible for the generation of charge carriers in semiconducting π-conjugated polymers when exposed to air, which modulates the electronic properties of the polymer. Analysis of P3HT OTFTs under increasing pressures of oxygen revealed that the formation of a CTC resulted in an increase in carrier concentration, an increase in conductivity, and a decrease in charge carrier mobility. With prolonged or higher intensity exposure to oxygen, the proper transistor behavior was lost [43]. The degree of carrier generation by oxidative doping upon air exposure is expected to increase as the ionization potential (IP) of the polymer is lowered. The *IP*, defined as the energy difference between the HOMO levels and a vacuum, is dependent on the effective π-conjugation lengths of the polymer. An extensive π-conjugation along the polymer chain leads to a lower IP for polythiophenes, and thus there is a greater tendency for them to be oxidatively doped. On the other hand, conjugated polymers with shorter effective π-conjugation lengths (such as small-molecule materials) possess higher IPs, and thus greater resistance against oxidative doping [13]. Therefore, organic semiconductors possessing low IPs and having an amorphous nature are predisposed to oxygen diffusion into the bulk, and are particularly susceptible to oxidative doping.

To reduce the degradation induced by atmospheric factors, fabrication and processing of these air-sensitive organic semiconductors can be done in an inert environment (e.g., vacuum, dry nitrogen). P3HT OTFTs fabricated in an inert environment tend to exhibit improved performance compared to those fabricated in air [16]. Moreover, most of the high performance solution-processed OTFTs reported in the literature involved fabrication and measurement in an inert environment, to reduce unwanted doping effects. Other reported techniques to address atmosphere-induced degradation involve treating the polymer film with a reducing agent (e.g., ammonia) to de-dope the polymer film, resulting in an increase in the on/off ratio [29]. Annealing is another technique to recover the polymer from degradation.

A more effective solution to address this material limitation is to develop air-stable organic materials; an excellent example is the PT derivative (PQT) developed by Ong *et al.* [13]. PQT-12 OTFTs fabricated and measured in air demonstrated excellent performance compared to other commercially available P3HT OTFTs fabricated in air. Merck Chemicals Ltd has also reported a thiophene polymer semiconductor with improved air stability. However, long term degradation of OTFT is inevitable; thus, passivation/encapsulation layers and other strategies must be considered to extend the device lifetime. More discussion on encapsulation techniques is presented in Section 2.4.5.

2.2 OTFT Operation and Characteristics

The TFT structure is well adapted to low conductivity materials, and is currently used in a-Si:H and polycrystalline silicon transistors [25]. There are a number of functional and structural differences between TFTs and the conventional

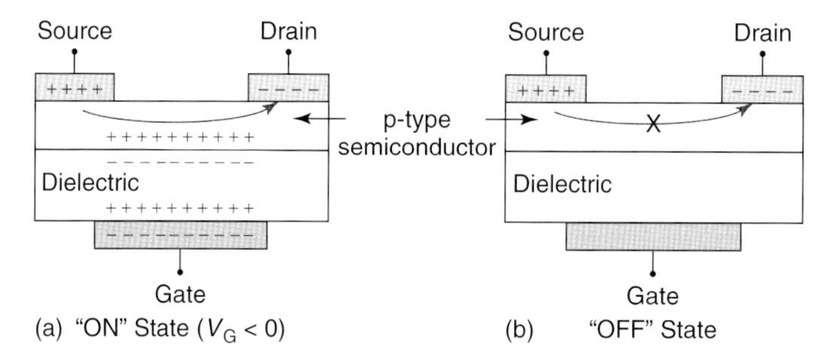

Figure 2.11 (a,b) Simplified illustration of the operation of a TFT with p-type semiconductor [7].

metal-oxide-semiconductor field-effect transistor (MOSFET) used in crystalline silicon ICs. First, there is no depletion region in TFTs to isolate the device from the substrate. Low off-currents (I_{OFF}) in TFTs are only guaranteed by the low conductivity of the semiconductor.[4] A second crucial difference is that the TFT operates in the accumulation regime while MOSFET operates in the inversion regime. This implies that the conduction channel in a TFT is formed by accumulating a majority of carriers at the gate dielectric and semiconductor interface, as illustrated in Figure 2.2a. In this figure, a TFT with a p-type semiconductor is turned "on" by applying a negative bias to the gate electrode. This sets up an electric field in the dielectric layer that induces positive charge carriers (i.e., holes) to accumulate in the semiconductor at the dielectric/semiconductor interface. This accumulation layer forms a conducting channel in the semiconductor, allowing charge carriers to be driven from the source to the drain by applying a bias to the drain electrode. Here, the source serves as the reference (grounded) electrode. When a TFT is in the "off" state (Figure 2.11b), the gate voltage is set so that there is no conducting channel in the semiconductor layer. This implies that, ideally, no current should flow between the source and drain electrode, even with an applied V_{DS} bias [7]. Figure 2.12 shows typical electrical characteristics of a p-type OTFT; the transition from the "off" state to the "on" state of the OTFT can be detected by the abrupt increase in drain current in the transfer ($I_D - V_{GS}$) characteristics near $V_{GS} = 0$ V in Figure 2.12a.

The TFT's gate-dielectric-semiconductor structure operates like a capacitor. Assuming the ideal case, application of a gate voltage induces an equal charge of opposite polarity in the semiconductor near the dielectric/semiconductor interface, as shown in Figure 2.11a. This charge forms a conducting channel if it is of the same type (polarity) as the majority charge carrier being injected into the semiconductor from the contacts. Conductance of the channel is proportional to the gate voltage.

4) Contrary to the standard silicon MOSFETs, OTFTs do not have a blocking p–n junction near the contacts to prevent the current from flowing between source and drain when the channel is not yet formed. As a result, the off-current (I_{OFF}) is relatively high in OTFTs compared to MOSFETs [26].

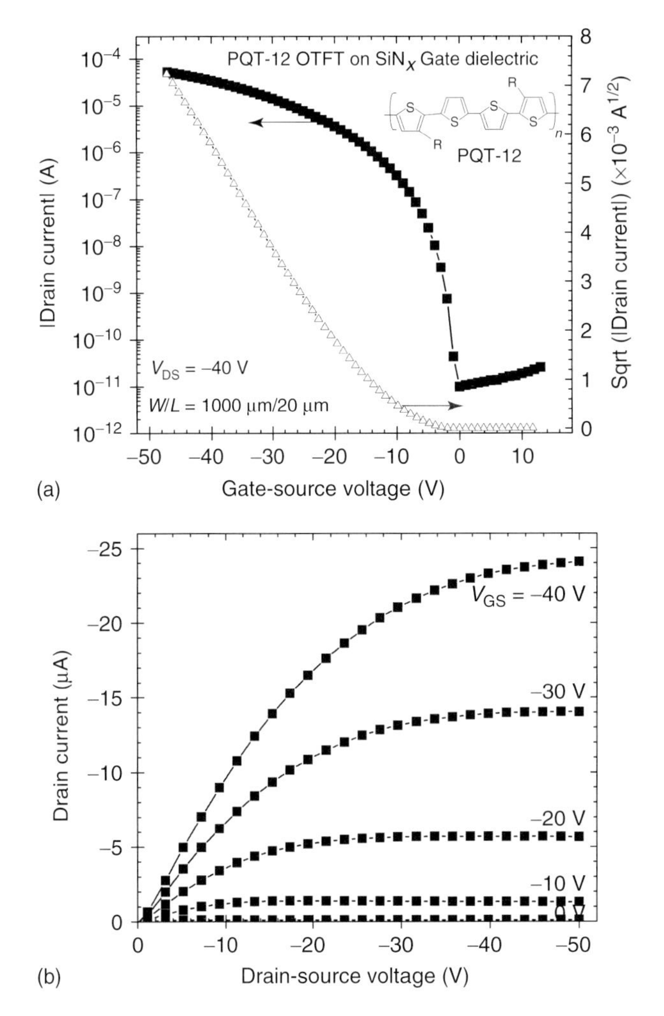

Figure 2.12 Typical electrical characteristics of a p-type OTFT. (a) Transfer curve (I_D–V_{GS}) in the saturation regime (V_{DS} = −40 V) and (b) output curves (I_D–V_{DS}) at different gate voltages. The device shown here is a PQT-12 OTFT on a PECVD SiN_x gate dielectric. Inset of (a) shows the molecular structure of PQT-12, which serves as the semiconductor in the device.

At low drain voltages, the current flowing between drain and source through the channel, I_{DS}, follows Ohm's law; I_{DS} is proportional to the drain and gate voltages. This operation mode is referred to as the *linear regime*. The linear current can be described by:

$$I_{DS,lin} = \mu_{FE} C_i \frac{W}{L} \left(V_{GS} - V_T - \frac{V_{DS}}{2} \right) V_{DS} \tag{2.3}$$

As the drain voltage increases and approaches the gate voltage, V_{DG} drops to zero and pinch-off of the channel occurs. At this point, the channel current (I_{DS}) becomes independent of the drain bias. This regime is called the *saturation operation*. Thus, for V_{DS} more negative than V_{GS}, I_{DS} tends to saturate owing to the pinch-off of the accumulation layer. The saturation current is approximated by:

$$I_{DS,sat} = \frac{1}{2}\mu_{FE} C_i \frac{W}{L} (V_{GS} - V_T)^2 \tag{2.4}$$

The transition from linear to saturation regime is observable in the output ($I_D - V_{DS}$) characteristics in Figure 2.12b. In these equations, μ_{FE} is the field-effect mobility, C_i is the capacitance per unit area of the dielectric layer, L is the channel length, W is the channel width, V_T is the threshold voltage, V_{GS} is the gate-source bias, and V_{DS} is the drain-source bias. V_T is usually defined as the gate voltage at which the channel conductance (at low V_{DS}) is equal to that of the whole semiconductor layer.

The operation and performance of OTFTs are generally governed by two key mechanisms: (i) charge transport in the organic semiconductor layer and (ii) charge injection and extraction at the source/drain contacts. Field-effect mobility (μ_{FE}) is often used to characterize the efficiency of charge transport in a device, and contact resistance (R_C) provides a measure of charge injection efficiency in an OTFT. The extraction of μ_{FE} and R_C, as well as other relevant device parameters, is discussed in the next section.

Note that μ_{FE} is commonly used as a figure of merit for reporting OTFT performance by the research community. However, the accurate extraction and modeling of μ_{FE} remain controversial, primarily due to incomplete understanding of the OTFT physics, artifacts in the extraction procedure (e.g., contact resistance, threshold voltage shifts, stress, geometry dependence), and the use of over-simplified models. As a result, the extracted value of μ_{FE} may not give a precise representation of OTFT performance. These concerns are not as critical in the analysis presented in this book because μ_{FE} is used to systematically compare device performance within various sets of experiments. Nonetheless, the absolute value of the extracted μ_{FE} should be analyzed with caution when it is assessed against data published in the literature.

2.2.1
OTFT Parameter Extraction

The current–voltage equations in Equations 2.3 and 2.4 provide a simple approach to approximating the OTFT characteristics, and are used to provide a basis for characterizing and comparing a variety of OTFT devices. The key device parameters of interest include: effective field-effect mobility (μ_{FE}, measured in $cm^2\,V^{-1}\,s^{-1}$), on/off current ratio (I_{ON}/I_{OFF}), threshold voltage (V_T, in V), and contact resistance (R_C). In particular, μ_{FE} describes how rapidly the charge carriers can move through the material, and is often used as a figure of merit for comparing the performance of various organic semiconductor materials. A high μ_{FE}, as well as high I_{ON}/I_{OFF}, are desirable qualities for OTFTs. To analyze the subthreshold behavior, the switch-on voltage (V_{SO}) and inverse subthreshold slope (S, in $V\,dec^{-1}$) are studied.

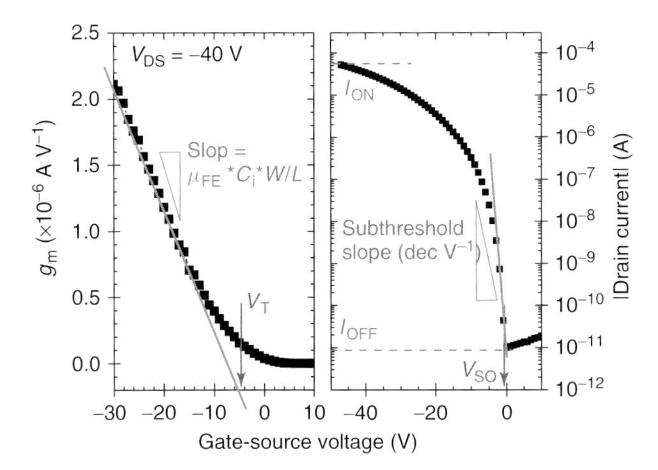

Figure 2.13 Extraction of OTFT device from electrical characteristics: (a) $g_m - V_{GS}$ plot for extracting μ_{FE} and V_T, and (b) $I_D - V_{GS}$ plot for deducing I_{ON}/I_{OFF} and subthreshold slope.

V_{SO} is defined as the gate voltage where the current starts to increase in the semi-logarithmic $I_D - V_{GS}$ plot (Figure 2.12a).

In this book, μ_{FE} and V_T are extracted from transconductance (g_m) measurements. In the linear regime, $g_{m,lin}$ is expressed as:

$$g_{m,lin} = \frac{\partial I_{D,lin}}{\partial V_{GS}} = \mu_{FE} C_i \frac{W}{L} V_{DS} \tag{2.5}$$

Accordingly, the effective mobility in the linear regime, $\mu_{FE,lin}$, can be extracted as follows:

$$\mu_{FE,lin} = \frac{g_{m,lin}}{C_i V_{DS}} \frac{L}{W} \tag{2.6}$$

In the saturation regime, $g_{m,sat}$ is expressed as:

$$g_{m,sat} = \frac{\partial I_{D,sat}}{\partial V_{GS}} = \mu_{FE} C_i \frac{W}{L} (V_{GS} - V_T) \tag{2.7}$$

By plotting $g_{m,sat}$ versus V_{GS} and performing linear curve fitting, μ_{FE} and V_T can be extracted from the slope and intercept, respectively, of the linearly fitted curve (see Figure 2.13a). The I_{ON}/I_{OFF}, V_{SO}, and subthreshold slope are easily deducible from the $I_D - V_{GS}$ curve (see Figure 2.13b). The calculation of contact resistance is discussed in Section 2.2.2.

In most cases, the current–voltage equations in Equations 2.3 and 2.4 can satisfactorily describe the actual OTFT curves; however, their derivation is based on several approximations that are not always fulfilled by organic semiconductors. These assumptions include [11]:

- Gradual channel approximation: the electric field along the channel is much lower than that across it, thus the voltages vary gradually along the channel from

the drain to the source. This assumption is valid when the distance between the drain and source is much larger than the thickness of the dielectric.
- The mobility, μ_{FE}, is constant.

Furthermore, this over-simplified model, derived based on concepts of silicon MOSFETs, overlooks many unique properties of OTFTs (e.g., the disordered nature of amorphous organic semiconductor materials, contact resistance, voltage-dependent mobility, the absence of an inversion layer). Thus, the extracted device parameters are subject to error. A more comprehensive model for TFTs based on disordered/amorphous semiconductors is outlined in Ref. [44]; this model can offer better insight on device behavior and provide more accurate extraction of device parameters. Improved models of OTFT based on different types of organic semiconductors continue to evolve through a growing understanding and stronger knowledge-base on OTFT physics.

2.2.2 Contact Resistance Extraction

Contact resistance (R_C) provides a means to evaluate the efficiency of charge injection. Devices with ohmic contacts and efficient charge injection are characterized by low contact resistance values. To investigate the role of contacts in device performance, a gated transmission line model is used for the extraction of contact resistance and channel resistance. The extraction procedure used in this book is adapted from the comprehensive TFT model for disordered semiconductors in Refs. [44, 45]. Output characteristics of TFTs with different channel lengths are used for extraction. The measured resistance of a TFT with unit width, denoted by $R_m W$, can be written as [44]:

$$R_m W = \frac{V_{DS}}{I_{DS,lin}} W = R_{DS} W + AL, \quad \text{where } A = f(R_{ch}) \tag{2.8}$$

Here, $R_m = V_{DS}/I_{DS,lin}$ is the measured resistance between drain and source electrodes, $R_{DS} W$ is the contact resistance for a unit width TFT, and A is a parameter related to channel resistance (R_{ch}). The total measured resistance, $R_m W$, is calculated at a small V_{DS} and plotted as a function of L for each V_{GS}, as depicted in Figure 2.14a. To facilitate evaluation of the linear component and the nonlinear component of contact resistance, Equation 2.8 is rewritten as:

$$R_m W = B + AL \tag{2.9}$$

where

$$B = R_{DS} W + A\Delta L, \text{ and } A = f(R_{ch}) \tag{2.10}$$

According to Equation 2.9, linear fitting of a $R_m W$ vs. L plot gives B and A as the coordinate intercept point and the slope, respectively. Sets of (A, B) values are obtained by plotting Equation 2.9 at various V_{GS}. Subsequently, plotting B as a function of A gives a line with $R_{DS} W$ and ΔL as the coordinate intercept point and

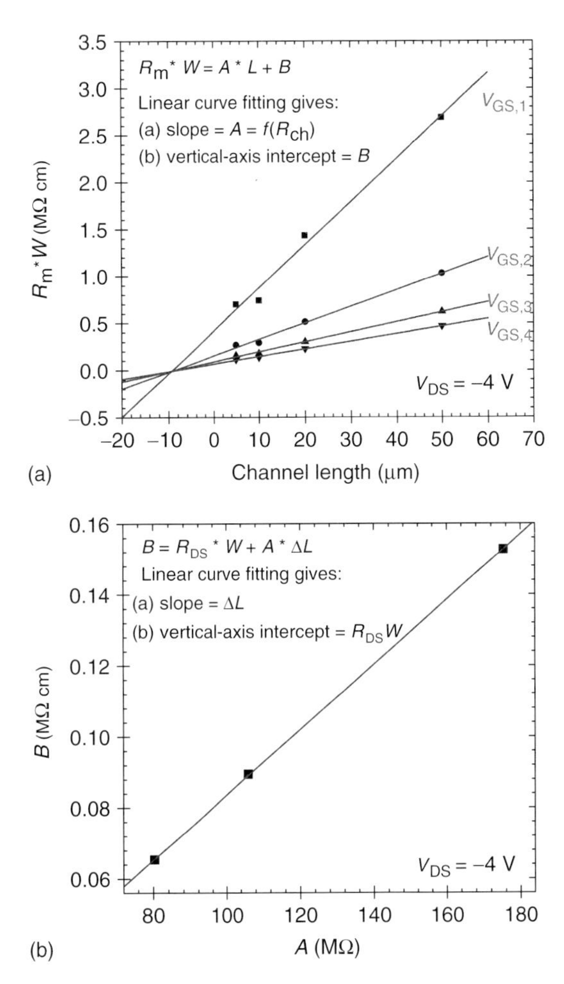

Figure 2.14 Contact resistance extraction. (a) Exemplary plot of $R_m W$ vs. L for different values of V_{GS} for an OTFT at a given V_{DS}, to extract A and B. (b) Exemplary plot of B vs. A at a given V_{DS}, to extract $R_{DS}W$ and ΔL.

slope, respectively. This is illustrated in Figure 2.14b. From these data, we extract two key parameters:

- $R_{DS}W$ (Ω cm) gives the constant part of the contact resistance, that is, the linear component;
- ΔL (μm) represents the voltage-dependent part of the contact resistance, that is, the non-linear component.

This extraction approach is applied in Chapter 6 to evaluate the effectiveness of contact interface treatments of OTFTs.

Of particular importance is the relative influence of the contact resistance R_C at the organic semiconductor/metal interface and the channel resistance R_{ch} formed at the organic semiconductor/dielectric interface. Depending on the charge carrier mobility in the semiconductor and the quality of the contact interface, contact resistances may dominate device operation and limit device performance. For devices with large contact resistance, where there is a significant portion of the total V_{DS} dropped across the contacts, the extracted field-effect mobility tends to be underestimated [46]. In addition, contact resistances may mask intrinsic bulk effects within the organic semiconducting layer, leading to misleading device parameter extraction. Therefore, it is important to minimize the contact effects in the operation of OTFTs. This necessitates a better understanding of the charge injection mechanisms, as well as a more comprehensive device model to account for contact resistance effects and correct extraction of mobility values. Overall, increased knowledge of charge transport and charge injection mechanisms is beneficial to assist in delivering enhanced device performance and in designing OTFT circuits for practical applications.

2.2.3
Desirable OTFT Characteristics

The preferable characteristics of a "good" OTFT typically include high field-effect mobility, a high on/off drain current ratio, low leakage current, minimal threshold voltage shift, sharp subthreshold slope, and low contact resistance. These general performance requirements are summarized in Table 2.2. In most cases, the performance of an OTFT is dictated by two key mechanisms:

1) Charge transport (related to μ_{FE}) in the semiconductor layer, which is dictated by the quality of the semiconductor layer and the dielectric/semiconductor interface;
2) Charge injection at the contacts (related to R_C), which is dictated by the energetic matching and compatibility between contact and semiconductor materials and the corresponding interface.

In this book, attempts to enhance charge transport (or channel quality) are made by optimizing the dielectric/semiconductor interface, as addressed in Chapters 4 and 5. Charge injection is examined in Chapter 6 by investigating contact/semiconductor interface treatment strategies.

2.3
OTFT Device Architecture

A number of OTFT structures can be obtained by varying the relative placement of the gate and source/drain electrodes with respect to the semiconductor layer. The resulting devices are described as top-contact, bottom-contact, top-gate, bottom-gate, or dual-gate structures. Each design possesses distinctive strengths

Table 2.2 General requirements for a good TFT [47].

Requirement	Descriptions
High field-effect mobility (μ_{FE})	Determined by quality, crystallinity, ordering, and microstructure of the semiconductor layer Requires a low defect density in the semiconductor layer and at interfaces Mobility influences switching speed in transistors
High on/off drain current ratio (I_{ON}/I_{OFF})	Depends on the quality of the semiconductor and gate dielectric layers Requires minimum leakage during "off-state" and maximum drain current during conduction
Low threshold voltage (V_T) and minimal threshold voltage shift (ΔV_T)	Low V_T signifies lower operation voltages. V_T depends on the bulk and interface states of the dielectric film Minimal ΔV_T is critical for stability reasons. ΔV_T is dependent on interfacial stresses, and on the creation of metastable defects in the semiconductor layer under a gate bias
Sharp subthreshold slope (S)	Sharp S (i.e., small value in units of V dec^{-1}) for faster switching of the TFT Requires a low density of deep gap states localized at and/or near the semiconductor/dielectric interface
Low leakage current (I_{leak})	When a TFT is off, I_{leak} needs to be small for higher retention of the stored charge, especially in display applications. I_{leak} is determined by the quality of the gate dielectric, the thermal generation current, the interface state density (ideally low), and interfacial stress
Low contact resistance (R_C)	Depends on the charge injection efficiency at the semiconductor/contact interface. High R_C limits current drive and switching speed

and weaknesses in terms of operating mode and ease of fabrication. Thus, each structure may find unique usage in specific applications/configurations.

2.3.1
Top-Contact and Bottom-Contact OTFTs

Depending on the arrangement of the source and drain contacts relative to the semiconductor layer, two configurations are possible, as depicted in Figure 2.15: a top-contact (or staggered) OTFT and a bottom-contact (or co-planar) OTFT. For top-contact OTFTs, the source and drain electrodes are placed on top of the semiconductor layer. For bottom-contact OTFTs, the organic semiconductor is deposited onto the gate dielectric and the prefabricated source and drain electrodes. The top-contact structure can have a performance advantage over bottom-contact

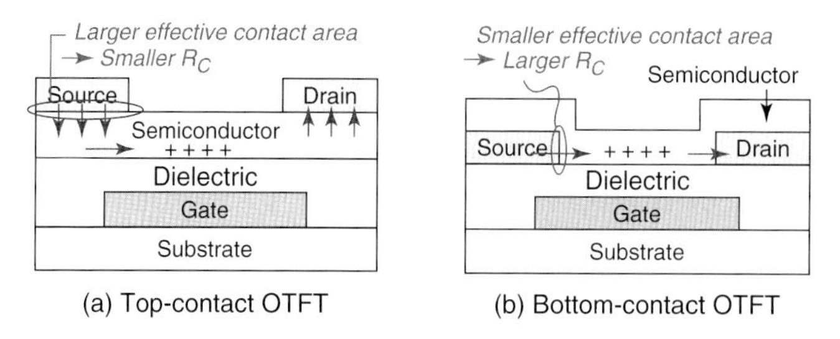

Figure 2.15 Cross-section of two bottom-gate OTFT configurations: (a) top-contact and (b) bottom-contact.

devices in terms of lower contact resistance. On the other hand, the bottom-contact structure enjoys simpler and more robust processing schemes over top-contact structures. These differences are explained next.

Top-contact and bottom-contact OTFTs often show different electrical performances. Considering bottom-gate devices, the application of a gate voltage induces a conductive channel at the interface between the organic semiconductor and the dielectric. For a bottom-contact OTFT, the carrier exchange between the channel and the source/drain contacts is limited to narrow lines along the contact edges (i.e., at the edge of the contacts along the width of the channel), where the contact and the induced-channel intersect (Figure 2.15b). Research has shown that this small contact area tends to generate large contact resistance in bottom-contact OTFTs [48]. In contrast, top-contact OTFTs usually have relatively larger areas for carrier injection at the contact/semiconductor interface, resulting in smaller contact resistance (Figure 2.15a). Due to more pronounced contact effects in bottom-contact designs, field-effect mobilities extracted from the electrical characteristics of top-contact OTFTs often exceed those calculated for bottom-contact devices, even in the case of identical materials, device dimensions, and process conditions [19]. This is due to artifacts in the empirical device models used for OTFT parameter extraction, where mobility often exhibits contact resistance dependence.

From a processing perspective, the bottom-contact design is preferable over the top-contact structure. This is because the deposition of the organic semiconductor constitutes the last fabrication step in bottom-contact OTFTs (assuming no encapsulation), and is not limited or affected by other processing steps. Thus, there is greater flexibility in selecting a patterning/deposition method for the various device layers. If an inorganic dielectric is used, the source/drain contacts can be patterned by a photolithography technique to deliver higher resolution features. On the other hand, the top-contact design is susceptible to physical damage of the organic layer from subsequent processing of source/drain contacts, or from metal–semiconductor reactions (which may influence the contact resistance). With the top-contact architecture, contacts are often deposited through shadow masks, with substantial loss of resolution. Moreover, the metal–organic interaction during metal deposition may lead to deterioration of device performance. For the results

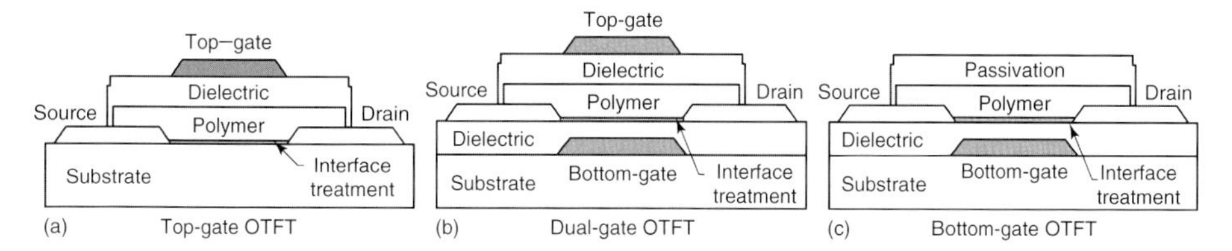

Figure 2.16 Cross-section of (a) top-gate, (b) dual-gate, and (c) bottom-gate OTFTs. Devices shown are in bottom-contact configuration.

reported in this book, fabrication processes are geared toward bottom-contact structures to ensure minimum process-induced disruption to the organic semiconductor layer.

2.3.2
Top-Gate, Bottom-Gate, and Dual-Gate OTFTs

Based on the arrangement of the gate electrode(s) relative to the semiconductor layer, OTFTs can be classified as bottom-gate, top-gate, or dual-gate, as illustrated in Figure 2.16. A feature of top-gate OTFTs is that the gate dielectric can provide encapsulation and protection of the organic semiconductor layer. Encapsulation is essential because most organic materials are chemically sensitive to environmental influence. However, the top-gate design poses some process integration challenges. First, the gate dielectric and gate electrode have to be deposited and structured on top of the organic semiconductor, and this process must preserve the organic material. Secondly, vertical interconnects and vias between the conductive layers have to be built through the organic semiconductor; this necessitates developing a compatible patterning/etching process for the organic layer. In our research, we demonstrated the first fully-encapsulated top-gate P3HT OTFT using silicon nitride (SiN_x) as the passivating gate dielectric, featuring a tailored etch process for patterning of the organic layer [49]; the results are presented in Section 3.3.

Comparatively, the fabrication of bottom-gate OTFTs is much simpler. Since the deposition of the organic semiconductor layer occurs after preparation of the gate and gate dielectric, there is a greater flexibility in the choice of materials and fabrication techniques. Most of the published research on OTFTs has adopted the bottom-gate design, and has demonstrated encouraging results. A relevant concern of this structure is the need for passivation of the organic layer. Passivation and encapsulation play an important role in prolonging the lifetime and stability of OTFTs, especially because most functional organic materials are very sensitive to environmental influences (including air, oxygen, moisture, water) and light, often leading to significant degradation or failure of devices [39–41].

In dual-gate OTFTs, the voltage bias on the bottom-gate has a distinct influence on the threshold voltage, subthreshold slope, on-current, and leakage current of the top-gate TFT. Similarly, there is a dependence of the bottom-gate TFT characteristics on the top-gate bias. This dual-gate arrangement offers the ability to control selected TFT parameters, making it attractive for circuit applications that demand high threshold voltage control and good reliability. The dual-gate OTFT also lends itself as a highly functional test structure for characterization of the density of states (DOSs) at the interfaces of the active organic and dielectric layers, to evaluate the interface integrity and provide insight into the underlying transport mechanisms [50]. Another promising application of dual-gate a OTFT is as an on-pixel circuit element for vertically integrated backplane electronics in active matrix displays and imagers. The use of a dual-gate OTFT can potentially improve the aperture ratio of the device and permit shielding of parasitic effects in the vertically integrated backplane electronics. (This concept is elaborated in Section 7.4.1.) Dual-gate OTFTs can be implemented by extending the fabrication procedure of top-gate OTFTs. Since processing must be done on the organic semiconductor layer, there is added processing complexity that requires careful consideration when designing a fabrication procedure for dual-gate OTFTs. In our research, we demonstrated one of the first dual-gate polythiophene OTFTs fabricated by photolithography methods [51]. The fabrication schemes that originated from this research to implement top-gate, bottom-gate, and dual-gate OTFTs are presented in Section 3.3.

2.4
OTFT Device Material Selection

As illustrated in the cross-sectional structure of OTFTs in Figure 2.1, the functional device layers are the semiconductor, the gate dielectric, and the electrodes. Since these layers are actively involved in the operation of OTFTs, the choice of materials and the compatibility between materials influence device performance. Figure 2.17 depicts the key interaction mechanisms between various device layers. There are two additional material layers that require consideration for OTFT fabrication: the substrate and the encapsulation. The substrate presents a platform for material deposition; in the case of organic electronics, there is an application-driven aspiration toward flexible substrates. Encapsulation provides protection/passivation for the functional device layers, which is particularly important for organic electronics due to the sensitivity of organic materials to environmental influences.

Table 2.3 summarizes the selection of device materials used for our OTFT research. The research focused on solution-processable p-type polymer semiconductors, inorganic gate dielectrics, and metallic electrodes. The various device layers are discussed in greater depth in this section. Well-established electronic materials were chosen to allow this research to focus on device integration and circuit development (instead of developing new materials). A solution-processable polymer semiconductor was selected specifically for this research because solution-based fabrication methods (e.g., inkjet printing, microcontact printing, and nanoimprinting) are projected to be the main workhorse for manufacturing organic electronics in the near future. Solution processing of OTFTs can eliminate the need for

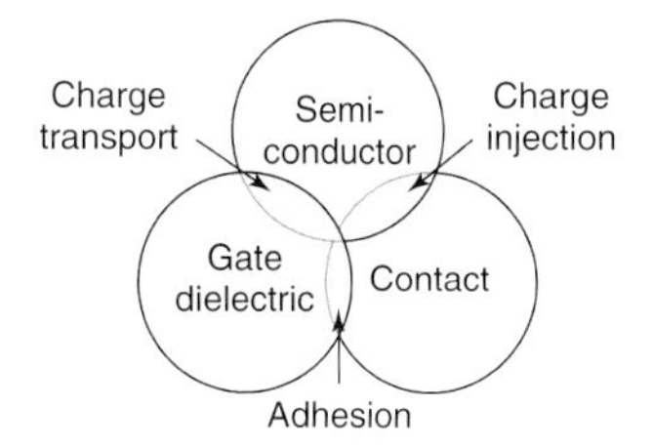

Figure 2.17 Interaction between the three key device layers in an OTFT.

Table 2.3 Device material consideration for the OTFT fabrication in this research.

OTFT device layers	Material choices
Semiconductor	Regioregular poly(3-hexylthiophene) (P3HT); *Aldrich* Poly(3,3‴-dialkyl-quaterthiophene) (PQT-12); *Xerox* Poly(2,5-bis(3-alkylthiophen-2-yl)thieno[3,2-*b*]thiophene) (PBTTT); *Merck*
Dielectric	Thermal silicon dioxide (SiO_2) PECVD silicon nitride (SiN_x) PECVD silicon oxide (SiO_x)
Electrodes	Source/drain contacts: Au, Cr Gate electrode: Mo, Al
Substrate	Silicon wafer Glass wafer Plastic substrate: Kapton, polyethylene naphthalate (PEN), poly(ethylene terephthalate) (PET)
Encapsulation	Parylene Photoresist PECVD silicon nitride (SiN_x)

expensive vacuum chambers and lengthy pump-down cycles, and can potentially reduce the cost and complexity of the fabrication processes [28].

2.4.1 Organic Semiconductor

The organic semiconductor is the core foundation of an OTFT. The semiconductor governs the charge transport in an OTFT. A high quality and high performance semiconductor is a pre-requisite for a high performance transistor. Motivated by the quest for low-cost manufacturing methods, this research focused on solution-processable organic polymer semiconductors. Initial experiments used RR-P3HT ($(C_{10}H_{18}S)_n$), from the Aldrich-Sigma Company, as the organic semiconductor layer. P3HT was chosen because it was commercially available and it had one of the highest reported mobilities among polymer semiconductor materials at that time [25, 26, 28]. However, our implementation of functional, high performance P3HT OTFTs was hindered by a number of challenges. First, many commercial organic semiconductors are very sensitive to air and moisture; thus, the lifetime and stability of the resulting OTFTs are a concern (especially as our processing and measurements were done in ambient, and nitrogen ambient for organic processing was unavailable). Secondly, the purity of the commercial-grade organic materials is questionable (despite claims of 99.9% purity). Impurities are manifested as defects in the organic semiconductor films, which can cause undesirable leakages and traps that obstruct proper device operation. Thirdly, the quality of commercial organic semiconductors tends to vary from batch to batch; this can have a negative impact on the uniformity of TFT characteristics and device yield. As a result, preliminary trials of P3HT OTFTs displayed high gate leakage currents, primarily due to impurities in the commercial-grade P3HT. Issues with impurities and instabilities of commercially available organic semiconductors are a well-known problem among researchers; however, these issues were not documented in the literature. Publications reporting high performance P3HT OTFTs typically employed ultra-high purity P3HT materials that were synthesized in-house or were subjected to additional purification steps (typically by thermal sublimation) to eliminate/reduce possible impurities and contaminants in the product; moreover, processing and testing were typically done in an inert (nitrogen) environment. Unfortunately, neither purification equipment nor nitrogen-purged chambers were accessible for this research; thus, limited device performance was measured for P3HT OTFTs.

A more effective and direct approach to attain OTFTs with improved performance and lifetime necessitates the development of stable polymer materials and an encapsulation strategy. The Xerox Research Centre of Canada has successfully developed a new family of PT materials, called poly(3,3‴-dialkylquarterthiophene) (PQT-12), with excellent stability and high mobility [13]. The core of this work was conducted using a PQT-12 polymer semiconductor [52]. Please refer to Figure 2.10a for the chemical structure of PQT-12.

More recently, Merck Chemicals introduced a new class of semiconducting liquid crystalline poly(2,5-bis(3-alkylthiophen-2-yl)thieno[3,2-*b*]thiophene) (PBTTT) polymers, featuring a fused ring structure (see Figure 2.10b). PBTTT OTFT demonstrated one of the highest mobilities and decent stability amongst solution-processed polymer semiconductors [31]. We recently demonstrated that high mobility can be achieved with PBTTT on a plasma enhanced chemical vapor deposited (PECVD) SiN_x gate dielectric with minimal surface treatment [53]; this is an important breakthrough as it signifies the potential to simplify OTFT fabrication process steps by clever selection of appropriate semiconductor–dielectric material pairs.

A number of factors can affect the quality and properties of the organic semiconductor layer, including solvent selection, deposition conditions, and surface properties. Different solvents produce films with varying degrees of molecular ordering, uniformity, and continuity [26]. A variety of solvents has been reported for P3HT, including chloroform, toluene, *p*-xylene, hexane, chlorobenzene, 1,1,2,2,-tetrachloroethane, and teterachloroethylene. Researchers observed variations in the mobility of P3HT OTFT by two orders of magnitude, depending on the choice of solvent [28]. OTFT device performance is also influenced by the deposition rate (e.g., spin coating speed of the polymer semiconductor layer) and temperature (substrate temperature), which affect the morphology of the semiconductor. Therefore, the choice of solvent, the concentration of the polymer solution, and the deposition parameters must be optimized to yield good quality films. In the case of the PQT polymer semiconductor, experiments have shown that a 1,2-dichlorobenzene solvent gave optimal device performance. Additional enhancements can be achieved by surface or interface treatment [14, 26, 27, 29] and post-deposition treatments such as annealing [54]; these processes can improve molecular ordering and contribute to improved device performance. Interface treatment techniques for OTFTs are examined in Chapters 5 and 6.

The main incentive for using polymer semiconductors is their compatibility with simple solution-based processing techniques. Spin-on methods are the simplest approach for depositing polymers uniformly across the entire substrate. The solvent is then dried off by evaporation after deposition. This technique can generate uniform films ($\sim$100 nm thick), as required in most organic light-emitting diode display (OLED) and OTFT devices. Although solution-based deposition methods can offer numerous advantages such as high-speed deposition over large substrate areas, concerns related to solvent interaction must be considered. The solvents used for one polymer layer may interact with previously applied polymer layers, thereby limiting the complexity and performance of the structures that can be achieved. Thus, the polymer solutions must be cautiously prepared and optimized to avoid chemical interactions in multilayer organic devices. Another critical drawback of simple spin-on deposition methods is the inability to locally pattern the electronic device. Patterning of the organic layer is needed for OLED color displays, to define the red (R), green (G), and blue (B) polymer OLED sub-pixels. The organic semiconductor layer of the OTFTs requires proper patterning for device isolation and circuit realization; these issues are discussed in Chapter 3.

2.4.2 Gate Dielectric

The gate dielectric is one of the most critical materials for organic transistor performance. It has an important function in establishing field-effect operation in TFTs. A general specification for the gate dielectric of TFTs is summarized as follows: the dielectric material has to withstand electric fields of about 2 MV cm^{-1} without breakdown, must have good insulating properties and low rate of charge trapping at lower electric fields, and it should form a high quality interface with the semiconductor layer [55]. In addition, the dielectric films should possess low trapping density at the surface, low surface roughness, low impurity concentration, and compatibility with organic semiconductors. Dielectric materials can be classified as inorganic or organic. Examples of inorganic gate dielectrics include silicon dioxide (SiO_2), silicon nitride (SiN_x), titanium oxide (TiO_2, $k = 41$), and aluminum oxide (Al_2O_3, $k = 8.4$) [56]. The OTFT research documented in this book focuses primarily on inorganic gate dielectrics; more specifically, PECVD silicon nitride (SiN_x) and silicon oxide (SiO_x) thin films. The key benefits of using inorganic dielectrics include their maturity, good dielectric strength and integrity, and the availability of high-quality films. The mature manufacturing processes and deposition technologies are capable of depositing a pinhole-free gate dielectric layer with a thickness of a few hundred angstroms [24]. The well-known dielectric characteristics of these materials significantly reduce process variability. PECVD gate dielectrics are investigated in depth in Chapter 4.

A high-k dielectric (e.g., Al_2O_3, TiO_2) is another candidate for the gate dielectric in OTFTs because its high-k value can increase the intrinsic gate capacitance ($C_i = k\varepsilon_0 A/d$) of a transistor, which in turn increases the current output (since $I_D \propto \mu C_i$). Thus, high-k dielectrics can partially compensate for the relatively low mobility (μ) of the organic semiconductors, and can enable device operation at lower drive voltages. However, a key shortcoming of high-k materials is their higher degree of disorder and surface roughness, which can create traps, reduce mobility, and lead to localization of carriers at the dielectric/semiconductor interface [56].

Although OTFTs with high quality inorganic gate dielectrics on rigid substrates have demonstrated noticeably improved performance over the past two decades, the insoluble and rigid nature of most inorganic dielectric materials has prompted researchers to investigate solution-processable organic dielectrics in order to fulfill the inherent merits of OTFTs for low-cost processing and mechanical flexibility. Examples of organic gate dielectrics for OTFTs include polyvinylphenol (PVP), polyvinylalcohol (PVA), polymethyl-methacrylate (PMMA), and polyimide [57–61]. Comparatively, organic dielectric materials are less matured; on-going research is needed to enhance material quality, dielectric strength, electrical integrity, and other properties. However, organic dielectrics with good solubility are ideal for low-cost solution-based processing methods (e.g., spin-coating, dip-coating, printing). Moreover, the low processing temperature of organic dielectrics is compatible with plastics substrates, making it suitable for flexible electronics applications.

Among the various polymer gate insulators, PVA appears promising because it has many merits such as a surface alignment effect by a rubbing process, high dielectric constant, and compatibility with solution-based processes [57]. However, PVA tends to absorb moisture easily due to the presence of a hydroxy group (−OH) in the material. A possible remedy is to use cross-linked PVA as the gate dielectric, which exhibits higher endurance against moisture than un-cross-linked PVA; moreover, the cross-linking process provides an opportunity for pattern definition by UV exposure. Insertion of a PMMA buffer layer on a cross-linked PVA gate insulator can provide an extra barrier against moisture penetration [57]. In addition, PMMA buffer layers on cross-linked PVA gate insulators have been shown to improve the ordering of the organic semiconductor layer owing to the excellent hydrophobicity of PMMA.

A more recent advancement is the development of very thin (2.3–5.5 nm) molecule-derived self-assembled organic dielectric multilayers that enable operation of OTFTs at very low voltages (sub-1 V) and without serious leakage currents [62]. Thin, nanostructurally ordered, pinhole-free, high-capacitance, and low leakage organic dielectrics can be formed by solution phase deposition methods. This chemically-tailored siloxane-crosslinked polymeric nanodielectric represents a new and promising approach to high-performance, low-power dissipation OTFTs, while providing excellent dielectric properties and efficient low-cost solution-phase processing characteristics.

Although there have been successful demonstrations of OTFTs with organic gate dielectrics, most of these devices exhibit a lower on/off ratio and higher leakage current than OTFTs with inorganic gate dielectrics. The surface roughness of the solution-processed organic dielectric is a key concern here, as the roughened dielectric/semiconductor interface can generate a higher density of interface traps, cause interface scattering, and degrade the mobility of bottom-gate OTFTs [61]. It is also important to ensure the solution-processable organic dielectrics do not chemically disrupt the organic semiconductor. This requires careful consideration for solvent selection and deposition conditions.

In addition to the bulk properties of the dielectric material, the interface properties between the dielectric and organic semiconductor have a profound effect on device performance. The molecular ordering of the organic semiconductor film, and hence the field-effect mobility of the device, is usually very sensitive to dielectric surface properties (e.g., surface roughness and surface energy). Interface modification methods, such as alkylsilane self-assembled monolayers (SAMs) treatment of the dielectric surface prior to semiconductor deposition, are often used to induce ordering of the polymer film (i.e., $\pi-\pi$ stacking between conjugated polymer molecules to form the pathway for charge carrier hopping from molecule to molecule). These interface treatments can improve device mobility by as much as an order of magnitude. Optimization of the gate dielectric and the interface properties form a critical part of this research; the results are presented in Chapters 4 and 5.

2.4.3 Electrodes/Contacts

The gate electrode and source/drain electrodes (also known as *contacts*) for OTFTs can be implemented using metals or organic conductors. Typically, the material specification for the gate electrode is relatively relaxed. Considerations for selecting gate material include high conductivity, compatibility with device layers, and low surface roughness in the case of bottom-gate structures. For example, molybdenum (Mo) is selected over aluminum (Al) for bottom-gate a-Si TFTs owing to molybdenum's smoother surface, which critically affects the quality of subsequent (or overlying) device layers.

In contrast, the choice of material for source/drain contacts is more stringent; conscientious energetic matching between contact and semiconductor is important

Table 2.4 Work function (Φ_M) of selected metals [63].

	Ag	Al	Au	Cu	Cr	Mo	Ni	Pd	Pt	Ti	W	ITO
Φ_M (eV)	4.26	4.28	5.1	4.65	4.5	4.6	5.15	5.12	5.65	4.33	4.55	4.8–5.2

to establish efficient charge injection in OTFTs. Ohmic contact is desirable, which requires the work function of the contact metal to match with the HOMO of the organic semiconductor in the case of a p-type OTFT, and with the LUMO in the case of an n-type OTFT. This implies a high work function metal for p-type OTFT and a low work function metal for n-type OTFT. The work function of various metals is listed Table 2.4. Gold (Au) and platinum (Pt) are favorable choices for their high work function, which has good electrical compatibility with the p-type organic semiconductors. Moreover, the environmentally stable nature of Au and Pt (in contrast to Ca and Al) ensures low contact resistance at the contact/semiconductor interfaces by shunning the formation of surface oxide and eliminating chemical interaction with most organic materials.

The interface between the source/drain contact and organic semiconductor requires low contact resistance, which is a function of both parasitic resistance and the energy barrier at the contact/semiconductor interface [24]. Low energy barriers necessitate matching the electrode work function (Φ_M) with the semiconductor ionization potential (IP_S). Figure 2.18 shows the energy scheme at the contact/semiconductor interface. Material should be selected so that the barrier of hole injection ($\varphi_B = IP_S - \Phi_M$) is minimized (for p-type OTFT), to allow for high charge injection efficiency at the contacts. In addition to energetic matching at the interface, the contact material must possess high conductivity for optimal OTFT performance [30]. Most metal/organic semiconductor junctions form a Schottky contact, indicating the presence of an injection barrier. Contact resistance is discussed in Chapter 6, along with techniques to improve the contact/semiconductor interface.

Metals are typically deposited by evaporation or sputtering; and thus require vacuum processing. To fulfill the inherent virtues of OTFTs such as low-cost processing and compatibility of flexible substrates, solution-processable organic conductors are favorable. Conductive polymers such as poly(3,4-ethylene dioxythiophene) doped with polystyrene sulfonic acid (PEDOT:PSS) and polyaniline (PANI) have been used as electrodes in OTFTs. Polymer conductors can be deposited by spin-coating, inkjet printing, screen-printing, or soft-lithography methods (e.g., microcontact printing) [30, 58, 59, 64]. More recently, solution-processable conductors based on Au/Ag/Cu nanoparticles, carbon nanotubes, and nanowires are emerging as promising candidates for electrode material [65–67].

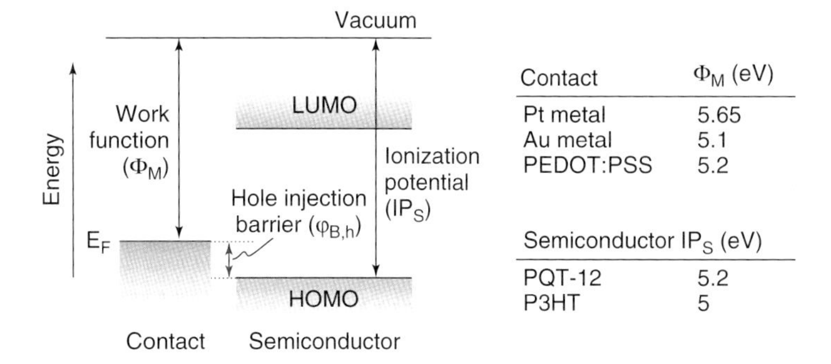

Contact	Φ_M (eV)
Pt metal	5.65
Au metal	5.1
PEDOT:PSS	5.2

Semiconductor	IP_S (eV)
PQT-12	5.2
P3HT	5

Figure 2.18 Energy band diagram at the contact/semiconductor interface, illustrating the concept of work function (Φ_M), ionization potential (IP), and injection barrier (φ_B) at the contact. Representative values for p-type polymer semiconductor and contact material are shown.

2.4.4 Substrate

Candidates for the substrate material for OTFTs include rigid substrates such as silicon wafers and glass substrates, and flexible substrates such as polyimide (or Kapton), poly(ethylene terephthalate) (PET), poly(ethylene naphthalate) (PEN), aluminum or stainless steel foil, paper, and fabric. The choice of substrate affects the OTFT performance. Rigid substrates are advantageous for high planarity, optimal surface smoothness, ease of processing, and capability of high temperature annealing of organic semiconductors [24]. Planarity and surface smoothness are particularly critical to OTFT performance because they affect the quality of the overlying thin film device layers. Large substrate surface roughness is expected to cause a greater density of trap states in the gate dielectric and at the dielectric/semiconductor interface for bottom-gate OTFTs, which can negatively impact the OTFT characteristics. For instance, Klauk *et al.* observed that pentacene OTFTs with a PVP gate dielectric on a PEN substrate have a slightly higher subthreshold swing and threshold voltage than OTFTs on a glass substrate. These differences are attributed to the larger surface roughness of the PEN substrate compared to glass (15 Å for PEN versus 8 Å for glass) [60].

Since organic electronics are envisioned to be flexible and lightweight, the ability to fabricate organic electronics on flexible substrates (e.g., plastic or foil) is of considerable interest for hand-held devices and military applications. The low material cost, opportunity for using high-speed roll-to-roll printing processes for reduced manufacturing costs, and the flexible form factor are all marketable features of organic electronics. Criteria for selecting flexible substrates used in OTFTs include thermal, chemical, and mechanical stability, surface roughness, and the cost of the films. A number of flexible substrate candidates are listed

Table 2.5 Properties of flexible substrate material candidates [70].

Flexible Substrate Material	Maximum Process Temperature (°C)	Characteristics
Steel foil	900	☑ Moderate chemical resistance, moderate coefficient of thermal expansion (CTE) ☒ Opaque, poor surface finish
Polyimide (Kapton®)	275	☑ Good chemical resistance, highest processing temperature for plastic substrates ☒ Orange color, higher cost, high moisture absorption
Polyetheretherketone (PEEK)	250	☑ Good chemical resistance, low moisture absorption ☒ Amber color, expensive
Polyethersulfone (PES)	230	☑ Clear, good dimensional stability, inexpensive, moderate moisture absorption ☒ Poor solvent resistance, expensive
Polyetherimide (PEI)	200	☑ Strong ☒ Hazy-colored, expensive
Polyethylene-napthalate (PEN)	150	☑ Clear, good chemical resistance, inexpensive, moderate CTE, moderate moisture absorption
Polyester (PET)	120	☒ Clear, good chemical resistance, inexpensive, moderate CTE, moderate moisture absorption

in Table 2.5, along with their maximum process temperatures and key material characteristics.

However, existing plastic substrates exhibit a number of shortcomings related to permeability, dimensional distortion, and temperature tolerance. Most plastics are permeable to water and oxygen, which may lead to rapid deterioration in organic device performance. Thus, it is vital to perform proper passivation of the plastic substrate prior to device fabrication. Also, inexpensive plastic materials typically cannot withstand high processing temperatures. Researchers are investigating various processing techniques to maintain the substrate temperature below 100–150 °C. New plastic substrate materials that can withstand higher temperatures (e.g., up to 300–350 °C) are also under development; however, these temperature-tolerant plastic substrates will likely involve higher costs [68]. Dimensional distortion (e.g., shrinking) of plastic substrates due to thermal effects can affect the stability of organic electronics.

To improve the dimensional stability of flexible plastic substrates for device applications, plastic substrates can be subjected to thermal treatment prior to device fabrication. For example, a PEN substrate can be preshrunk in a vacuum oven at 200 °C and at a rate of 3 °C min^{-1}, followed by cooling to room temperature at about 1 °C min^{-1}. Klauk *et al.* [60] observed that OTFTs on an untreated PEN exhibited dimensional distortions greater than 0.5%, but OTFTs on a pre-treated PEN substrate had dimensional distortion of less than 0.02%. These observations suggested that thermal treatment of the plastic substrate prior to device processing can help to secure sufficient dimensional stability in the resulting organic device.

Using a fabric or textile as a device substrate for organic electronics has recently gained increased visibility. Wearable electronics, such as clothing with sensors that monitor critical body health parameters, are beginning to be promoted by companies such as Infineon and Philips [69].

2.4.5
Encapsulation Strategies

Because many organic semiconductors and conductive polymers are easily influenced by water, oxygen, and other environmental elements present in ambient conditions, the lifetime, and reliability of organic electronics are a critical concern. This demands an optimum strategy to protect organic electronics from environmental damage. Encapsulation presents a viable solution to protect organic electronic devices from the environment which helps to maintain device performance and prolong device lifetime. The proper encapsulation method for OTFTs should meet these basic criteria [71]: low-temperature deposition ($<$80 °C) to enable direct deposition on temperature-sensitive organic devices, intrinsically defect-free films to provide a reliable barrier against moisture and oxygen, compatibility with large-area substrate manufacturing systems (e.g., for displays), minimal effect on device performance, low cost, and good durability to provide long-term protection from moisture, oxygen, alkali ion, or other contaminants.

Various encapsulation strategies have been investigated for organic devices, including metal/glass lid encapsulation for OLEDs, photoresist-based encapsulation, PVA-based encapsulation, and multilayer coating. For example, the current OLED technology uses monolithic encapsulation with metal or glass lids to protect the OLED devices. However, this approach is regarded only as a short-term expedient. Improved materials are needed as sealants and as dessicants to minimize/eliminate water and oxygen penetration into the device [68]. Photoresist materials have been investigated as an encapsulation material for polymer devices. Lu *et al.* [72] encapsulated their P3HT OTFTs with a thin layer (l μm) of the photoresist (AZ1312 SFD) using a spin-casting technique. The mobility of the encapsulated OTFT remained quite stable over a 60-day period compared to an unencapsulated device, implying enhanced electrical stability of the device in the long term. However, a decrease in mobility by ~60% was observed after the encapsulation step, suggesting undesirable interaction between the photoresist and the organic semiconductor film [72]. Therefore, a more reliable and robust solution is required.

A similar resist-based encapsulation was demonstrated by Qui *et al.* [73], in which a 200 μm thick UV curable resin was coated by a doctor blade method on the top of a pentacene OTFT for encapsulation. For an unencapsulated OTFT, the field-effect mobility decreased by 30% and the on/off current ratio decreased to 20% of the original value after storing the OTFT in the atmosphere for 500 h [73]. In contrast, encapsulated OTFTs showed reasonable electrical stability. Degradation of the organic material was not reported after the coating process, implying that this UV curable resin may offer a feasible encapsulation solution for OTFTs.

Encapsulation of OTFTs using PVA has been studied [74]. Although PVA was able to passivate the OTFT from environmental effects, some groups have reported degradation in OTFT electrical characteristics after the PVA coating (formed by a wet process). The decrease in mobility is proposed to be due to the migration of PVA solution into the organic semiconductor layer, and/or due to the sheer stress by the viscosity of PVA when it was coated and dried.

To eliminate interaction between the solution-based chemicals, encapsulation based on multilayer coatings has been introduced. Lee *et al.* reported an encapsulation method using an adhesive multilayer formed by a conventional lamination process. The adhesive multilayer consisted of polyacrylate-based adhesive (15 μm thick) and Al (185 μm thick). For OTFT encapsulation, the Al film is adhered onto the semiconductor layer in a dry nitrogen atmosphere using a proper adhesive [74]. The passivated OTFTs showed no degradation after the encapsulation procedure and demonstrated reasonable stability in air over time. This adhesive multilayer coating technique provides an excellent barrier against moisture penetration with a very low water vapor transmission rate, and is suitable for use as a flexible and light-weight encapsulation for flexible OTFT devices [74]. Multilayer coatings based on alternating layers of inorganic and polymer materials are also under development [68]. These coatings are especially useful to compensate for the high porosity of plastic films, and are used as barrier layers to withstand moisture and oxygen penetration.

Inorganic gas barriers such as tin oxide (SnO_2), aluminum oxide (AlO_x), and silicon oxide (SiO_x) have been considered as encapsulation for OTFTs. Kim *et al.* reported the long term stability of pentacene OTFTs encapsulated with transparent SnO_2 thin film. Although the field effect mobility degraded from 0.62 to 0.5 $cm^2\,V^{-1}\,s^{-1}$ with encapsulation, enhanced long-term device stability and lifetime was observed [75]. Caution should be exercised when using these inorganic barrier layers due to the potential for process-induced damage to the organic semiconductor layer; the damage may arise from exposure to energetic ions, X-rays, electron beams, and high temperatures during deposition of the inorganic barrier layers.

Inorganic material such as SiN_x is commonly used as passivation layers in silicon ICs. We have demonstrated P3HT OTFTs with SiN_x encapsulation [49, 50]. However, a reduced mobility is observed after depositing SiN_x on top of the P3HT layer by PECVD. The reduction is believed to stem from plasma-induced material degradation, and can be circumvented by the incorporation of a protective layer (e.g., an organic dielectric such as parylene) to eliminate direct plasma exposure of the organic semiconductor layer during the PECVD process. Parylene is ideal for this purpose as it is routinely used as a conformal coating for many commercial and industrial applications [76]. Successful demonstrations of OTFT devices and circuits with parylene passivation are reported in Chapters 3 and 7, respectively.

2.5 Summary

This chapter presented a background overview of OTFT technology. The fundamental properties of organic semiconductors and the basic operation of OTFT were reviewed. A procedure for the extraction of OTFT device parameters was explained to provide a basis for evaluation of OTFT characteristics. The device architecture and material selection considered for OTFT fabrication in this research were also specified. The discussion presented in this chapter lays a foundation for subsequent discussions and investigations reported in this book.

References

1. Morrison, R.T., Boyd, R.N., and Boyd, R.K. (1992) Organic Chemistry, 6th edition (Benjamin Cummings).
2. Cantatore, E., Gelinck, G.H., and de Leeuw, D.M. (2002) Polymer electronics: from discrete transistors to integrated circuits and active matrix displays, *IEEE Proceedings of the Bipolar/BiCMOS Circuits and Technology Meeting*, p. 167.
3. Shaw, J.M. and Seidler, P.F. (2001) Organic electronics: introduction. *IBM J. Res. Dev.*, **45** (1), 3.
4. Norden, B. and Krutmeijer, E. (2000) The Noble Prize in Chemistry, 2000: Conducting polymers (Advanced Information). Noble Foundation, *http://nobelprize.org/nobel_prizes/chemistry/laureates/2000/adv.html* (accessed 15 Jan 2011).
5. Ito, T., Shirakawa, H., and Ikeda, S. (1974) Simultaneous polymerization and formation of polyacetylene film on the surface of concentrated soluble Ziegler-type catalyst solution. *J. Polym. Sci. Chem. Ed.*, **12** (1), 11.
6. Chiang, C.K., Fincher, C.R., Park, Y.W., Heeger, A.J., Shirakawa, H., Louis, E.J., Gau, S.C., and MacDiarmid, A.G. (1977) Electrical conductivity in doped polyacetylene. *Phys. Rev. Lett.*, **39**, 1098.
7. Higgins, S.J., Eccelston, W., Sedgi, N., and Raja, M. (2003) Plastic Electronics, Education in Chemistry. *http://www.rsc.org/lap/educatio/eic/2003/higgins_may03.htm* (accessed 2008).
8. Cameron, C.G. (2000) Enhanced rates of electron transport in conjugated-redox polymer hybrids. PhD. thesis, Department of Chemistry, Memorial University of Newfoundland, Canada. *http://mmrc.caltech.edu/colin/science/past/PhD/html-thesis/node8.html* (accessed 2008).
9. Dimitrakopoulos, C.D. and Malenfant, P.R.L. (2002) Organic field-effect transistors for large area electronics. *Adv. Mater.*, **14** (2), 99.
10. Vissenberg, M.C.J.M. and Matters, M. (1998) Theory of the field-effect mobility in amorphous organic transistors. *Phys. Rev. B*, **57**, 12964.
11. Horowtiz, G. (2006) Organic transistors, in *Organic Electronics: Materials, Manufacturing and Applications* Chapter 1 (ed. H. Klauk), Wiley-VCH Verlag GmbH, Weinheim, 3–32.
12. Ong, B.S., Wu, Y., Liu, P., and Gardner, S. (2005) Structurally ordered polythiophene nanoparticles for high-performance OTFTs. *Adv. Mater.*, **17**, 1141.

13. Ong, B.S., Wu, Y., Liu, P., and Gardner, S. (2004) High-performance semiconducting polythiophenes for organic thin-film transistors. *J. Am. Chem. Soc.*, **126** (11), 3378.
14. Sirringhaus, H., Brown, P.J., Friend, R.H., Nielsen, M.M., Bechgaard, K., Langeveld-Voss, B.M.W., Spiering, A.J.H., Janssen, R.A.J., Meijer, E.W., Herwig, P.T., and de Leeuw, D.M. (1999) Two-dimensional charge transport in self-organized, high-mobility conjugated polymers. *Nature*, **401**, 685.
15. Park, Y.D., Lim, J.A., Lee, H.S., and Cho, K. (2007) Interface engineering in organic transistors. *Mater. Today*, **10**, 46–54.
16. Ficker, J., Ullmann, A., Fix, W., Rost, H., and Clemens, W. (2003) Stability of polythiophene-based transistors and circuits. *J. Appl. Phys.*, **94** (4), 2638.
17. Schon, J.H., Berg, S., Kloc, C., and Batlogg, B. (2000) Ambipolar pentacene field-effect transistors and inverters. *Science*, **287**, 1022.
18. Kelley, T. (2006) High performance pentacene transistors, in *Organic Electronics: Materials, Manufacturing and Applications* Chapter 2 (ed. H. Klauk), Wiley-VCH Verlag GmbH, Weinheim, 35–57.
19. Jackson, T.N., Lin, Y.-Y., Gundlach, D.J., and Klauk, H. (1998) Organic thin-film transistors for organic light-emitting flat-panel display backplanes. *IEEE J. Sel. Top. Quantum Electron.*, **4** (1), 100.
20. Bao, Z., Dadabalapur, A., Katz, H.E., Raju, R.V., and Rogers, J.A. (2001) Organic semiconductors for plastic electronics. *IEEE-TAB New Technology Directions Committee (NTDC), 2001 IEEE Workshop on New and Emerging Technologies.*
21. Afzali, A., Dimitrakopoulos, C.D., and Breen, T.L. (2002) High-performance, solution-processed organic thin film transistors from a novel pentacene precursor. *J. Am. Chem. Soc.*, **124**, 8812.
22. Kawasaki, M., Ando, M., Imazeki, S., Sekiguchi, Y., Hirota, S., Sasaki, H., Uemura, S., and Kamata, T. (2005) Printable organic TFT technologies for FPD applications. *Proc. SPIE*, **5940**, 59400Q1–5940010.
23. Cantatore, E. (2000) Organic materials: a new chance for electronics? *Proceedings of the SAFE/IEEE Workshop*, November 27.
24. Chason, M., Brzis, P.W., Zhang, J., Kalyanasundaram, K., and Gamota, D.R. (2006) Printed organic semiconducting devices. *Proc. IEEE.*, **93** (7), 1348.
25. Horowitz, G. (1998) Organic field-effect transistors. *Adv. Mater.*, **10** (5), 365.
26. Bao, Z., Dodabalapur, A., and Lovinger, A.J. (1996) Soluble and processable regioregular poly(3-hexylthiophene) for thin film field-effect transistor applications with high mobility. *Appl. Phys. Lett.*, **69**, 4108.
27. Bao, Z., Feng, Y., Dodabalapur, A., Raju, V.R., and Lovinger, A.J. (1997) High-performance plastic transistors fabricated by printing techniques. *Chem. Mater.*, **9**, 1299.
28. Dimitrakopoulos, C.D. and Mascaro, D.J. (2001) Organic thin-film transistors: a review of recent advances. *IBM J. Res. Dev.*, **45** (1), 111.
29. Sirringhaus, H., Tessler, N., and Friend, R.H. (1998) Integrated optoelectronic devices based on conjugated polymers. *Science*, **280**, 1741.
30. Sirringhaus, H., Kawase, T., Friend, R.H., Shimoda, T., Inbasekaran, M., Wu, W., and Woo, E.P. (2000) High-resolution inkjet printing of all-polymer transistor circuits. *Science*, **290**, 2123.
31. McCulloch, I., Heeney, M., Bailey, C., Genevicius, K., MacDonald, I., Shkunov, M., Sparrowe, D., Tierney, S., Wagner, R., Zhang, W., Chabinyc, M.L., Kline, R.J., McGehee, D., and Toney, M.F. (2006) Liquid-crystalline semiconducting polymers with high charge-carrier mobility. *Nat. Mater.*, **5**, 328–333.
32. Bao, Z., Lovinger, A.J., and Brown, J. (1998) New air-stable n-channel organic thin film transistors. *J. Am. Chem. Soc.*, **120**, 207.
33. Katz, H.E., Johnson, J.J., Lovinger, A.J., and Li, W. (2000) Naphthalenetetracarboxylic diimide-based n-channel transistor semiconductors: structural

variation and thiol-enhanced gold contacts. *J. Am. Chem. Soc.*, **122**, 7787.
34. Babel, A. and Jenekhe, S.A. (2002) n-channel field-effect transistors from blends of conjugated polymers. *J. Phys. Chem. B*, **106** (24), 6129.
35. Anthopoulos, T.D., Singh, B., Marjanovic, N., Sariciftci, N.S., Ramil, A.M., Sitter, H., Colle, M., and de Leeuw, D.M. (2006) High performance n-channel organic field-effect transistors and ring oscillators based on C_{60} fullerene films. *Appl. Phys. Lett.*, **89**, 213504.
36. Tiwari, S.P., Namdas, E.B., Ramgopal Rao, V., Fichou, D., and Mhaisalkar, S.G. (2007) Solution-processed n-type organic field-effect transistors with high on / off current ratios based on fullerene derivatives. *IEEE Electron Dev. Lett.*, **28**, 880.
37. Anthopoulos, T.D., Setayesh, S., Smits, E., Colle, M., Cantatore, E., de Boer, B., Blom, P.W.M., and de Leeuw, D.M. (2006) Air-stable complementary-like circuits based on organic ambipolar transistors. *Adv. Mater.*, **18**, 1900.
38. Meijer, E.J., de Leeuw, D.M., Setayesh, S., van Veenendaal, E., Huisman, B.-H., Blom, P.W.M., Hummelen, J.C., Scherf, U., and Klapwijk, T.M. (2003) Solution-processed ambipolar organic field-effect transistors and inverters. *Nat. Mater.*, **2**, 678.
39. Han, J.I., Kim, Y.H., Park, S.K., Moon, D.G., and Kim, W.K. (2004) Stability of organic thin film transistors. *Mater. Res. Soc. Symp. Proc.*, **814**, I4.3.1.
40. Chabinyc, M.L., Endicott, F., Vogt, B.D., DeLongchamp, D.M., Lin, E.K., Wu, Y., Liu, P., and Ong, B.S. (2006) Effects of humidity on unencapsulated poly(thiophene) thin-film transistors. *Appl. Phys. Lett.*, **88**, 113514.
41. Chabinyc, M.L., Street, R.A., and Northrup, J.E. (2007) Effects of molecular oxygen and ozone on polythiophenes-based thin-film transistors. *Appl. Phys. Lett.*, **90**, 123508.
42. Allport, P.P., Booth, P.S.L., Casse, G., Eccleston, W., Higgins, S., Marsland, J., and Sdghi, N. (2003) Ionizing Radiation Detection Using Semiconducting Conjugated Polymers. *http://hepwww.rl.ac.uk/pprp/Experimental_reports/ExpRep_2002/polymer.pdf* (accessed 2008).
43. Abdou, M.S.A., Orfino, F.P., Son, Y., and Holdcroft, S. (1997) Interaction of oxygen with conjugated polymers: charge transfer complex formation with poly(3-alkylthiophenes). *J. Am. Chem. Soc.*, **119**, 4518.
44. Servati, P., Striakhilev, D., and Nathan, A. (2003) Above-threshold parameter extraction and modeling for amorphous silicon thin-film transistors. *IEEE Trans. Electron Devices*, **50**, 2227.
45. Servati, P. (2004) Amorphous silicon TFTs for mechanically flexible electronics, PhD thesis. University of Waterloo.
46. Hamadani, B.H., Gundlach, D.J., McCulloch, I., and Heeney, M. (2007) Undoped polythiophene field-effect transistors with mobility of 1 $cm^2 V^{-1} s^{-1}$. *Appl. Phys. Lett.*, **91**, 243512.
47. Nathan, A. (2001) *Large Area Thin Film Electronics*, University of Waterloo, Waterloo, ON.
48. Halik, M., Klauk, H., Zschieschang, U., Schmid, G., Radlik, W., Ponomarenko, S., Kirchmeyer, S., and Weber, W. (2003) High-mobility organic thin-film transistors based on α,α-didecyloligothiophenes. *J. Appl. Phys.*, **93**, 2977.
49. Li, F.M., Koul, S., Vygranenko, Y., Sazonov, A., Servati, P., and Nathan, A. (2005) Fabrication of RR-P3HT-based TFTs using low-temperature PECVD silicon nitride. *Mater. Res. Soc. Symp. Proc.*, **871E**, I9.3.1.
50. Servati, P., Karim, K.S., and Nathan, A. (2003) Static characteristics of a-Si:H dual-gate TFTs. *IEEE Trans. Electron Devices*, **50**, 926.
51. Li, F.M., Koul, S., Vygranenko, Y., Servati, P., and Nathan, A. (2005) Dual-gate SiO_2/P3HT/SiN_x OTFT. *Mater. Res. Soc. Symp. Proc.*, **871E**, I9.4.1.
52. Li, F.M., Wu, Y., Ong, B.S., and Nathan, A. (2007) Organic thin-film transistor integration using silicon nitride gate dielectric. *Appl. Phys. Lett.*, **90**, 133514.

53. Li, F.M., Dhagat, P., Jabbour, G.E., Haverinen, H.M., McCulloch, I., Heeney, M., and Nathan, A. (2008) Polymer TFT without surface pre-treatment on silicon nitride gate dielectric. *Appl. Phys. Lett.* **93** (7), 073305.
54. Joung, M.J., Kim, C.A., Kang, S.Y., Baek, K.H., Kim, G.H., Ahn, S.D., You, I.K., Ahn, J.H., and Suh, K.S. (2005) The application of soluble and regioregular poly(3-hexylthiophene) for organic thin-film transistors. *Synth. Met.*, **149**, 73.
55. Sazonov, A., Striakhilev, D., Nathan, A., and Bogomolova, L.D. (2002) Dielectric performance of low temperature silicon nitride films in a-Si:H TFTs. *J. Non Cryst. Solids*, **299–302**, 1360.
56. Wang, G., Moses, D., Heeger, A.J., Zhang, H.M., Narasimhan, M., and Demaray, R.E. (2004) Poly(3-hexylthiophene) field-effect transistors with high dielectric constant gate insulator. *J. Appl. Phys.*, **59**, 316.
57. Jin, S.H., Yu, J.S., Kim, J.W., Lee, C.A., Park, B.G., Lee, J.D., and Lee, J.H. (2003) PMMA buffer-layer effects on electrical performance of pentacene OTFTs with a cross-linked PVA gate insulator on a flexible substrate. *Society of Information Display (SID) Sympoisum 2003 DIGEST*, p. 1088.
58. Garnier, F., Hajlaoui, R., Yassar, A., and Srivastava, P. (1994) All-polymer field-effect transistors realized by printing techniques. *Science*, **265**, 1684.
59. Halik, M., Klauk, H., Zschieschang, U., Schmid, G., Radlik, W., and Weber, W. (2002) Polymer gate dielectrics and conducting-polymer contacts for high-performance organic thin-film transistors. *Adv. Mater.*, **14**, 1717.
60. Klauk, H., Halik, M., Zschieschang, U., Eder, F., Schmid, G., and Dehm, C. (2003) Pentacene organic transistors and ring oscillators on glass and on flexible polymeric substrates. *Appl. Phys. Lett.*, **82**, 4175.
61. Kang, G.W., Park, K.M., Song, J.H., Lee, C.H., and Hwang, D.H. (2005) The electrical characteristics of pentacene-based organic field-effect transistors with polymer gate insulators. *Curr. Appl. Phys.*, **5**, 297.
62. Facchetti, A., Yoon, M.H., and Marks, T.J. (2005) Gate dielectrics for organic field-effect transistors: new opportunities for organic electronics. *Adv. Mater.*, **17**, 1705.
63. Barbalace, K. Periodic Table of Elements. EnvironmentalChemistry.com 1995–2010. Last accessed on-line: 2010. *http://environmentalchemistry.com/yogi/periodic/*.
64. Kawase, T., Sirringhaus, H., Friend, R.H. *et al.* (2003) Inkjet printing of polymer thin film transistors. *Thin Solid Films*, **438**, 279.
65. Wu, Y., Li, Y., Ong, B.S., Liu, P., Gardner, S., and Chiang, B. (2005) High-performance organic thin-film transistors with solution-printed gold contacts. *Adv. Mater.*, **17**, 184.
66. Gamerith, S., Klug, A., Scheiber, H., Scherf, U., Moderegger, E., and List, E.J.W. (2007) Direct ink-jet printing of Ag-Cu nanoparticle and Ag-precursor based electrodes for OFET applications. *Adv. Funct. Mater.*, **17**, 3111.
67. Zhang, Y.Y., Shi, Y., Chen, F., Mhaisalkar, S.G., Li, L.J., Ong, B.S., and Wu, Y. (2007) Poly(3,3-didodecylquarterthiophene) field effect transistors with single-walled carbon nanotube based source and drain electrodes. *Appl. Phys. Lett.*, **91**, 223512.
68. Bardsley, J.N. (2004) International OLED technology roadmap. *IEEE J. Sel. Top. Quantum Electron.*, **10** (1), 3.
69. Marculescu, D., Marculescu, R., Park, S., and Jayaraman, S. (2003) Ready to wear. *IEEE Spectrosc.*, **40**, 28.
70. Wagner, W., Gleskova, H., Sturn, J., and Suo, Z. (2000) in *Technology and Applications of Amorphous Silicon*, Chapter 5 (ed. R.A. Street), pp. 222–251.
71. Symmorphix Inc. (2004) Flat Panel Displays. *www.symmorphix.com/index.asp?pgid=16* (accessed 2008).
72. Lu, X., Xie, Z., Abdou, M.S.A., Deen, M.J., and Holdcroft, S. (1993) Studies of polymer-based field effect transistors. *IEEE Can. Conf. Electr. Comput. Eng.*, **2**, 814.
73. Qiu, Y., Hu, Y., Dong, G., Wang, L., Xie, J., and Ma, Y. (2003) H_2O effect on the stability of organic thin-film

field-effect transistors. *Appl. Phys. Lett.*, **83**, 1644.

74. Lee, J.H., Kim, G.H., Kim, S.H., Lim, S.C., Yang, Y.S., Oh, J., Youk, J.H., Jang, J., and Zyung, T. (2005) The novel encapsulation method for organic thin-film transistors. *Curr. Appl. Phys.*, **5**, 348.
75. Kim, W.J., Koo, W.H., Jo, S.J., Kim, C.S., Baik, H.K., Lee, J., and Im, S. (2005) Passivation effects on the stability of pentacene thin-film transistors with SnO_2 prepared by ion-beam-assisted deposition. *J. Vac. Sci. Technol. B*, **23** (6), 2357.
76. Parylene Conformal Coatings Specifications and Properties, Technical notes, Specialties Coating Systems, *http://www.scscookson.com/parylene_knowledge/specifications.cfm* (accessed 2010).

3
OTFT Integration Strategies

As described in the previous chapter, the availability of high performance organic semiconductor material systems is, undeniably, a principal ingredient for successful realization/implementation of organic electronics. Of similar importance is the establishment of a robust and compatible fabrication infrastructure that can capitalize on these material systems for practical applications. Without a dependable fabrication process and integration strategy, researchers cannot reliably reproduce organic thin film transistor (OTFT) devices/circuits for proper characterization studies; as a result, the advancement, maturation, and deployment of OTFT technology would be hindered.

The development of robust processing schemes for manufacturing OTFT circuits is a challenge due to the sensitivity of organic electronic materials to standard microelectronic fabrication methods and material compatibility limitations/issues. The device layout, architecture, and material selection have a strong bearing on the choice of processing techniques and the overall construct of the fabrication procedure. This chapter explores various aspects of OTFT fabrication. The technological challenges underlying OTFT fabrication are addressed in Section 3.1. Deposition methods and patterning techniques for organic semiconductor materials are described in Section 3.2, along with integration strategies in Section 3.3. To take advantage of the solution-processability of organic materials, this research focuses primarily on OTFTs based on soluble organic semiconductors. For a list of device materials used, please refer to Table 2.3.

3.1
Technological Challenge in OTFT Integration

Although the OTFT device structures (as presented in Chapter 5) appear relatively straightforward, implementation, and fabrication of these devices can be difficult. One of the major challenges when developing an OTFT fabrication process is to address the sensitivity of the functional organic layers. More specifically, the intricacy lies in formulating compatible methods to pattern the organic semiconductor layer in order to enable the production of fully integratable OTFT circuits and systems. In addition, compatibility between various device material layers, to

Organic Thin Film Transistor Integration: A Hybrid Approach, First Edition. Flora M. Li, Arokia Nathan, Yiliang Wu, and Beng S. Ong.

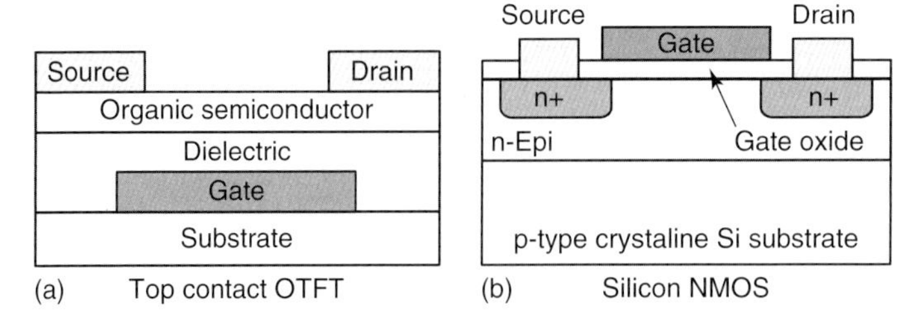

Figure 3.1 Cross-sectional diagram of (a) top-contact OTFT and (b) silicon n-type MOSFET (NMOS).

ensure minimum chemical interaction and process-induced material degradation, must not be overlooked.

Most of the early demonstrations of OTFT devices used a continuous/unpatterned layer of organic semiconductor [1, 2]; this device configuration is predisposed to higher leakages and parasitic effects than silicon metal-oxide-semiconductor field-effect transistor (MOSFET) devices. Figure 3.1 illustrates the structural difference between an OTFT and a silicon n-channel MOSFET (NMOS). MOSFET has lower leakages and lower off-currents than OTFT owing to the presence of a depletion layer. In the case of an n-channel MOSFET (NMOS), a sufficiently high positive V_G results in the formation of an inversion layer at the dielectric/semiconductor interface. This inversion layer provides a conducting channel between the source and the drain, which turns the device on. One of the main advantages of the MOSFET structure is that the depletion region between the p-type substrate and the n-type channel provides isolation from other devices fabricated on the same substrate; this effectively suppresses parasitic leakage and cross-talk between individual transistors. Moreover, the n+ regions below the source/drain contacts form reverse-biased p-n junctions with the p-type substrate. These reverse-biased junctions are responsible for impeding leakage and generating very low off-currents.

In contrast, an OTFT operates by accumulation, and there is no reverse-biased junction to isolate the source and the drain when the device is turned off. Consequently, a leakage current always flows between source and drain through the semiconductor bulk, and thus, OTFTs are liable to higher off-currents and parasitic leakages than MOSFETs. Moreover, since many existing OTFT fabrication processes do not incorporate local isolation between adjacent OTFT devices, lateral electrical cross-talk is an issue for OTFTs fabricated on the same substrate.

To improve the reliability of OTFTs, minimize parasitic leakage and reduce cross-talk, it is crucial to develop appropriate isolation schemes and patterning techniques for the organic semiconductor. The semiconductor should be confined to the active channel region to isolate each transistor from neighboring devices. There are three key benefits to patterning the semiconductor: reduction of cross-talk, improvement of the on/off ratio, and removal of material from the optical path (as in the case of a backlit display, the backplane should be as clear as possible for maximum brightness) [3]. By removing the semiconductor from the non-active transistor area, non-gated current-carrying pathways (which contribute to a constant leakage current) can be removed; this decreases the off-current while making no significant changes to the on-current. Our experiments also showed that isolating the device via patterning of the organic semiconductor layer led to a reduction in gate leakage current. This suggests a possible contribution of gate leakage from the interaction between the organic semiconductor (impurities or solvents) and gate dielectric; more research is needed to clarify this speculation. In addition, a means to integrate reverse-biased p–n junctions in OTFT would be ideal, to permit further reduction in leakages and parasitic effects.

Accordingly, patterning of the organic semiconductor layer is important from a device performance perspective as well as an integrated circuit (IC) fabrication perspective. Unfortunately, developing compatible methods to achieve patterned organic films has been a major bottleneck. Most organic films, particularly soluble polymers, are very sensitive to solvents and chemical attack, so they are typically not amenable to conventional photoresist-based photolithographic patterning approaches (e.g., photoresist coating, developing, film etching, and photoresist removal). In addition, the structure of organic materials is generally optimal at deposition and most post-treatments tend to degrade the material properties [3]. Exposure to liquid solvents, perhaps during a wet etching/patterning process, can degrade the charge transport properties of organic thin films. Therefore, most types of wet processing are typically unsuitable for patterning organic semiconductor materials.

The most desirable method for patterning the organic layer involves direct printing of the active materials, in which deposition and patterning occurs simultaneously in one single step, eliminating any etching steps. For example, in inkjet printing (IJP), the functional organic material is printed only in the desired area, thus reducing material cost/waste and eliminating the need for subsequent patterning. Printing tools are the subject of extensive research and development, and factors such as resolution, printing speed, and so on, need to be refined before large-scale manufacturing can occur [4, 5]. An all-printed fabrication process, hence the term "*printed electronics*", is envisioned for the organic electronics industry in the long term.

Shadow masks also permit generation of patterned organic layers; however, this approach is limited to vacuum-deposited organic materials and is not applicable to solution-processed polymers. Other emerging patterning processes for organic semiconductors include laser ablation, selective deposition by substrate treatment, mold printing/patterning, microcontact printing, and soft lithography techniques [4, 6–10].

For the current OTFT research work, we developed an organic-compatible photolithography fabrication approach for patterning the organic semiconductor layer, which involved investigating a number of passivation/masking materials and strategically tailoring the patterning steps to ensure minimum process-induced damage to the organic layer. A more versatile/pliable approach based on a hybrid photolithography–IJP fabrication scheme was also conceived/formulated to enable robust and direct patterning of the organic semiconductor layer. Before presenting the details of these OTFT integration strategies in Section 3.3, Section 3.2 provides an overview of various patterning techniques.

3.2 Overview of Processing and Fabrication Techniques

To produce electronic devices, the existence of a foolproof/impeccable fabrication process is as critically important as the availability of high-performance materials. With regards to the OTFT fabrication schemes (see Section 3.3), the device processing steps can be grouped into four categories: deposition, patterning, etching/removal, and interface/surface modification.

- Deposition refers to depositing/coating a thin film of material onto the substrate. For this research, relatively standard thin-film deposition techniques are used [11, 12], including plasma enhanced chemical vapor deposition (PECVD) for the gate dielectric layer, sputtering and thermal evaporation for the electrode layer, spin-coating, or IJP for the solution-processed organic semiconductor layer, and chemical vapor deposition (CVD) for the polymer passivation layer.
- Patterning refers to shaping or altering the existing shape of the deposited materials, or directly depositing materials into a selected region to form a film with patterned structures. Three patterning approaches are considered: shadow mask, photolithography, and IJP.
- Etching/removal refers to removing material in selective regions from the wafer. Once an image is transferred from a mask to a wafer, one has to remove or etch material from selected regions to form the final device. Specially designed wet or dry etching processes allow selective removal of materials once the resist has been patterned. For the photolithography fabrication schemes outlined in Section 3.3, inorganic device layers are patterned by wet-etching. Organic layers are dry-etched in oxygen plasma to avoid chemical/solvent interactions between wet-etchants and organic device layers. Metal source/drain contacts are patterned by lift-off processes.
- Interface/surface modification refers to functionalizing the device interfaces for performance enhancements. For OTFTs, and other transistor technologies alike, the interfaces between device layers play a significant role in establishing functionality and performance. Interface modification has been a vital step for OTFT fabrication; it improves interfacial interaction and controls the quality/ordering/structure of subsequently deposited organic layers. Modifications are typically done on the dielectric surface and contact surface prior to deposition of the organic semiconductor layer, in the case of bottom-gate bottom-contact devices. Chapters 5 and 6 are dedicated to interface treatment and analysis.

Table 3.1 summarizes the deposition and patterning techniques used to process the various device layers for the OTFT fabrication schemes discussed. Particulars of the processing sequence are given in Section 3.3.

Table 3.1 Deposition and patterning techniques employed for OTFT fabrication in this research.

Device layer	Deposition techniques	Patterning techniques
Gate metal (Mo, Al)	Sputtering	Photolithography, wet etching with PAN[a]
Gate dielectric (silicon nitride (SiN_x), silicon oxide (SiO_x))	Plasma enhanced chemical vapor deposition (PECVD)	Photolithography, wet etching with BHF[b]
Source/drain contacts (Au, Cr)	Thermal evaporation	Photolithography and lift-off Shadow mask
Organic semiconductor (PQT-12, P3HT, PBTTT)	Spin-coating Inkjet printing	Photolithography, dry etching with O_2 plasma Inkjet printing
Passivation or capping layers (SiN_x, parylene)	SiN_x: PECVD Parylene: vapor deposition	Photolithography, dry etching

[a] PAN is a mixture of phosphoric acid (H_3PO_4), acetic acid ($H_4C_2O_2$), and nitric acid (HNO_3), for etching metallization (e.g., Al).
[b] BHF is hydrofluoric (HF) acid buffered with NH_4F, for etching silicon oxide or silicon nitride.

One of the key challenges in OTFT fabrication lies in patterning the device layers. Due to the sensitivity of most organic thin films to the chemicals/solvents used in standard photolithographic processes, careful/special consideration must be given when developing a compatible fabrication process for organic electronic devices. During the early stages of OTFT research, where material development was the main driving force, most researchers adapted a very simple bottom-gate OTFT device configuration[1] with a single-patterning step on source/drain contacts and a continuous organic semiconductor layer (see Figure 3.2). This procedure delivers a very simple and convenient platform to develop prototype devices and evaluate different organic material systems, and circumvents complications associated with device patterning. For preliminary material characterization, an unpatterned semiconductor layer in a single/discrete transistor configuration would suffice, since the key interests are device mobility and interface properties. However, this approach is inadequate for creating highly integrated devices and robust circuits. Functional circuitries require individually-addressable gate electrodes, via holes through the dielectric layer to make electrical interconnection between different metal layers, and a patterned semiconductor layer to define the transistor's active

1) The device configuration is a bottom-gate bottom-contact OTFT using a highly doped silicon substrate as a bottom gate, a continuous layer of gate dielectric, defining source/drain electrodes through a shadow mask or by lithographic means, and depositing a continuous organic semiconductor film by spin-coating or vacuum evaporation.

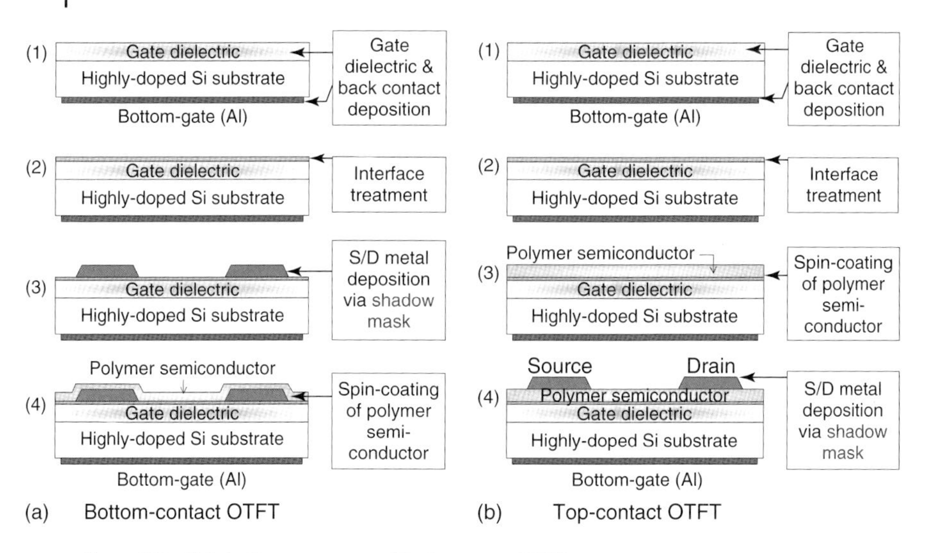

Figure 3.2 Fabrication sequence of bottom-gate OTFT by the shadow mask patterning approaches, for (a) bottom-contact and (b) top-contact configuration.

region. Therefore, patterning of the organic semiconductor active layer is crucial to fulfill various performance, functionality, and integration objectives:

- Patterning the active layer reduces cross-talk between adjacent devices, reduces parasitic resistance, and limits leakage current (off-current). As a result, the devices have better performance characteristics (e.g., smaller I_{OFF} and higher I_{ON}/I_{OFF} ratio for OTFTs).
- Patterning is more important when OTFTs share a common gate, in which case the leakage through the gate dielectric may become significant.
- For circuit implementation, patterning the semiconductor layer is even more critical because transistors must be isolated from one another and have individually addressable gates to achieve circuit functionality.
- From an integration standpoint, a patterned polymer layer is required in applications such as color or backlit polymer/organic light emitting diode displays (OLED), where the OTFT pixel circuit should only consume a small footprint to get a high-fill factor active matrix organic light-emitting diode (AMOLED) system with OTFT backplane.

Three patterning techniques are considered here. The first is the shadow-mask technique, which provides a simple method to form patterns of the source and drain contacts for OTFTs. The second approach is the photolithography technique, which enables precise device definition and higher integration complexities. Lastly, IJP enables direct deposition of patterned layers, which is envisioned to be an important work-horse for future generations of plastic or printed electronics. Each of these techniques is studied in greater detail in this section. The focus of this section is primarily on deposition (Section 3.2.1) and patterning (Sections 3.2.2–3.2.4) techniques for organic semiconductor materials. On the other hand, the technologies used for processing metals and silicon-based inorganic semiconductor/dielectric materials are relatively standard; please consult the appropriate references for details [11, 12].

3.2.1 Deposition Methods for Organic Semiconductors

The two most common deposition methods for organic materials are vacuum evaporation and solution processing techniques (e.g., spin-coating, printing). These two methods are discussed briefly in this section.

3.2.1.1 Vacuum Evaporation

Small-molecule and oligomer organic semiconductors are typically deposited by vacuum evaporation, which consists of heating the material under reduced pressure. The process is conducted in a very high or ultra-high vacuum chamber. The organic material is put into a metal boat, which is heated by Joule effects or with an electron gun, and the substrate is placed a few centimeters above the boat [1]. This technique is usually not applicable for polymers, because polymers tend to decompose by cracking at high temperatures. The main advantages of vacuum evaporation are the easy control of the thickness and purity of the deposited film, and the ability to realize highly ordered films by monitoring the deposition rate and the substrate temperature [1]. Patterned films can be realized by evaporation through a shadow mask. The primary drawback is the need for sophisticated vacuum-based instrumentation, which is in contrast to the simplicity of spin-coating or other solution-processing techniques.

3.2.1.2 Solution-Processed Deposition

Various techniques are available for processing soluble polymer semiconductor films from the liquid phase; these include spin-coating, casting, printing, and soluble precursor conversion. One of the most simple and effective ways to realize a nice polymer film is by spin-coating. When the technique is well handled, it allows the production of very homogeneous films with precise control of their thickness over relatively large areas. Control parameters of a spin-coating process include spin-coating speed, ramping rate, choice of solvent for the polymer solution, drying temperature, drying time, and so on; these parameters need to be optimized to achieve a film of desired thickness and uniformity. Spin-coating is the principal technique we have employed to deposit organic semiconductors. The spin-coating environment affects the performance of air-sensitive organic materials. For instance, for extremely air-sensitive n-type organic materials, such as phenyl-C61-butyric acid methyl ester (PCBM) or C_{60} [1, 13, 14], device processing, and measurements must be done in an inert environment (e.g., nitrogen) in order to attain functioning OTFTs with good mobility. Fortunately, the poly(3,3‴-dialkylquarterthiophene) (PQT-12) polymer semiconductor is relatively stable in air; thus we have spin-coated PQT-12 in ambient conditions for the experiments considered.

Spin-coating requires polymers with good solubility. The solubility of conjugated polymers can be tuned by grafting solubilizing groups to the polymer backbone. An alternative approach is to use a soluble precursor polymer, which can undergo subsequent chemical reaction to convert the material to the desired conjugate oligomer or polymer. The precursor method has been applied for the preparation of pentacene [15] and polythienylenevinylene (PTV) films [16]. This precursor route technique typically produces polymer films with lower mobilities than evaporated films, due to a higher degree of molecular disorder in the film [16]. In general, solution-based processing techniques are more pertinent to polymers than small molecules. Polymers have higher viscosity, which is essential to obtain uniform and sufficiently thick films. However, the mobility of polymer semiconductors is inferior to that of small-molecule semiconductors due to the more random orientation of molecular units or the relatively short conjugation length in polymer backbones [17].

A limitation of spin-coating is the lack of patterning capability. Spin-coating forms a continuous film across the substrate. This shortcoming can be circumvented by direct printing techniques, which are discussed in Section 3.2.4. Alternative solution-based processing/patterning methods, including microcontact printing, screen printing, and nanoimprinting, also demonstrate excellent potential for organic electronics fabrication.

3.2.2 Patterning by Shadow Mask

Patterning by depositing vaporized materials through shadow masks was one of the earliest and simplest routes for making OLEDs and OTFTs. In this process, metals or low molecular weight organic molecules are evaporated from a source in a physical vapor deposition system and travel through openings in masks placed near the surface of the substrate. The deposition typically occurs under high vacuum (10^{-8}–10^{-6} Torr) such that the mean free path of the evaporated species exceeds the distance between the source and the substrate [18]. When this condition is satisfied, the evaporated material travels in a directional manner through the gaps in the mask and onto the substrate [12]. This technique is additive at the substrate level, which enables sequential deposition of multiple layers of different materials.

The shadow mask technique is attractive for OTFT fabrication because it can guarantee a simple, robust, and low cost source/drain contact definition process without damage to underlying organic materials if present. A shadow mask can also provide a constructive tool to deposit patterned low-molecular weight organic films, to define the active semiconductor of the OTFT. Since we focus on solution-processable polymer semiconductors, the shadow mask is used only to define source/drain contacts.

Figure 3.2 displays the fabrication sequence of bottom-gate OTFT patterned by the shadow-mask approach in this research work. Preliminary samples are fabricated on thermally oxidized silicon wafers, where the SiO_2 serves as the gate dielectric and the highly doped Si substrate serves as the gate electrode. Other gate dielectrics tested include PECVD silicon nitride and PECVD silicon oxide. The dielectric surface is pre-treated with an octyltrichlorosilane self-assembled monolayer (OTS SAM), to promote proper alignment of the polymer semiconductor molecules and to improve the quality of the dielectric/semiconductor interface. Following the deposition of the polymer semiconductor layer, the source/drain contact metal (typically gold) is deposited through a (stainless steel) shadow mask to define the transistor geometries for top-contact structures (see Figure 3.2a). A similar process can be applied to fabricate bottom-contact OTFTs (see Figure 3.2b), but with the source/drain electrodes deposited directly on top of the dielectric surface and followed by spin-coating of the polymer semiconductor layer as the last step. Discrete OTFTs with different W and L have been fabricated using this shadow mask approach.

As demonstrated above, the shadow mask fabrication process is quite simple. It provides a convenient approach to producing discrete OTFTs for prototyping and device characterization purposes. However, the above-described scheme has one major setback: it is not well-suited for fabrication of circuits with high complexity due to its inability to perform gate patterning. For the approach outlined in Figure 3.2, all transistors on the substrate share a common gate. To achieve proper circuit functionality, individually-addressable gates are required to drive/bias individual transistors in a circuit. Moreover, an unpatterned gate electrode results in considerable overlap/parasitic capacitances, which limit the circuit's speed performance. Figure 3.3 illustrates a more practical device configuration, with individual gates for each transistor. To extend the capability of the shadow mask process for circuit fabrication, several adaptations are necessary, including:

- Developing a method for patterning the gate electrodes, in order to realize circuits;
- Developing a method for patterning the polymer semiconductor layer, in order to reduce parasitic leakages; and
- Establishing a method to allow accurate alignment (e.g., fiducial marks) of the various masks/layers.

Multilevel registration with shadow masks has been reported by the use of physical mounting brackets, mask translating fixture, and alignment systems [19].

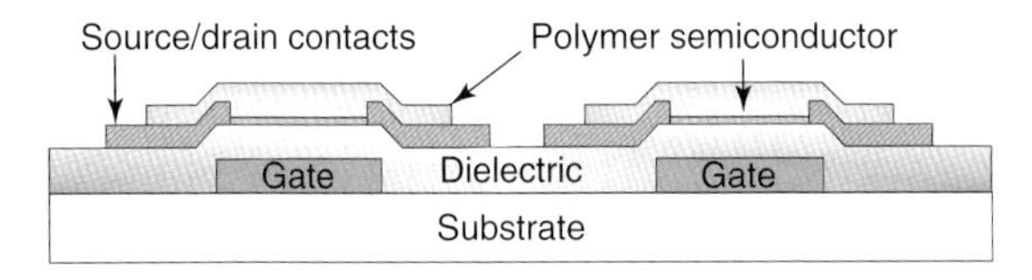

Figure 3.3 Illustration of individually addressable gates on the same substrate, which is a basic requirement to enable circuit implementation.

Multiple organic light-emitting layers and electrodes can be patterned sequentially by successive shadow masks, and several companies have demonstrated organic devices, such as full-color OLED displays (15 in^2 active matrix OLED by Kodak-Sanyo, 13 in^2 active matrix OLED by Sony, Pictiva OLED display by Osram, etc.) using this shadow mask approach [20, 21]. However, the processing demands are more stringent for the fabrication of OTFT circuitries, thus parameters such as multilayer registration, alignment, and feature resolution still pose significant challenges for the shadow mask approach.

Another shortcoming of the shadow mask approach is its limited resolution compared to other patterning techniques. Conventional metal shadow masks, fabricated by photochemical machining or laser beam machining, have inherent limits on the process resolution and aspect ratio. The resolution of such masks is typically of the order of $\sim$50 μm, and must be of the same order as the mask thickness [22]. Practical resolution limits are set by (i) sizes of openings that can be generated in masks that retain sufficient rigidity (i.e., thickness) to be mechanically stable, (ii) mask-to-substrate separation distances that can be reproducibly achieved without unwanted physical contact, and (iii) levels of directionality in the material flux. New technology to fabricate higher resolution shadow masks (since smaller L is preferred for OTFTs) and/or adoption of alternate mask materials are needed. One research group reported a new shadow mask fabrication procedure, combining micro-electro-discharge machining (micro-EDM) and electrochemical etching (ECE), to produce a high-aspect-ratio and high-resolution stainless steel shadow mask [23]. OTFTs with $W = 150$ μm and $L = 3.6$ μm were achievable with this new shadow mask. A novel use of a set of polymeric shadow masks to fabricate pentacene-based organic circuits was also demonstrated [24].

Although alternative patterning technologies (e.g., high-resolution rubber stamping, microcontact printing, soft lithography) can offer resolution better than 10 μm, each of these methods has fundamental drawbacks, including more complicated process steps, limited manufacturability for large area applications, and brittleness of the mask [23]. Therefore, by developing higher resolution shadow masks with precise multilayer alignment capability, shadow masks may be adaptable for OTFT circuit fabrication and can potentially compete with other patterning technologies.

3.2.3
Patterning by Photolithography

Even though the industry forecasts printing technologies/processes to be a key manufacturing standard for organic electronics, there are a number of applications in organic electronics that can benefit from a robust photolithography process at this time. Photolithography is a well-established method, commonly used in the semiconductor industry today. It enables circuit integration, through patterning and etching the multiple device layers, and creating vias and interconnects with modest alignment tolerances. Photolithography facilitates device fabrication that incorporates more than one function, for instance, integrating capacitors, transistors, and diodes in the same device flow. The ability to implement higher resolution devices, as well as complex OTFT structures and ICs, are other advantages of photolithography compared to the shadow masking process. The reported development of a large area, high-resolution lithography system with high volume, roll-to-roll production capability exemplifies the potential and practicality of the lithography-based patterning approach for organic electronics [5]. Our primary motivation for using photolithography is to provide a more immediate fabrication solution for organic electronics, as well as to bridge the technological transfer and development to an all-printed process for the near future. The basic photolithography process is reviewed in Section 3.2.3.1. The challenges/concerns related to adaptation of the photolithography process for organic electronics are considered in Section 3.2.3.2. The photolithography approaches developed in this research for OTFT fabrication are presented in Section 3.3.

3.2.3.1 Photolithography Basics

Photolithography is a process used in microfabrication to selectively remove parts of a thin film (or the bulk of a substrate). It uses light to transfer a geometric pattern from a photomask to a light-sensitive chemical (photoresist) on the substrate. A series of chemical treatments then engraves the exposure pattern into the material underneath the photoresist. The basic photolithography process steps are outlined in Figure 3.4. It typically begins with coating the substrate with photoresist, whose properties can be modified by exposure to ultraviolet (UV) light. UV light is passed through a photomask that manipulates the phase and/or amplitude to achieve patterned exposure on the photoresist layer. The photomask can be in direct contact with (contact mode) or in proximity to (proximity mode) the photosensitive material. Alternatively, imaging optics can magnify or demagnify the mask image and project it onto the substrate (projection mode). UV exposure

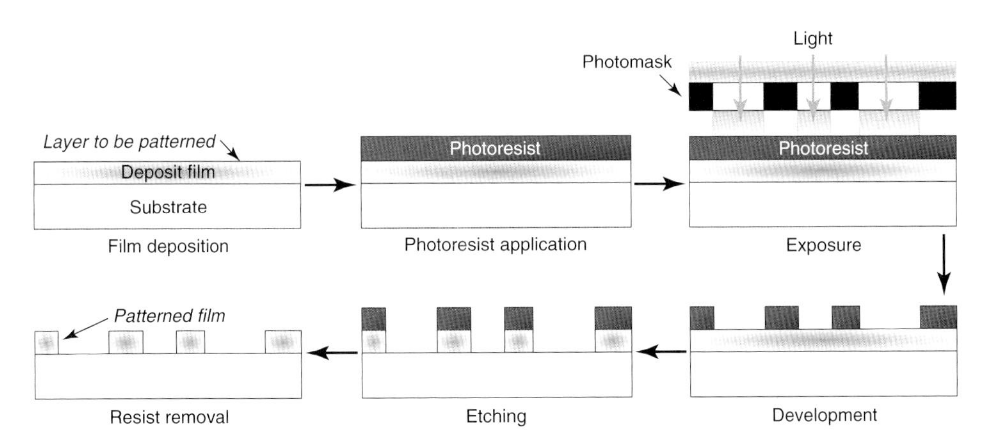

Figure 3.4 Basic photolithography process: the exposure and etching process that allows one to transfer a pattern to the film/wafer.

changes the chemical structure/solubility of the photoresist in a selected region. Subsequent development of photoresist (in a developer solution) removes/dissolves the more soluble regions, leaving behind a patterned layer of photoresist. This patterned photoresist acts as an etch mask for the underlying film, where areas uncovered by photoresist are removed by wet- or dry-etching processes. Finally, photoresist is stripped and the image from the photomask is replicated on the film/substrate.

3.2.3.2 **Photolithography Considerations for OTFTs**

Photolithography is the industry's workhorse for manufacturing inorganic microelectronics and optoelectronics. The high speed, parallel patterning capability, and high resolution features of (projection mode) photolithography also make it attractive for applications for organic devices. However, straightforward implementations can be difficult due to (i) the incompatibility of photoresists, solvents, developers, and UV light exposure with many organic active materials, (ii) challenges in resolution and registration caused by rough, and often dimensionally unstable, plastic substrates, and (iii) cumbersome implementations needed for large-area patterning [18]. Despite these challenges, photolithography still offers many attractive advantages for organic circuit integration, provided that we can meticulously and strategically design the processing sequence to avoid process-induced degradation of device properties. One possible consideration is to execute the majority of the photolithographic steps prior to deposition of the active organic layers; however, this approach would limit device structures to a bottom-gate bottom-contact configuration, such as the one shown in Figure 7.4.1c.

Alternatively, a more versatile strategy is to incorporate a passivating/capping layer (e.g., SiN_x, AlO_x, parylene, polyvinylalcohol (PVA)) on top of an active organic semiconductor to provide protection against chemical attack during photolithography processing and from deposition of subsequent device layers. This concept is illustrated in Figure 3.5. The benefits of this approach are several-fold: enabling post-deposition patterning of the organic layer to realize a greater variety of OTFT architectures, incorporating passivation for the organic active layer, and,

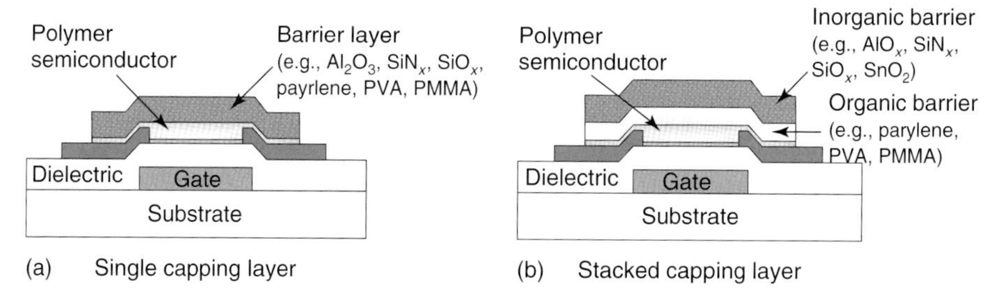

Figure 3.5 Cross-sectional illustration of OTFTs with (a) single capping layer and (b) stacked inorganic–organic capping layers.

more importantly, generating a fully-patterned device for implementation of ICs. Inorganic capping layers such as SiO_2, SiN_x, AlO_x have been used. Since these films have low permeability against solvents, oxygen, and moisture, they serve as excellent barriers to protect the underlying organic semiconductors from solvent processing. However, deposition of these inorganic barrier films may induce damage to the organic layer (e.g., thermal heating during evaporation, plasma exposure during PECVD, or physical damage from energetic ions during sputtering) [1, 25–27]. These concerns can be eliminated with the use of organic-based capping layers (e.g., parylene, PVA) with better processing compatibility with the underlying organic semiconductor layer. Parylene is a vapor-deposited polymer, deposited at room temperature. It is widely used as a tough and conformal coating, and is one of the materials used in our process [28]. Another organic capping material is water-soluble PVA. A concern with organic-based passivation films is their higher permeability to oxygen, moisture, and solvents, when compared to inorganic candidates. A more comprehensive solution is delivered by a stacked inorganic–organic capping layer structure, where an inorganic barrier is deposited on top of the organic buffer. The inorganic barrier can effectively prevent diffusion of chemical/solvent into the organic semiconductor layer, while the organic buffer can minimize process-induced damage to the active organic layer.

Photolithography patterning can be executed on the capping layers to define the active channel region. The capping layers then serve as an etch mask to etch the uncovered regions of the underlying organic semiconductor layer. Oxygen plasma generated by a reactive ion etcher (RIE) is used to etch organic-based films in this research; dry etching is preferable over wet etching because it provides improved control and anisotropy.

This photolithography approach with a capping layer has been successfully applied for the fabrication of OTFTs and circuits (please refer to results in Chapter 7) [25–27]. However, reduced device performance is typically observed. Jia *et al.* studied the source of degradation for their poly(3-hexylthiophene) (P3HT) OTFT with parylene/Al_2O_3 capping layers [5]. For their process, the degradation occurred primarily during the atomic layer deposition (ALD) of Al_2O_3 and etching of the capping layer. There was 30% degradation in mobility, a nearly twofold reduction in drive current, and an increase in threshold voltage after the ALD Al_2O_3 deposition. In the capping layer etching, a near 50% degradation in mobility was observed. The patterned devices displayed mobility of $0.02\,cm^2\,V^{-1}\,s^{-1}$ [5]. For PQT-12 OTFT on SiN_x, devices patterned with parylene had mobility of $0.04–0.12\,cm^2\,V^{-1}\,s^{-1}$, which is very comparable to unpatterned devices $0.05–0.14\,cm^2\,V^{-1}\,s^{-1}$. This shows the reliability of our process. Nonetheless, careful handling and process optimization is required to minimize any process-induced degradation using this approach.

Despite the maturity and effectiveness of the photolithography process for circuit fabrication, it is not the optimal solution for organic electronics in the long run, due to its relatively high costs and process complexity. Many research groups are assertively exploring innovative fabrication techniques (e.g., IJP, nanoimprinting, microcontact printing) that can capitalize on the solution-processability offered by organic materials [7, 9].

3.2.4
Patterning by Inkjet Printing

Inkjet printing is a precision micro-fluid dispensing technology for creating prints and images via non-contact printing onto a substrate. The IJP system works by producing a fine jet of ink, called "*ink-jet*," through a printing nozzle of up to 50 μm in diameter, and positioning these ink droplets as required on the substrate [29]. Inkjet printing, conventionally a popular consumer desktop publishing technology, is now an emerging and revolutionizing technique for fabrication of organic electronics and optical devices (e.g., OLEDs, OTFTs, organic solar cells). The rising field of "printed electronics" is receiving a great deal of attention. Plastic Logic, a Cambridge-based UK company whose core technology is in printed plastic electronics manufacturing, acquired large sums of investment in early 2007 to build the first factory to manufacture plastic electronics on a commercial scale [30]. This announcement further exemplifies the promising future of IJP for electronic fabrication.

Inkjet printing, along with other solution-based processing methods (e.g., spin-coating, imprinting, stamping), can reduce fabrication costs and lead to large area, cost-effective reel-to-reel production of organic electronics. The advantages of IJP technology over the photolithography process for silicon ICs include low capital investment, large area capability, elimination of high-temperature and vacuum deposition processes, compatibility with flexible substrates, ease of customization, quick cycle and turnaround times, robustness, and an environmentally friendly production process [31]. Because custom images can be generated using a software program, IJP presents a low-cost and convenient method to create prototype organic devices. Figure 3.6 illustrates the simplicity of the IJP fabrication process over the photolithography process, where printing combines deposition and patterning all in one step; thus, reducing processing steps and material wastage [16]. Table 3.2 summarizes the key advantages of using IJP for OTFT fabrication.

Despite the many advantages, there are some challenges and limitations associated with IJP organic electronics material. Because the requirements of printing electronic functions are very different from those of printing visual images, the adaptation of inkjet systems for processing organic devices will require careful consideration of many factors, including print-head performance (e.g., nozzle geometry, ink delivery, reliability, drive electronics); solvent compatibility with polymer material and its functionality (e.g., solubility, viscosity); print-head and liquid compatibility; polymer liquid–substrate interactions (e.g., surface tension vs. substrate energy); polymer solidification (e.g., adhesion, mechanical integrity, functionality); and placement accuracy [32]. More specifically, when using IJP for OTFT fabrication, technological concerns pertain to film continuity, multilayer registration/alignment, device resolution, and ink formulation. Some of these issues are explained below.

- *Continuous and pinhole-free films*: A feature of IJP (and other non-contact printing techniques) is the utilization of discrete dots to form surface images, and the image resolution is measured by the number of dots per inch (dpi) [33]. However,

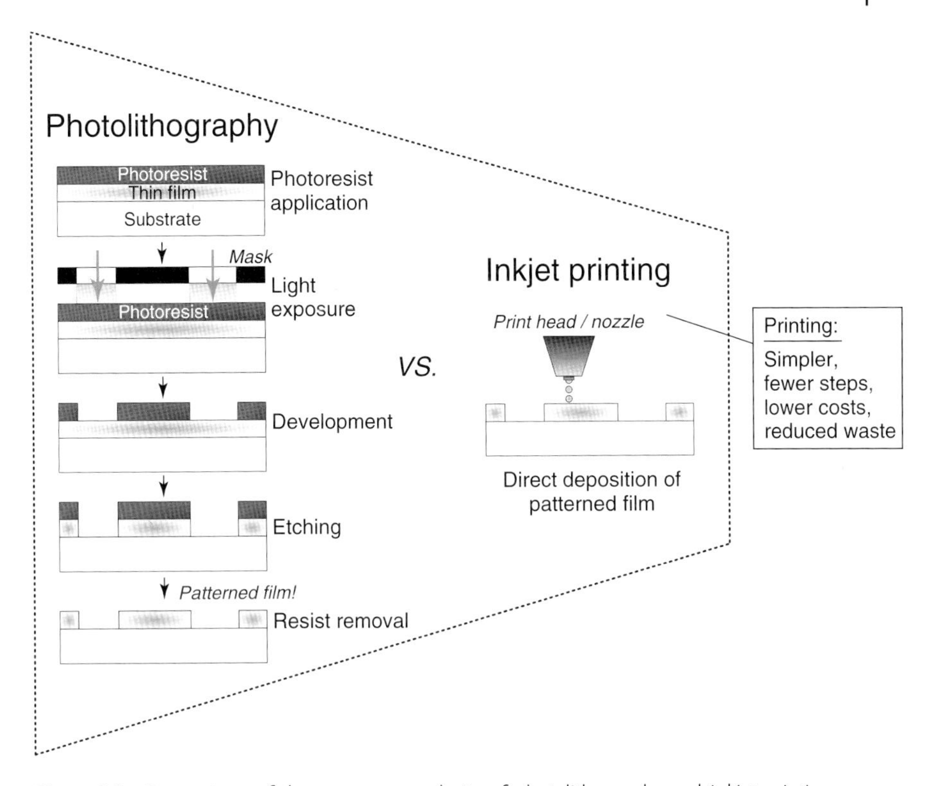

Figure 3.6 Comparison of the process complexity of photolithography and inkjet printing.

this attribute can be a concern for electronic device fabrication, because the discrete dots must have electrical continuity for functional devices. Also, it must be possible to print pinhole-free layers to avoid shorting of devices. These requirements impose a challenge when designing printing systems for device fabrication.

- *Printing multi-layer devices*: IJP should have the capability to print multiple layers such that the various layers form discrete unmixed layers, or they mix and react to form a single material, depending on the specific process goals [31].
- *Higher image resolution*: Typical image resolution for IJP is ~25 μm [34]. Higher device resolution is desirable for fabricating OTFTs, because smaller channel lengths can result in higher output currents and faster switching speeds (refer to Equations 2.3 and 2.4). The resolution of most direct printing techniques is often limited by difficulties in controlling the flow and spreading of the liquid inks on surfaces. One approach to overcoming these resolution limitations is by "surface-energy-assisted printing," where the functional ink is deposited onto a substrate containing a predefined surface-energy pattern that is able to steer the deposited ink droplets into place. OTFTs with channel length of 500 nm

Table 3.2 Key advantages of the inkjet printing process for the fabrication of organic electronics [29, 31].

Feature of inkjet printing	Description
Additive patterning technique	Allows simultaneous patterning and deposition of material, simplifying the patterning process Reduces consumption of material and energy Minimal ink/chemical wastage Circumvents use of chemicals such as photoresist or developer, which can adversely deteriorate functional organic layers Different kinds of material can be deposited by simply changing solutions in an inkjet head
Non-contact patterning method	No direct contact between substrate and equipment Free from mechanical damage or contamination to a device that could arise from a direct contact of a device surface with a tool Compatibility with a variety of substrates, irregular shapes, or surfaces Provides accurate alignment with patterns already on a substrate to create multilayered devices
Computer-controlled process	Excellent flexibility and adaptability Designs can be generated and modified quickly on the computer Ideal for small volumes or complex designs On-site design and printing functionality will enable new applications such as personalized ICs or electronic tickets
Sufficiently high print speed	Ideal for high-throughput production Higher print speed is achievable by increasing the jetting frequency and the number of nozzles on a print head

have been demonstrated with this surface-energy-assisted IJP method [35]. More discussion on the dewetting technique is provided in this section.

- *Ink formulation and printable materials*: Solution-processable organic electronic material must be formulated into printable inks, with the right viscosity, surface tension, concentration, and solvent. Compatibility of the organic ink with the print-head, nozzle size (to avoid clogging), and other specifications of the printer system is also important. Ink formulation and printing parameters strongly influence the performance of a printed device. Good knowledge of material processing conditions and material properties is vital for developing a set of suitable, printable, and electrically functional materials systems for OTFT fabrication [33].

Researchers are actively addressing these technological challenges in order to develop a versatile IJP technology that can gain further momentum in the field of organic or printed electronics.

3.2.4.1 **Inkjet Printing of OTFTs**

The relevance of IJP to organic electronics is made evident by vast reports in the literature. The earliest demonstrated applications are inkjet-printed OLEDs using organic electroluminescent materials [31, 36, 37]. Investigations have also been directed to the IJP of conducting lines as electrodes for OLEDs and OTFTs [35, 38–40]. One of the key objectives here is to improve printing resolution for channel length reduction. A significant breakthrough was made with the introduction of a selective dewetting technique [6, 35], which enabled patterning of submicron channels. There is also ongoing research focused on improving homogeneity, morphology, and molecular ordering of the inkjet-printed semiconductor film for OTFT fabrication [40–42]. In all applications, the printed media (i.e., inks) must be carefully adapted to the printing process, the print heads, and the underlying structures.

Inkjet printing of OTFTs has been demonstrated by various research groups [4, 6, 29, 39, 43]. For example, Kawase *et al.* reported IJP of a top-gate OTFT structure, using solution-processable polymer materials [6]. The source/drain electrodes were inkjet printed in air from an aqueous dispersion of water-soluble poly(3,4-ethylenedioxythiophene) doped with polystyrene sulfonic acid (PEDOT:PSS) on a glass substrate. The active semiconductor layer was a fluorene-bithiophene copolymer (poly(9,9-dioctylfluorene-co-bithiophene) (F8T2))[2] spin-coated from xylene solution. The gate insulator, poly(vinylphenol) (PVP), was spin-coated from isopropanol solution. A gate electrode was printed on the gate dielectric to cover the channel. Different types of solvent were used to deposit those four layers to avoid dissolving or damaging the underlying layers or structures [6]. The inkjet-printed F8T2 OTFT has a channel length (L) of 50 μm and a channel width (W) of 2 mm and demonstrated a relatively high on/off current ratio of $>10^5$ [6]. However, the inkjet-printed thin film transistor (TFT) has a relatively low drain current (I_D), of the order of 0.4 μA with V_{DS} and V_{GS} biased at −40 V [6].

The channel length of 50 μm may be too large to obtain sufficient drain current at a lower voltage. It is desirable to reduce L to submicrometer dimensions to obtain sufficient output current at a lower voltage, since $I_D \propto W/L$, and to allow faster switching speeds. However, the device yield of the IJP process is compromised by a decreasing channel length due to the fluctuation in the flight direction of the inkjetted droplets [6]. In addition, inhomogeneous wettability of a substrate surface leads to irregularities in printed patterns and short circuit formation between source and drain electrodes. To suppress the uncontrolled spreading of

2) F8T2 is a nematic liquid crystalline conjugated polymer semiconductor. It can be preferentially oriented by rubbed polyimide PI layers.

the ink droplets, a novel surface-energy-assisted IJP technique was demonstrated; this approach involves pre-patterning the substrate with a "dewetting pattern" of hydrophobic lines, which confines the organic ink to specific (hydrophilic) regions on the surface [34, 41].

3.2.4.2 Improved Resolution by Surface-Energy Assisted Inkjet Printing

Sirringhaus and coworkers incorporated this "dewetting" prepatterning approach in their inkjet-printed OTFT [4]. The prepattern was prepared with polyimide (PI) by photolithography and by etching with O_2 plasma. The surface of the patterned PI strip is hydrophobic and the surface of the etched regions is hydrophilic. The geometry of the PI strip defines the channel of the polymer TFT. When a droplet of a PEDOT aqueous dispersion (for source and drain electrodes) is inkjetted onto the etched region along the PI strip (see Figure 3.7a), the droplet spreads up to the edge of the PI strip and is confined in the hydrophilic etched region. The atomic force microscopy (AFM) image in Figure 3.7b shows that the dried PEDOT is successfully deposited up to the edge of the PI strip (or channel edge) and the spreading of the droplet is effectively controlled. The success of this approach relies heavily on the wettability contrast between the two regions. Transistors with a channel length as small as 2 μm have been reported with this technique; the resolution is notably better than the standard inkjet process (with resolution around 25 μm). The transfer characteristics of an OTFT with $L = 5\,\mu m$ and $W = 3\,mm$ exhibited an on/off switching ratio $>10^5$. The drain current was also higher than 10 μA (with V_{DS} and V_{GS} biased at -40 V). The mobility (μ_{FE}) was $\sim 0.02\,cm^2\,V^{-1}\,s^{-1}$, which is two to three times larger than the mobility of a non-prepatterned inkjet-printed polymer TFT [6]. More recently, devices with a 500 nm channel length have been demonstrated with this surface-energy-assisted IJP method with stringent control over the flow of liquid ink droplets and optimization of the wetting/dewetting mechanisms [35].

One shortcoming of this prepatterning inkjet method that might limit it from full-scale manufacturing deployment is that it requires a photolithography or

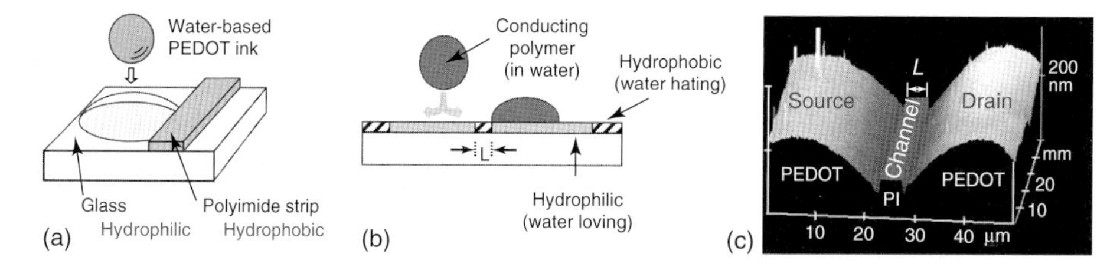

Figure 3.7 Surface-energy-assisted inkjet printing to improve image resolution. (a) Schematic view of inkjet deposition of PEDOT solution along the hydrophobic PI strip. (b) Control of ink spreading by a "dewetting pattern" of hydrophobic lines that confines the organic ink to specific hydrophilic regions on the surfaces. (c) Topographic AFM image of the transistor channel defined by PI, and the PEDOT source and drain electrodes. The width of the strip (channel length) is 5 μm (adapted from [6]).

e-beam lithography step to create the high resolution dewetting pre-patterns. Photolithography might not be the most cost-effective option for low-cost production of surface energy patterns in the long term. Other non-lithographic approaches for modifying the surface chemistry to get wettability contrast are under investigation; examples include soft lithographic stamping, embossing, and laser patterning.

3.2.4.3 Printing Peripheral Circuit: Vias and Interconnects

Additional criteria to make IJP fully adaptable for IC fabrication include the ability to form inter-layer electrical connections (i.e., implementation of vias and interconnects), and the capability to print other circuit elements including resistors, capacitors, and pad electrodes. An IJP process for creating via-holes based on the local dissolution of materials by inkjet deposition of a suitable solvent has been demonstrated [4, 6]. For example, to fabricate a via-hole through a polymer insulating layer of PVP, a droplet of isopropanol is first inkjetted at the appropriate location. The isopropanol dissolves the PVP locally, and excavates a hole that can be subsequently filled with a conducting polymer (e.g., PEDOT:PSS) to form the via-hole interconnect. Resistors can also be inkjet printed by using a low conductivity polymer material such as PEDOT diluted with PSS. The resistance value can be adjusted over a wide range by varying the concentration of PEDOT in PSS, the length of the resistor structure, or the density of drop depositions [4]. The realization of inkjet-printed polymer TFTs, via-holes, and resistors has permitted the fabrication of simple inverter circuits, which are the basic building blocks of a logic circuit. The first inkjet-printed inverters were reported by Sirringhaus *et al.* [4]. They fabricated an enhancement-load inverter using two p-type inkjet-printed F8T2 polymer transistors, where the drain and gate of the load transistor are connected together through a via-hole interconnect. An inkjet-printed resistance-load inverter was also demonstrated, where a printed resistor was used as the load element. The inkjet-printed polymer transistors used in these inverters are fabricated using the prepattern-dewetting approach, to attain channel lengths of 5 μm. Both circuit configurations demonstrated clean inverter action. These inkjet-printed inverters can be switched at frequencies up to a few hundred Hz [4]. Further improvements are expected with improving mobility of the polymer semiconductor and conductivity of the PEDOT electrodes, as well as decreasing the channel length and the source/drain-to-gate overlap capacitance. The performance of these inkjet-printed all-polymer TFT circuits is believed to be adequate for applications such as active-matrix displays or identification tags [4].

3.2.5 Microcontact Printing

Microcontact printing, which was introduced by Whitesides and coworkers [44], is one of the most investigated techniques for patterning micro- and nano-structures. This common soft lithography method uses an elastomeric stamp, generally made of poly(dimethylsiloxane) (PDMS), with replicas from master structures fabricated in materials such as silicon by conventional lithographic processes. A typical strategy is to coat the surface of the PDMS stamp with an "ink" comprising the materials to be transferred. When bringing the PDMS stamp into contact with a substrate, nanoscale interactions between the substrate and the materials on the stamp enable transfer of materials from the stamp to the substrate. Microcontact printing has been used to pattern a wide range of materials, including biological molecules, organic electronic molecules, and metals. It has been extensively reviewed [45], and we focus here solely on the use of microcontact printing in the fabrication of OTFTs.

Microcontact printing could be used to pattern electrodes for the fabrication of OTFTs. The first implementation of microcontact printing in fabricating source and drain electrodes for OTFTs was reported by Rogers and coworkers [46]. Small channel lengths down to 2 μm were achieved, which is comfortably smaller than the typical channel length requirement for many applications. The high printing resolution is one of the advantages of microcontact printing. It enables the demonstration of radio frequency identification tags and flexible display [47]. In order to reduce macroscopic distortion, a rigid carrier microcontact printing technique to print interdigitated electrodes for OTFTs was employed by Leufgen *et al.* [48]. Variations of printing pressure, printing time, and the concentration of molecular ink in the printing process were systematically studied, resulting in OTFTs having a mobility of $0.01\ cm^2\ V^{-1}\ s^{-1}$ with a vacuum evaporated α,α'-dihexylquaterthiophene semiconductor. The IBM group has also developed a microcontact printing process for gate electrodes [49]. A microcontact printing phosphonic acid self-assembled monolayer (SAM) on aluminum was used as the etch resist, and the SAM layer functioned as a thin gate dielectric for the same device, enabling a low operating voltage of 3 V [50]. All the above studies are based on printing a monolayer such as thiol SAM as the etch resist, or so-called mask, on a pre-deposited metal film, followed by etching the metal into electrode features. Alternatively, microcontact printing was also used to print a hydrophobic silane layer on a flexible substrate, followed by electroless plating of Ni as the gate electrode only at the region without the silane layer [51]. Recently, metal nanoparticles such as gold and silver nanoparticles have been developed as printable conductive materials for OTFTs [52, 53]. Direct deposition of gold nanoparticles using a microcontact printing technique was attempted by Wu *et al.* [54]. The printed electrodes from gold nanoparticles functioned as well as vacuum-evaporated electrodes in OTFTs.

Microcontact printing has also been used to pattern a solution-processable organic semiconductor layer for OTFT. Similar to patterning an electrode, hydrophilic and hydrophobic patterns could be achieved through microcontact printing molecules with hydrophobic tail groups on a gate dielectric such as silicon oxide. Subsequently, semiconductor was selectively deposited on the hydrophilic regions only [55]. Another approach is to print semiconductor material directly into the patterned semiconductor layer [56, 57]. For example, Park and coworkers patterned a P3HT semiconductor for a transistor array on a plastic substrate, and obtained devices with a mobility of $0.02\,cm^2\,V^{-1}\,s^{-1}$ and current on/off ratio of about $10^3 - 10^4$ [56].

Besides the aforementioned high resolution, microcontact printing offers other advantages such as low capital cost, and great capability to print on curved, flexible, or structured substrates. However, the challenges remain in registration for multiple layer integration, and mechanical properties of the elastomeric stamp to prevent distortions. In addition, fabrication of the master structures still needs conventional photolithography.

3.2.6
Other Deposition Methods

Other deposition methods, such as conventional screen printing, gravure printing, and offset printing have been used to fabricate OTFTs or parts of OTFTs. Screen printing is in use for printing circuit boards. Recently, it has been utilized to print OTFTs. A pattern can be formed by squeezing a specially formulated ink through a screen mask onto a substrate surface. Silver and gold nanoparticle inks have been screen printed as source and drain electrodes for OTFTs [52, 58]. The printed devices showed similar performance to evaporated metal electrodes. Other layers, such as dielectric and semiconductor, can also be printed through a screen. For example, Bao and coworkers used screen printing to fabricate all the active layers for OTFTs [59]. The device showed similar performance to those made by photolithographically defined Au electrodes. Screen printing is usually limited to a smallest feature size of about $75\,\mu m$. However, a recent study demonstrated screen printed patterns as small as $30\,\mu m$ by optimizing the fabrication conditions [60].

Gravure printing uses a cylinder bearing engraved patterns as an image carrier to "pick up" ink materials and transfer the materials directly onto the substrate by an impression cylinder. Gravure printing has also been used to print different layers for OTFTs. For example, gate, source, and drain electrodes were fabricated by gravure printing using a patterned PDMS stamp and silver paste on a plastic substrate. A field effect mobility up to $0.1\,cm^2\,V^{-1}\,s^{-1}$ was achieved using bis(triisopropyl-silylethynyl) pentacene [61]. The semiconductor layer was printed via the gravure method, showing similar performance to other printing technologies [62].

Laser thermal transfer printing should be noted as well, since it has been used to print a variety of materials, including polymers, and polymer nanoparticle composites, and shows great potential for large-area plastic electronic devices. A donor sheet with the transfer material is used for this process. Upon heating from the back side of the donor sheet with a laser, the material can be transferred from the heated region to a receiver substrate. Conducting polymers have been printed as electrodes for large-area and high resolution OTFTs [63]. Since it is a dry process, one of the advantages of laser thermal transfer printing is avoidance of the solvent compatibility issue for OTFT integration.

3.3
OTFT Fabrication Schemes

The OTFT integration strategies considered here are presented in this section. The key patterning techniques used are photolithography and IJP. The processing sequence, device layout, and photomask designs are varied to produce a variety of OTFT architectures.

The mask sets used for OTFT fabrication were designed to facilitate systematic characterization of OTFT's electrical behavior. These masks consist of OTFT designs with different geometries, with $W = 20–6000\,\mu m$ and $L = 5–400\,\mu m$ in discrete transistor or transistor array formats. Testing of these devices enables accurate extraction of device parameters (e.g., effective mobility, threshold voltage, contact resistance, off current), which are valuable for studying device physics and for device modeling. The masks also include simple OTFT circuits (e.g., inverters, current mirrors, source followers, ring oscillators), display pixel circuits, imaging pixel circuits, and phototransistors. Functional OTFT circuits were fabricated, and preliminary results from circuit characterization are presented in Chapter 7.

Device fabrication was primarily conducted using PQT-12 OTFTs; please refer to Table 2.3 for a list of device materials used for this research. The one-mask bottom-gate OTFT process was also used to fabricate solution-processed poly(2,5-bis(3-alkylthiophen-2-yl)thieno[3,2-*b*]thiophene) (PBTTT) OTFTs [64]. It also provides a convenient platform to demonstrate and evaluate solution-processed nanocomposite TFTs based on spin-coated single-wall carbon nanotubes (SWCNTs), silicon nanowires (SiNWs), zinc oxide (ZnO) nanowires, and zinc oxide tetrapods; this work was done in collaboration with the University of Cambridge. Our hybrid photolithography–inkjet printing approach led to one of the first demonstrations of IJP carbon nanotube TFTs and PQT-12/SiNW nanocomposite TFTs [65, 66]. These results demonstrate the versatility of our integration strategies. The fabrication schemes and mask designs described can be easily extended to fabrication of TFTs based on other novel organic or nanocomposite thin-film material systems, thus providing a convenient platform for initial prototyping of new devices.

3.3.1
Basic One-Mask Processing Scheme for Bottom-Gate OTFT

The fabrication sequence of a bottom-gate bottom-contact OTFT on highly doped Si substrate as gate, employing a single patterning step (i.e., one-mask process), is depicted in Figure 3.8. The source/drain contacts can be defined by a shadow mask or by photolithography. Although this common-gate configuration may not deliver devices with the best overall performance (e.g., overlap capacitance will limit speed) and may not be highly compatible for circuit integration (e.g., lack of individually addressable gates), this design provides the simplest method to evaluate material systems and interfaces. Furthermore, it has the least number of processing variables and the lowest possibility of process-induced damage, thus

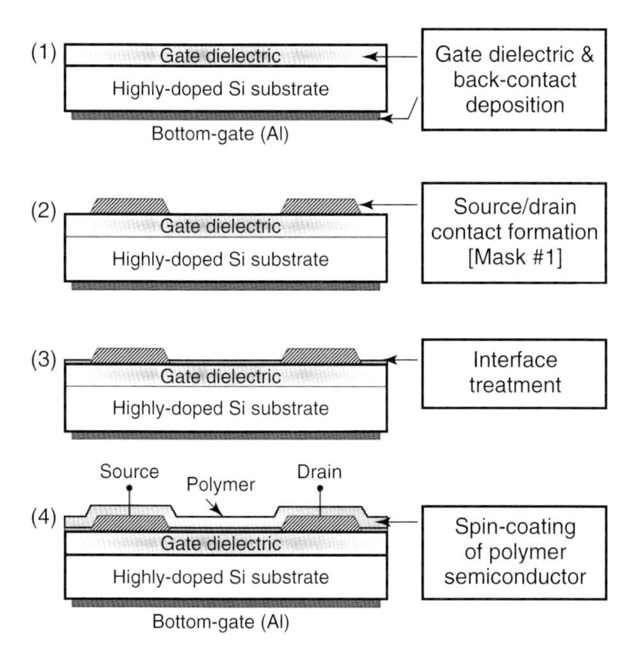

Figure 3.8 Fabrication sequence of bottom-gate bottom-contact OTFT on highly-doped Si substrate with a one-mask patterning step to define the source/drain contacts (by photolithography or shadow mask).

one can focus strictly on evaluation and/or characterization studies of new material systems. This configuration and fabrication scheme serves as the basic platform for the dielectric and interface investigations reported in Chapters 4–6.

Initial stages of this research used a shadow mask to define the source/drain contacts. However, the image resolutions produced by evaporation through the shadow mask were poor (the smallest channel that can be reliably reproduced is 90 μm). As discussed in Section 3.2.2, manufacturing high-resolution shadow masks is an industrial-wide challenge. Furthermore, the edges of the resulting source/drain contacts exhibited uneven (zigzag) ridges; these ridges were inherited from fabrication of the actual shadow mask. Due to the shadowing effect during deposition, the geometry of the deposited contacts often deviates slightly from the dimensions of the objects on the shadow mask. In contrast, photolithography provides a much cleaner, sharper, and well-controlled source/drain contact definition than the shadow-mask approach. Consequently, photolithography was the technique-of-choice for this one-mask OTFT fabrication scheme. Bottom-gate OTFTs with photolithographically-defined bottom-contacts form the basis for the experimental work reported in Chapters 4–6. One exception was the fabrication of OTFTs on plastic substrates (in Section 4.4.5), where the shadow mask was used to minimize potential damage to the thin film on a flexible substrate (e.g., cracking) from handling or photolithography processing steps.

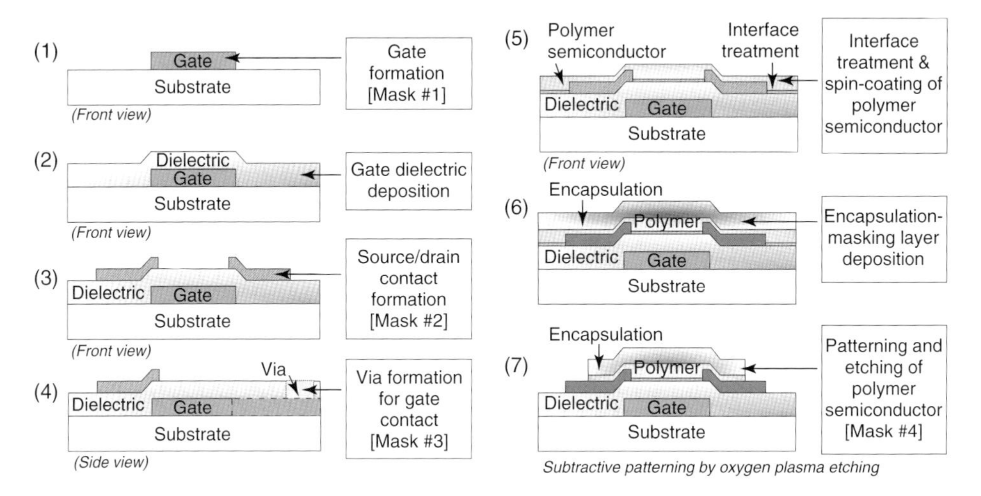

Figure 3.9 Fabrication sequence of fully-patterned fully-encapsulated bottom-gate bottom-contact OTFT by a four-mask photolithography process.

3.3.2 Photolithography Scheme for Fully-Patterned and Fully-Encapsulated Bottom-Gate OTFT

Figure 3.9 shows the processing scheme for a photolithographically-defined fully-patterned and fully-encapsulated bottom-gate bottom-contact OTFT, developed in this research. First, the bottom-gate metal is deposited and patterned (mask 1), followed by deposition of the bottom-gate dielectric across the wafer. A second metal layer is deposited on the dielectric surface, and is subsequently patterned by photolithography to define the source/drain contacts (mask 2). Prior to deposition of the organic semiconductor, interface treatment is performed on the dielectric and/or contacts to improve the molecular ordering and quality of the organic semiconductor film, for enhanced device performance. (The role of interface treatment is analyzed in depth in Chapters 5 and 6.)

The most critical step in this process is patterning of the organic semiconductor layer. Two approaches were investigated:

1) **Direct patterning**, where the standard photolithography process was performed directly on the organic semiconductor layer. Photoresist was processed directly on the semiconductor, and was preserved on the final device to serve as passivation. Figure 3.10a illustrates a cross-section of the directly-patterned structure with photoresist passivation.
2) **Indirect patterning**, where a buffer layer was used to isolate the organic semiconductor from the photoresist. Parylene was used as the buffer material in our process. Photoresist was processed on the buffer layer, and was later preserved on the device after processing to provide an additional level of

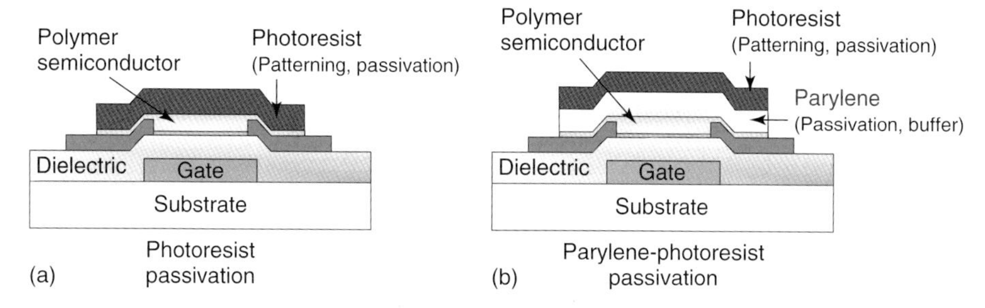

Figure 3.10 Two photolithographically-defined bottom-gate OTFT structures with (a) photoresist passivation and (b) parylene-photoresist passivation.

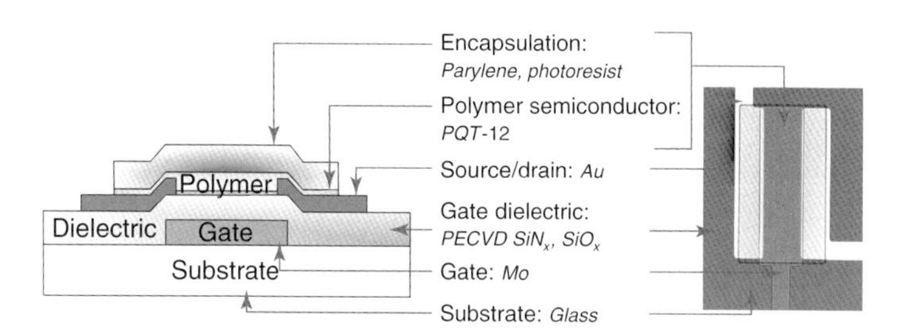

Figure 3.11 Cross-sectional illustration and photograph of a fully-patterned OTFT fabricated using four-mask photolithography scheme with parylene-photoresist passivation.

passivation. Figure 3.10b shows a cross-section of the final structure with parylene-photoresist passivation.

Figure 3.11 displays a cross-sectional diagram and actual microphotograph of a fully-patterned parylene-passivated PQT-12 OTFT fabricated using the scheme presented in Figure 3.9; the microphotograph shows that well-defined device layers are produced using the photolithography approach. OTFT circuits were successfully fabricated using this scheme; more details are presented in Chapter 7. Details of the direct and indirect patterning approaches are explained below.

3.3.2.1 **Directly Patterned OTFTs**

With the direct patterning approach, spin-coating a uniform photoresist layer on the polymer semiconductor surface (PQT-12 in this experiment) across the entire 3 in wafer can be a challenge. This can be due to inhomogeneity of the PQT-12 thin film surface and adhesion issues of the photoresist on PQT-12. Imperfect photoresist coverage degrades the device yield across a single 3 in wafer. To gain better control over the patterning process and to improve adhesion/uniformity of the photoresist on the PQT surface, it is recommended to fine-tune the processing parameters of the photoresist (e.g., spin rate, baking temperature), and to optimize

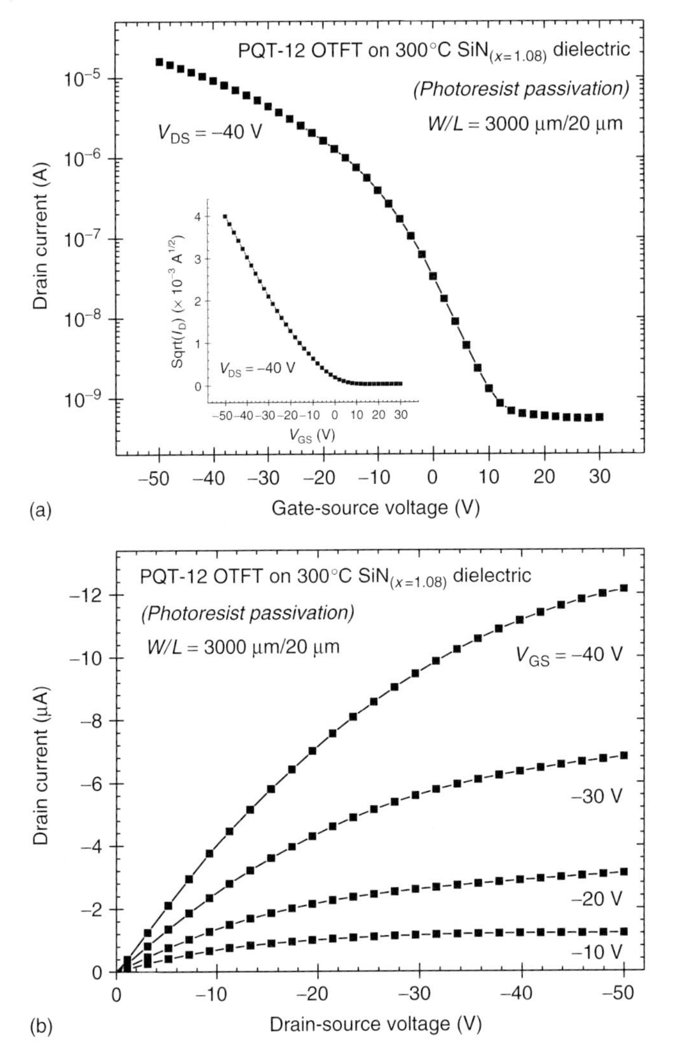

Figure 3.12 (a) Transfer and (b) output characteristics of a fully-patterned bottom-gate bottom-contact PQT-12 OTFT with SiN_x gate dielectric on a glass substrate. The polymer semiconductor layer was directly patterned photolithographically, with direct deposition of photoresist on the PQT-12 layer.

the surface energy of the photoresist (e.g., by modifying the photoresist formulation or pre-treating the semiconductor surface).

Following spin-coating, the photoresist layer was patterned by the standard photolithography sequence: soft-baking, UV exposure, developing, and hard-baking. The organic semiconductor was carefully etched using oxygen plasma in RIE mode, where the photoresist pattern served as a masking layer.

Functional OTFTs and circuits were produced using this direct patterning approach. The electrical characteristics of a photoresist-passivated directly-patterned OTFT are displayed in Figure 3.12. However, the resulting PQT-12 OTFTs displayed rather low mobility, in the range 0.003–0.008 $cm^2\ V^{-1}\ s^{-1}$; a 10-fold reduction compared to the indirect patterning approach with a parylene buffer layer, discussed next. These results indicate that direct patterning using photoresist may induce unfavorable changes to the organic semiconductor layer. Damage may arise from chemical interactions with photoresist, developer solution, or oxygen plasma during etching. Nevertheless, the demonstration of functional OTFTs is a breakthrough, since it is generally reported that the standard photolithography process is destructive to OTFTs. Therefore, with careful control and planning, concerns related to photolithography can be overcome, giving a compatible process for organic electronics fabrication.

3.3.2.2 **Indirectly Patterned OTFTs**

Indirect patterning presents a more reliable approach, where the organic semiconductor is encapsulated with a buffer layer prior to photolithographic patterning. Parylene-C film (200–300 nm thick), deposited using room temperature CVD (Cookson CVD deposition system), was used as the buffer/passivation layer in our process. Parylene is an attractive masking/capping/dielectric/passivation material owing to its ability to form a conformal pinhole-free coating, excellent dielectric strength (2.2 MV cm^{-1}, $\varepsilon_r = 2.95–3.15$), temperature-independent electrical properties, high molecular weight (~500 000), high melting temperatures (290 °C), superior barrier properties (i.e., gas and moisture permeability) compared to other polymeric materials, good resistance to chemical attack at room temperature, low solubility in organic solvents up to 175 °C, and good adhesion to a wide variety of substrates (where surfaces can be treated with silane to improve adhesion) [28]. More information on the material properties and processing procedure of parylene is available in Reference [28].

Patterning of the organic semiconductor with a parylene buffer layer is relatively straightforward. Spin-coating of photoresist on parylene results in a relatively uniform film across the wafer. Regular exposure and development steps were carried out to pattern the photoresist. Oxygen plasma RIE is used to transfer the pattern onto the parylene and the underlying organic semiconductor layer; photoresist served as an etch mask for the parylene and semiconductor layers. Again, careful timing of the etching in oxygen plasma is extremely critical to avoid over-etching and to suppress/avoid any possible process-induced degradation of the organic semiconductor layer.

A processing concern in this approach is the possible penetration of water or solvent under the parylene layer (through film edges) during photoresist developing and rinsing, which can affect the quality/properties of the underlying organic semiconductor layer. Possible remedies include the use of a thicker parylene layer, improving the adhesion of the parylene on the organic semiconductor via interface treatment, and cautious device processing/handling. Despite these minor imperfections, parylene-passivated PQT-12 OTFTs on glass substrates displayed

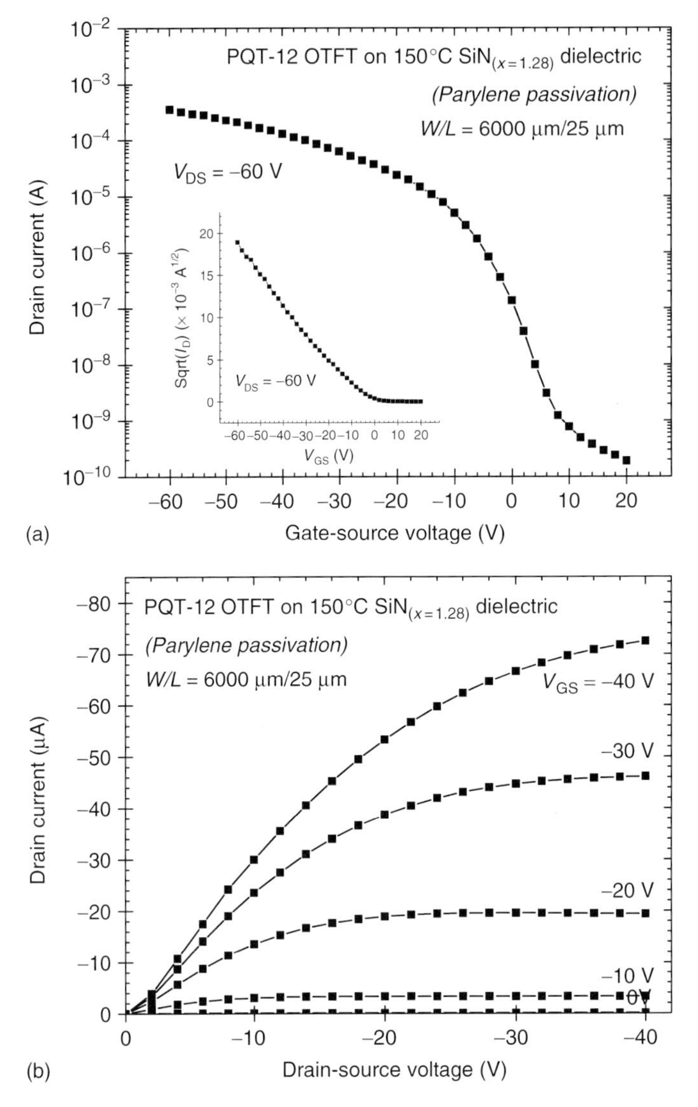

Figure 3.13 (a) Transfer and (b) output characteristics of a fully-patterned bottom-gate bottom-contact PQT-12 OTFT with SiN_x gate dielectric on a glass substrate. The polymer semiconductor layer was photolithographically patterned with parylene passivation layer.

a mobility of 0.04–0.12 $cm^2\,V^{-1}\,s^{-1}$; this is comparable to unpatterned PQT-12 OTFTs fabricated with the same SiN_x gate dielectric on a Si wafer using the scheme reported in Section 3.3.1. The ability to maintain comparable device mobility suggests successful implementation of this parylene-passivated approach for a photolithographically-defined fully-patterned and fully-encapsulated bottom-gate bottom-contact OTFT. Figure 3.13 illustrates the electrical characteristics of a parylene-passivated OTFT.

As an extension of this fabrication scheme, further process improvements can be achieved by incorporating an inorganic passivation layer (e.g., SiN_x, SiO_x, AlO_x) on top of parylene, to provide an additional level of protection against moisture, oxygen, or solvent penetration. This multilayer passivation approach, as illustrated in Figure 3.5b, is expected to minimize the environmental sensitivity of the organic material, and improve the stability and lifetime of OTFTs.

3.3.3 Hybrid Photolithography–Inkjet Printing Scheme for Fully-Patterned Bottom-Gate OTFT

We developed a hybrid photolithography-inkjet printing fabrication scheme to enable non-disruptive patterning of the organic semiconductor layer via printing, while providing the necessary feature resolution, patterned gate structures, and via-interconnect structures by photolithographic means. Since an all-printed process is envisioned for organic electronics fabrication in the future, this hybrid process can help bridge the transfer to an all-printed fabrication process while providing an immediate solution for OTFT circuit integration. The fabrication sequence for our hybrid photolithography–inkjet printing approach to produce fully patterned bottom-gate OTFTs is depicted in Figure 3.14. Figure 3.15 displays a cross-section diagram and actual microphotograph of a fully-patterned PQT-12 OTFT fabricated using this hybrid IJP scheme; the microphotograph shows that the inkjet printed polymer semiconductor has higher non-uniformity and inhomogeneity than the photolithographically-defined polymer layer (in Figure 3.11).

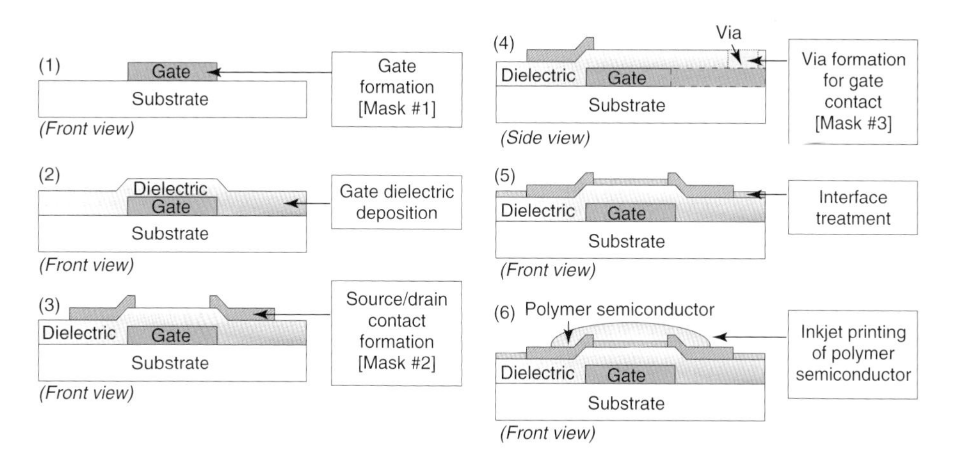

Figure 3.14 Hybrid photolithography–inkjet printing fabrication scheme for fully-patterned bottom-gate bottom-contact OTFT.

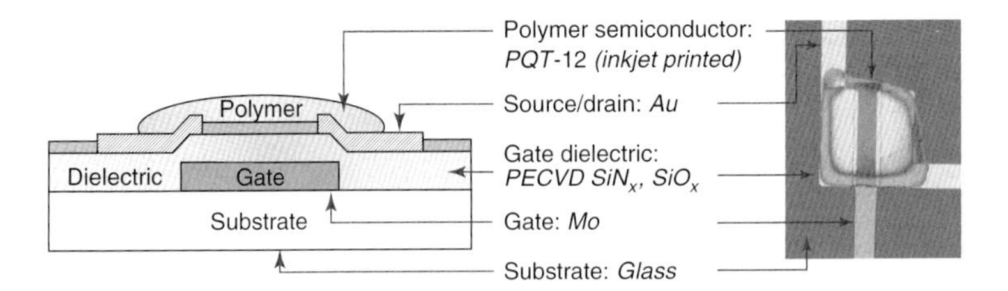

Figure 3.15 Cross-sectional illustration and photograph of a fully-patterned OTFT fabricated using the hybrid photolithography–inkjet printing scheme.

PQT-12 OTFT circuits (including inverters, ring oscillators, pixel circuits) were successfully fabricated using this approach; the results are documented in Chapter 7.

The gate electrode, source/drain contacts, and via/interconnect structure are defined by photolithography techniques. An organic semiconductor (PQT-12 in dichlorobenzene solvent) is inkjet printed onto the pre-patterned substrates using a Dimatix Materials Printer to define the transistor's active region [67]. Inkjet printed PQT-12 OTFTs on a SiN_x gate dielectric displayed effective mobility in the range $0.005–0.03\,cm^2\,V^{-1}\,s^{-1}$. Typical electrical characteristics are shown in Figure 3.16. The effective mobility of OTFTs with an inkjet printed PQT-12 is lower (by 5 to 10 times) than OTFTs with spin-coated PQT-12; this is attributed to non-uniformity and possibly to discontinuity of the inkjet printed organic layer, as shown in Figure 3.17. Device variation across a wafer was also more pronounced with the IJP approach. These observations indicate that better control over the IJP process is important to achieve better device performance.

Experiments showed that the resolution, quality, and performance (e.g., mobility) of the inkjet printed device are sensitive to ink formulation and a variety of printing parameters, including printing direction (see Figure 3.18), drop spacing, substrate and ink cartridge temperature, frequency and amplitude (or voltage) of inkjet driving waveform, and so on. For instance, the choice of solvent and ink concentration affects the evaporation rate of the ink droplet, which in turn influences the configuration of the printed film (i.e., dot-like or ring-like). A low evaporation rate yields small dot-like film and a fast evaporation rate results in wide ring-like film [68]. Ink formulation also determines the spreading of the ink droplet on the substrate, which affects film uniformity and image resolution. In addition, printing parameters or settings must be carefully controlled to deliver a continuous and uniform inkjet-printed organic layer needed for high performance OTFTs. Table 3.3 summarizes experimental observations on the impact of selected printing parameters on the quality of inkjet printed PQT-12 organic semiconductor thin film.

Researchers at Palo Alto Research Center (PARC) reported jet printed polythiophene OTFTs with mobility ($0.1\,cm^2\,V^{-1}\,s^{-1}$) comparable to spin-coated devices [43]. Thus, with technological optimization and precise control of the printing parameters, high performance inkjet printed OTFTs can be delivered. Moreover, the

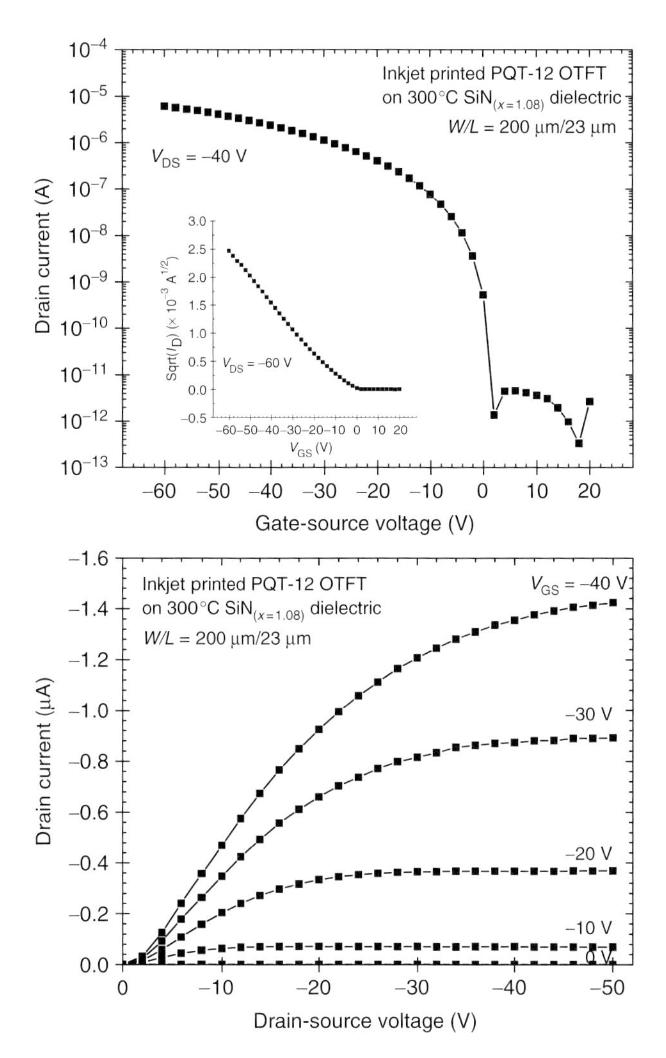

Figure 3.16 (a) Transfer and (b) output characteristics of a fully-patterned bottom-gate bottom-contact PQT-12 OTFT with SiN_x gate dielectric on a glass substrate. The organic semiconductor layer was deposited by inkjet printing.

ease of patterning compared to other approaches make IJP particularly promising and versatile for the fabrication of OTFTs for low cost electronics.

3.3.4 Photolithography Scheme for Top-Gate and Dual-Gate OTFTs

The fabrication scheme for a photolithograhically-defined dual-gate OTFT based on a five-mask process is outlined in Figure 3.19. This procedure can be adapted

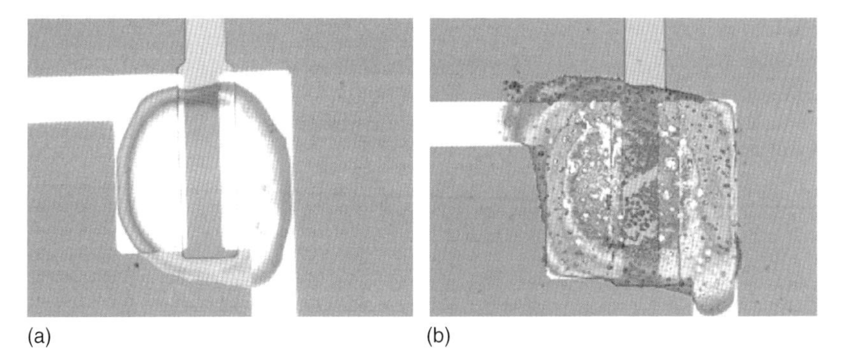

Figure 3.17 Photograph of OTFTs with inkjet printed organic semiconductor showing (a) a thin uniform layer and (b) a thicker non-uniform layer.

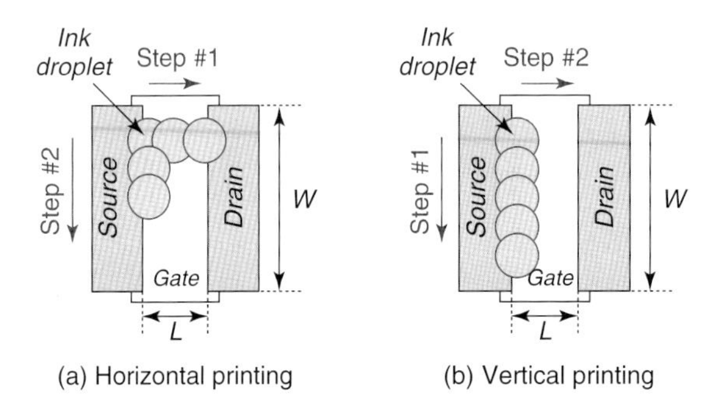

Figure 3.18 Illustration of inkjet printing direction: (a) horizontal and (b) vertical printing.

for fabrication of a top-gate OTFT, with the omission of Step (1) for bottom-gate deposition and Steps (8–9) for via/interconnect. Observe that a bottom-gate OTFT can also be realized with Steps (1–6) in Figure 3.19. Dual-gate and top-gate P3HT OTFTs have been demonstrated using this fabrication scheme [25, 27]; device characteristics are presented later in this section. The advantages of top-gate and dual-gate structures are discussed in Section 2.3.2.

Our earliest attempts to fabricate fully-patterned and fully-encapsulated photolithograhically defined top-gate and dual-gate OTFTs utilized PECVD SiN_x deposited at 75 °C as the top gate dielectric and masking/passivation layer. Low temperature SiN_x was selected deliberately to ensure thermal compatibility with the organic semiconductor. The optimized processing conditions and material characterization of the 75 °C PECVD SiN_x were reported in Ref. [69]. For top-gate and dual-gate OTFT designs, the deposition of the SiN_x layer proceeds in two stages. First, a thin SiN_x layer (50 nm thick) is formed and serves as a masking layer for the organic semiconductor film. Then, the organic semiconductor film is patterned by dry etching to define active regions of the semiconductor. A second SiN_x layer (150 nm thick) is deposited over the entire wafer and encapsulates

Table 3.3 Experimental summary of the impact of selected printing parameters on the quality of inkjet printed PQT-12 organic semiconductor thin film.

Inkjet printing parameter	Impact on quality of inkjet printed organic semiconductor layer
Printing direction (see Figure 3.18)	Vertical printing produces better device performance and less coffee ring effect compared to horizontal printing
Nozzle temperature	Increasing nozzle temperature (~60 °C) can prevent precipitation of polymer solution and avoid blockage of cartridge nozzles
Substrate temperature	Higher substrate temperature improves stability of ink after printing and enhances quality of the printed line. (60 °C was used)
Printing amplitude (voltage)	Amplitude needs to be adjusted to deliver large and stable drops Optimal jetting voltage varies with ink formulation Higher amplitude needed for more viscous fluids and higher concentration inks Higher amplitude increases drop volume, jetting speed, line-width
Printing frequency	Lowest jetting frequency gave stable drops with good registration Higher frequency leads to misalignment of the printed layer
Drop spacing	Drop spacing (i.e., center to center distance between adjacent drops) should be slightly smaller than drop size to ensure continuity of the printed line or pattern

the organic semiconductor layer. The top SiN_x layer is selectively etched to open source/drain contact windows. Finally, top metallization is deposited to define the top-gate electrode and contact pads.

3.3.4.1 Top-Gate OTFT

Electrical characteristics of a photolithographically-defined top-gate P3HT OTFT on glass are presented in Figure 3.20. The glass substrate surface was treated with OTS SAM prior to deposition of the P3HT semiconductor layer. Low-temperature PECVD SiN_x served as the top gate dielectric while providing a simultaneous solution for encapsulation of the organic semiconductor layer. The device demonstrated p-type field-effect transistor behavior and gate leakage current below 5 pA. However, the top-gate OTFT exhibited a low drain current, limited mobility and low on/off current ratio. Possible causes of the limited device performance are analyzed in Section 3.3.4.3.

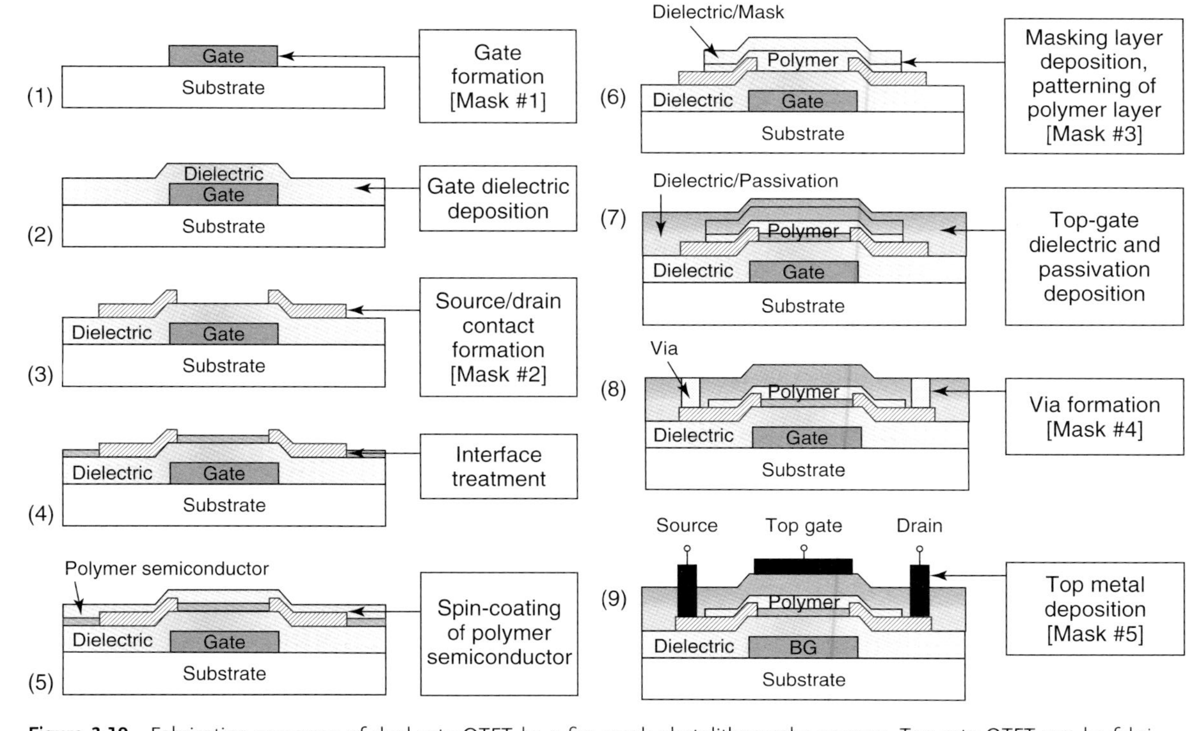

Figure 3.19 Fabrication sequence of dual-gate OTFT by a five-mask photolithography process. Top-gate OTFT can be fabricated by excluding steps (1), (8), and (9).

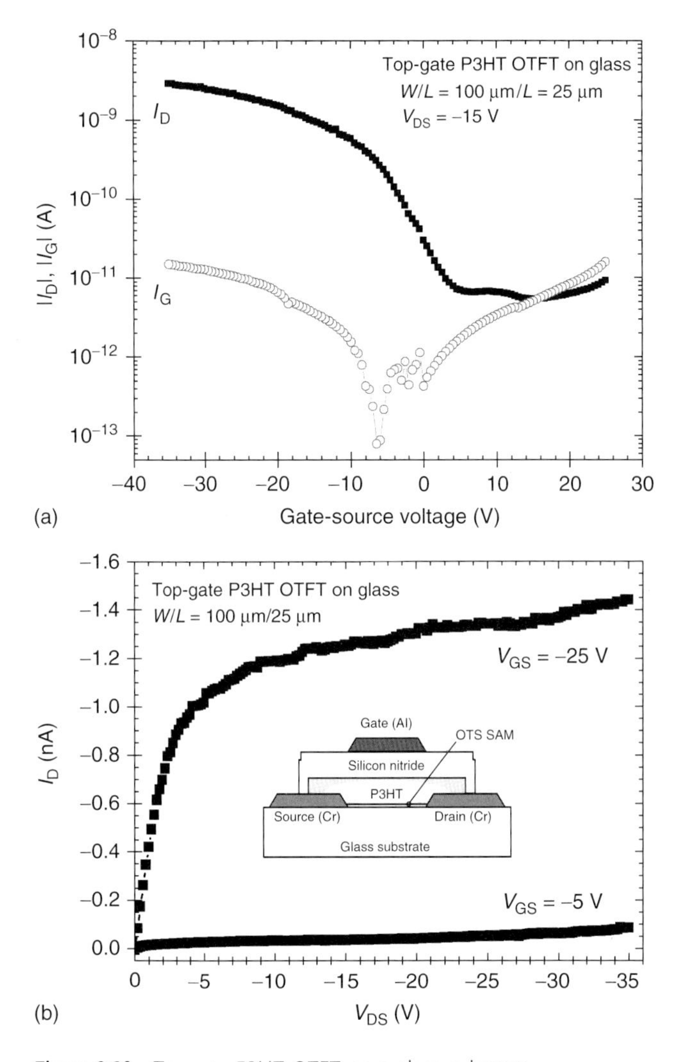

Figure 3.20 Top-gate P3HT OTFT on a glass substrate, fabricated by photolithography. (a) Transfer characteristics (I_D–V_{GS}). (b) Output characteristics (I_D–V_{DS}).

An advantage offered by the top-gate structure is that the source/drain contact layer is the first layer to be defined on the substrate. This first layer is generally the easiest to pattern since it is not inhibited by and does not need to contend with the topology of other device layers; thus, there is more flexibility in selecting an appropriate patterning technique to achieve a small L for enhancing μ_{FE}. Another feature of a top-gate OTFT is that the gate dielectric can provide encapsulation and protection of the organic semiconductor layer. Encapsulation is essential because most organic materials are chemically sensitive to environmental influences. Nevertheless, there are a number of process integration challenges associated with the top-gate OTFT design. First, the gate dielectric and gate electrode have to be deposited and structured on top of the organic semiconductor layer, and this process must preserve the organic material. Secondly, vertical interconnects and vias between the conductive layers necessitate the development of a compatible etching process for the organic layer. These issues were addressed in our process, in which low temperature SiN_x was selected as a passivating gate dielectric to ensure thermal conformity with the organic semiconductor, and a tailored etch recipe was developed for patterning the polymer layer with excellent compatibility.

3.3.4.2 **Dual-Gate OTFT**

This section describes one of the first dual-gate OTFTs reported in the literature. Our initial dual-gate OTFTs were prepared on an oxidized Si wafer, with a thermal SiO_2 layer as the bottom-gate dielectric and PECVD SiN_x as the top-gate dielectric and passivation [27]. The fabrication process of this SiO_2/P3HT/SiN_x dual-gate OTFT is similar to that outlined in Figure 3.19, except these OTFTs consisted of a uniform bottom-gate provided by the highly doped Si substrate; thus, the bottom-gate deposition step is neglected. Figure 3.21a shows the transfer characteristic of a dual-gate OTFT as a function of the top-gate voltage (V_{TG}), with the bottom-gate (V_{BG}) biased at 10 V. The output characteristics of the same device at various top-gate voltages are displayed in Figure 3.21b. The device demonstrated p-type transistor behavior, with a field effect mobility (μ_{FE}) of $10^{-5}\,cm^2\,V^{-1}\,s^{-1}$, a threshold voltage ($V_T$) of −1.9 V, an on/off current ratio (I_{ON}/I_{OFF}) of 10^3, and an inverse subthreshold slope (S) of 1.8 V dec^{-1}. The top gate leakage current was in the range 10^{-14}–10^{-12} A, indicating good electrical characteristics of the 75 °C SiN_x passivation dielectric. On the other hand, the bottom-gate component displayed poor field-effect transistor behavior, and a large leakage current was observed through the bottom-gate SiO_2 dielectric. Further assessments indicated that comparable dielectric leakage was present in OTFTs with SiN_x as the bottom-gate dielectric, and that the leakage current became significant only after polymer deposition on the bottom-gate dielectric. A similar dielectric leakage phenomenon was reported by Raja *et al.* [70, 71]. The large increase in dielectric leakage currents as a result of polymer deposition was attributed to the displacement of dopant ions or impurities present in the polymer film, as well as impurity contamination of the oxide from the polymer-chloroform solution. (A brief discussion of the limitations of commercially available polymer semiconductor is presented in Section 2.4.1) More in-depth investigation of this matter is needed. It is believed that purification of the commercially-purchased polymer material prior to deposition and selection of an appropriate solvent are necessary to alleviate this leakage problem.

A valuable/interesting property of dual-gate OTFTs is that the voltage bias on the bottom-gate has a distinct influence on the threshold voltage, subthreshold slope, on-current, and leakage current of the top-gate TFT. A larger negative V_{BG} resulted in a larger $|I_D|$ and positive shift in V_T. Likewise, there is a dependence of the bottom-gate TFT characteristics on the top-gate bias. Similar dual-gate effects were reported in Ref. [72]. With the ability to control selected TFT parameters, the

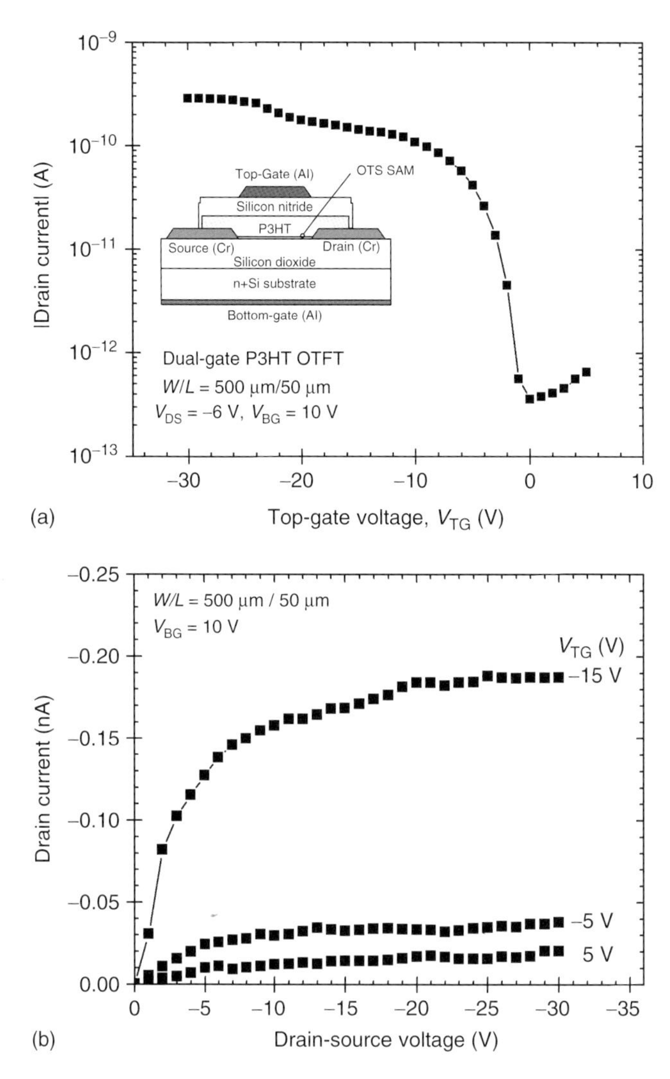

Figure 3.21 Dual-gate P3HT OTFT on a Si wafer. Bottom-gate dielectric is SiO_2, and top-gate dielectric is PECVD SiN_x. (a) Transfer characteristics (I_D-V_{TG}) of the dual-gate OTFT as a function of top-gate voltage. (b) Output characteristics (I_D-V_{DS}) of the dual-gate OTFT at various top-gate voltages.

dual-gate OTFT structure is attractive for circuit applications that demand high threshold voltage control and good reliability. The dual-gate OTFT also lends itself as a highly functional test structure for characterization of the density of states at the top and bottom interfaces of the active organic and dielectric layers, to evaluate the interface integrity and provide insight into the underlying transport mechanisms. Another promising application of the dual-gate OTFT is seen in AMOLED display circuits where vertical integration is desirable to achieve a high aperture ratio and a high on-pixel integration density. Another promising application of the dual-gate OTFT is to implement a vertically stacked active-matrix backplane for displays or imagers; vertical integration is desirable to achieve a high aperture ratio and a high on-pixel integration density. This concept is discussed in Section 7.4.1.

3.3.4.3 **Analysis**

The limited performance (i.e., low current drive, mobility, current on/off ratio) of these preliminary photolithographically-defined top-gate and dual-gate OTFTs can be attributed to various factors, including environmental sensitivity of the polymer semiconductor, process-induced degradation, non-optimized dielectric/semiconductor interfaces, and large contact resistance. First, the poor field-effect mobility and on/off current ratio, relative to other P3HT OTFTs reported in the literature [73, 74], are likely linked to the degradation of P3HT due to exposure to air. P3HT is sensitive to oxygen and water molecules, and becomes p-type doped upon absorption of atmospheric oxygen. Consequently, the material experiences an increase in free carrier density and a higher conductivity [75]. These changes translate to higher off-current, lower on/off current ratio, and negligible field-effect and gate-modulation effects in OTFT. Our observations are consistent with the data reported by Han *et al.*, where they noted a decrease in mobility from 10^{-3} to $10^{-5}\,cm^2\,V^{-1}\,s^{-1}$ and a decay in on/off current ratio from 10^4 to 10 following exposure of the P3HT OTFTs to air for only 15 min [75]. The rapid degradation of P3HT in air, during processing and measurement, also accounts for the reduced performance of our devices. Processing in an inert environment, as well as purification of the polymer materials, should lead to enhanced OTFT performance.

Other potential factors contributing to the limited OTFT performance include process-induced degradation and inferior quality of the top P3HT/SiN_x interface. Process-induced device degradation can occur during deposition of the PECVD SiN_x films, where ion bombardment during the initial deposition stage can adversely impact the P3HT/SiN_x interface quality, and emission of UV radiation from plasma processing can alter the P3HT film properties. These observations are substantiated by the results reported in Ref. [76]. The effects of stress in the SiN_x layer can also undermine the device performance. These concerns can be circumvented by selecting alternative top passivation dielectric materials with improved compatibility with the organic semiconductor layer. As demonstrated in Section 3.3.2, parylene is a viable candidate. An evaporated SiO_2 layer and a solution-processed benzocyclobutene (BCB) layer are other possible options for the passivation dielectric in photolithographically-defined OTFTs.

Contact effects constitute another source of the poor OTFT performance. Cr was used as the source/drain contacts in these preliminary dual-gate and top-gate OTFTs, due to the unavailability of higher work function metals (e.g., Au) during early stages of the research. The lower work function of Cr may cause an injection barrier at the Cr/P3HT interface, thus restricting device performance. Application of a higher work function contact material and optimization of the contact/polymer

Table 3.4 Summary of strengths and weaknesses of the various OTFT integration approaches developed in this research. The approaches are grouped by the technique used for depositing/patterning organic layer: photolithography and inkjet printing.

Approaches to pattern polymer semiconductor	Advantages	Challenges/ drawbacks	Reliability
Photolithography:			
Direct patterning of polymer semiconductor (PQT-12) using photoresist	One less deposition step (compared to indirect patterning)	Adhesion issues between PQT-12 and photoresist; solvent processing may affect PQT-12	Inconsistency in reproducibility
Indirect patterning of polymer semiconductor using parylene as buffer/ passivation layer	Better processing control and photoresist handling than direct patterning; highest mobility amongst the schemes reported	Water/solvent may penetrate under parylene during photolithography processing	Good
Indirect patterning of polymer semiconductor using SiN_x as buffer/ passivation layer	SiN_x is a good/stable passivation material	PECVD deposition of SiN_x may induce damages in polymer layer	Limited
Inkjet printing:			
(Hybrid photolithography–inkjet printing scheme)	Easiest patterning method for organic material; reduced process complexity; robust method	Resolution limitation; uneven spreading of ink may cause non-uniform semiconductor in the TFT channel	Good

interface (by appropriate interface treatment) are expected to generate enhanced device characteristics in future experiments.

3.3.5 Fabrication Scheme Comparisons

The strengths and limitations of the above-described fabrication schemes and patterning techniques are summarized in Table 3.4. Table 3.5 compares the performance of OTFTs fabricated by the various integration schemes. The photolithography approach using indirect patterning with a parylene buffer layer delivered the highest mobility, and, thus, is currently the most robust method overall. However, IJP presents a larger technological impact owing to its ease of

Table 3.5 Comparing device performance of fully-patterned OTFTs fabricated by various approaches. Best values of field-effect mobility and on/off current ratio measured are shown.

Approach	Mobility ($cm^2 V^{-1} s^{-1}$)	On/off current ratio	Remark
Photolithography: direct patterning with photoresist	0.003–0.008	2.89×10^4	Low mobility and on/off current ratio
Photolithography: indirect patterning with parylene buffer layer	0.04–0.12	1.89×10^6	Highest mobility
Hybrid photolithography–inkjet-printing	0.005–0.033	1.84×10^7	Highest on/off current ratio

patterning for the various device layers. Nonetheless, extensive optimization of printing parameters and processing conditions is needed to excel the performance of inkjet printed OTFTs.

3.4 Summary and Contributions

The establishment of robust and reliable OTFT integration strategies is a prerequisite for full-scale deployment of organic electronics manufacturing. The key challenge in this area is finding a reliable method to pattern functional organic materials. This chapter presents a number of OTFT fabrication schemes, which involved tailoring the photolithography process for organic materials and exploring the IJP technique to enable robust and effective direct patterning of polymers. The objective here is to exploit photolithography to enable immediate fabrication of highly integrated and high resolution OTFT circuits, allowing timely evaluation of device behavior and demonstration of integrated organic electronics. Inkjet printing is gradually incorporated into the fabrication process as the IJP technology gains maturity and strength. The development of a hybrid photolithography–inkjet printing scheme serves as a technological bridge toward an all-printed fabrication process for the near future.

Key contributions highlighted in this chapter include:

- Development of integration schemes that enable fabrication of fully-patterned fully-encapsulated OTFTs and circuits.

Table 3.6 Comparison of the strengths and drawbacks of the fabrication/patterning techniques considered.

Patterning techniques	Strengths	Drawbacks
Shadow mask (Figure 3.2)	☑ Simple, robust process ☑Enabled patterning of evaporated source/drain contacts and organic films ☑Cost-effective, large-area ☑Direct deposition on the surface	☒Limited feature resolution ☒Difficult to pattern multilayers and circuits ☒Lack of alignment mechanism ☒Not applicable to solution-processed polymers
Photolithography (Figures 3.9 and 3.19)	☑High feature resolution ☑Precise device definition ☑Ease of integration, standard IC compatible process ☑Enables higher complexity devices/circuits ☑Mature technology – provides a good starting point for building and studying OTFT circuits	☒Higher process complexity ☒Issues with solvent compatibility ☒Requires a compatible and robust masking layer ☒Etching solutions and developers may cause damage to polymer films
Inkjet printing	☑Non-contact, additive printing process ☑Direct patterning ☑Minimum material consumption ☑Low cost, fewer steps ☑Custom circuit design	☒Typical system resolution: ~25 μm, limited by spreading of ink droplet on the substrate ☒Film non-uniformity and discontinuity ☒Requires precise control of viscosity and careful choice of solvents
Hybrid photolithography-inkjet printing (Figure 3.14)	☑Non-disruptive patterning of polymer semiconductor layer by inkjet printing ☑High resolution by photolithography ☑Bridges the transfer to an all-printed fabrication process	☒Reliant on photolithography as a temporary solution for defining critical dimensions of the transistor

- Design and formulation of an organic-friendly photolithography fabrication process, providing solutions that address concerns of chemical attacks and other process-induced damage to the organic layer.
- Demonstration of a hybrid photolithography–inkjet printing approach to allow robust fabrication of high-resolution OTFT devices. This development helps bridge the technological transfer to an all-printed process in the longer term.
- Realization of a variety of OTFT architectures using these integration schemes, including bottom-gate, top-gate and dual-gate OTFTs.

- Demonstration of the applicability and expandability of these integration strategies to other material systems (e.g., nanocomposite TFTs), providing a convenient platform to prototype and evaluate a wide range of new/novel material systems.

Table 3.6 presents an overall summary of the strengths and weaknesses of the patterning techniques considered. A comparison of the pros and cons of the various techniques suggests that IJP is the most promising method for patterning organic materials. However, additional research and development is needed to advance the IJP technique to facilitate well-controlled mass-production of high performance organic devices. Before IJP technology can reach its full potential and maturity for electronics production, the hybrid photolithography–inkjet printing approach provides an appealing interim solution, as it enjoys the high resolution and registration advantages of photolithography, while benefiting from the convenient and robust deposition of a patterned organic layer by IJP. This hybrid scheme delivers an integration strategy with workable manufacturing yields while lowering costs compared to conventional processes.

References

1. Horowitz, G. (1998) Organic field-effect transistors. *Adv. Mater.*, **10**, 365.
2. Dimitrakopoulos, C.D. and Mascaro, D.J. (2001) Organic thin-film transistors: a review of recent advances. *IBM J. Res. Dev.*, **45** (1), 111.
3. Kymissis, I., Dimitrakopoulos, C.D., and Purushothaman, S. (2002) Patterning pentacene organic thin film transistors. *J. Vac. Sci. Technol. B*, **20**(3), 956.
4. Sirringhaus, H., Kawase, T., Friend, R.H., Shimoda, T., Inbasekaran, M., Wu, W., and Woo, E.P. (2000) High-resolution inkjet printing of all-polymer transistor circuits. *Science*, **290**, 2123.
5. Jia, H., Gross, E.K., Wallace, R.M., and Gnade, B.E. (2007) Patterning effects on poly(3-hexylthiophene) organic thin film transistors using photolithographic processes. *Org. Electron.*, **8**, 44.
6. Kawase, T., Shimoda, T., Newsome, C., Sirringhaus, H., and Friend, R.H. (2003) Inkjet printing of polymer thin film transistors. *Thin Solid Films*, **438**, 279.
7. Jabbour, G.E., Radspinner, R., and Peyghambarian, N. (2001) Screen printing for the fabrication of organic light-emitting devices. *IEEE J. Sel. Top. Quantum Electron.*, **7** (5), 769.
8. Rogers, J.A., Bao, Z., Meier, M., Dodabalapur, A., Schueller, O.J.A., and Whitesides, G.M. (2000) Printing, molding, and near-field photolithographic methods for patterning organic lasers, smart pixels and simple circuits. *Synth. Met.*, **115**, 5.
9. Rogers, J.A., Bao, Z.N., and Raju, V.R. (1998) Nonphotolithographic fabrication of organic transistors with micron feature sizes. *Appl. Phys. Lett.*, **72**, 2716.
10. Chabinyc, M.L., Wong, W.S., Salleo, A., Paul, K.E., and Street, R.A. (2002) Organic polymeric thin-film transistors fabricated by selective dewetting. *Appl. Phys. Lett.*, **81**, 2460.
11. Sze, S.M. (1988) *VLSI Technology*, McGraw-Hill, New York.
12. Sze, S.M. (1985) *Semiconductor Devices, Physics and Technology*, John Wiley & Sons, Inc., New York.
13. Anthopoulos, T.D., Singh, B., Marjanovic, N., Sariciftci, N.S., Ramil, A.M., Sitter, H., Colle, M., and de Leeuw, D.M. (2006) High performance n-channel organic field-effect transistors and ring oscillators based on C_{60} fullerene films. *Appl. Phys. Lett.*, **89**, 213504.

14. Tiwari, S.P., Namdas, E.B., Ramgopal Rao, V., Fichou, D., and Mhaisalkar, S.G. (2007) Solution-processed n-type organic field-effect transistors with high on/off current ratios based on fullerene derivatives. *IEEE Electron Dev. Lett.*, **28**, 880.
15. Afzali, A., Dimitrakopoulos, C.D., and Breen, T.L. (2002) High-performance, solution-processed organic thin film transistors from a novel pentacene precursor. *J. Am. Chem. Soc.*, **124**, 8812.
16. Bao, Z., Dadabalapur, A., Katz, H.E., Raju, R.V., and Rogers, J.A. (2001) Organic semiconductors for plastic electronics. *IEEE-TAB New Technology Directions Committee (NTDC), 2001 IEEE Workshop on New and Emerging Technologies.*
17. Sirringhaus, H., Brown, P.J., Friend, R.H., Nielsen, M.M., Bechgaard, K., Langeveld-Voss, B.M.W., Spiering, A.J.H., Janssen, R.A.J., Meijer, E.W., Herwig, P.T., and de Leeuw, D.M. (1999) Two-dimensional charge transport in self-organized, high-mobility conjugated polymers. *Nature*, **401**, 685.
18. Menard, E., Meitl, M.A., Sun, Y., Park, J.U., Shir, D.J., Nam, Y.S., Jeon, S., and Rogers, J.A. (2007) Micro- and nanopatterning techniques for organic electronic and optoelectronic systems. *Chem. Rev.*, **107**, 1117.
19. Tian, P.F., Bulovic, V., Burrows, P.E., Gu, G., Forrest, S.R., and Zhou, T.X. (1999) Precise scalable patterning of vacuum-deposited organic light emitting devices. *J. Vac. Sci. Technol., A*, **17**, 2975.
20. Bartic, C., Jansen, H., Campitelli, A., and Borghs, S. (2002) Ta_2O_5 as gate dielectric material for low-voltage organic thin-film transistors. *Org. Electron.*, **3**, 65.
21. Forrest, S.R. (2004) The path to ubiquitous and low-cost organic electronic appliances on plastic. *Nature*, **428**, 911.
22. Ling, M.M. and Bao, Z.N. (2004) Thin film deposition, patterning, and printing in organic thin film transistors. *Chem. Mater.*, **16**, 4824.
23. Yi, S.M., Jin, S.H., Lee, J.D., and Chu, C.N. (2005) Fabrication of a high-aspect-ratio stainless steel shadow mask and its application to pentacene thin-film transistors. *J. Micromech. Microeng.*, **15**, 263.
24. Baude, P.F., Ender, D.A., Kelley, T.W., Muyres, D.V., and Theiss, S.D. (2004) Pentacene-based RFID transponder circuitry. *62nd Device Research Conference (DRC) Digest*, pp. 227–228.
25. Li, F.M., Vygranenko, Y., Koul, S., and Nathan, A. (2006) Photolithographically defined polythiophene OTFTs. *J. Vac. Sci. Technol. A*, **24**, 657.
26. Li, F.M., Koul, S., Vygranenko, Y., Sazonov, A., Servati, P., and Nathan, A. (2005) Fabrication of RR-P3HT-based TFTs using low-temperature PECVD silicon nitride. *Mater. Res. Soc. Symp. Proc.*, **871E**, pp. I9.3.1–I9.3.6.
27. Li, F.M., Koul, S., Vygranenko, Y., Servati, P., and Nathan, A. (2005) Dual-gate SiO_2/P3HT/SiN_x OTFT. *Mater. Res. Soc. Symp. Proc.*, **871E**, pp. I9.4.1–I9.4.6.
28. Parylene Conformal Coatings Specifications and Properties, Technical notes, Specialties Coating Systems. *http://www.scscookson.com/parylene_knowledge/specifications.cfm* (accessed 2010).
29. Hilord Chemical Corporation (2003) Ink-Jet Printing, *http://www.hilord.com/InkJet.htm* (accessed 2010).
30. (2007) (Jan 3: 2007) UK in Plastic Electronics Drive. BBC News. *http://news.bbc.co.uk/2/hi/business/6227575.stm* (accessed 2010).
31. Calvert, P. (2001) Inkjet printing for materials and devices. *Chem. Mater.*, **13**, 3299.
32. Micro Device Science (MDS) Ltd. (2003) Plastic Electronics. *http://www.mdslsolutions.com/plastronics.htm* (accessed 2010).
33. Zhang, J., Brazis, P., Chowdhuri, A.R., Szczech, J., and Gamota, D. (2002) Investigation of using contact and non-contact printing technologies for organic transistor fabrication. *Mater. Res. Soc. Symp. Proc.*, **725**, 6.3.1.
34. Bao, Z. (2004) Conducting polymers – fine printing. *Nat. Mater.*, **3**, 137.
35. Wang, J.Z., Zheng, Z.H., Li, H.W., Huck, W.T.S., and Sirringhaus, H.

(2004) Dewetting of conducting polymer inkjet droplets on patterned surfaces. *Nat. Mater.*, **3**, 171.
36. Shimoda, T., Morii, K., Seki, S., and Kiguchi, H. (2003) Inkjet printing of light-emitting polymer displays. *Mater. Res. Soc. Bull.*, **28**, 821–827.
37. Hebner, T.R., Wu, C.C., Marcy, D., Lu, M.H., and Sturm, J.C. (1998) Ink-jet printing of doped polymers for organic light emitting devices. *Appl. Phys. Lett.*, **72**, 519.
38. Kim, D., Jeong, S., Lee, S., Park, B.K., and Moon, J. (2007) Organic thin film transistor using silver electrodes by the ink-jet printing technology. *Thin Solid Films*, **515**, 7692.
39. Sirringhaus, H., Kawase, T., and Friend, R.H. (2001) High-resolution ink-jet printing of all-polymer transistor circuits. *Mater. Res. Soc. Bull.*, **26**, 539.
40. Burns, E.S., Cain, P., Mills, J., Wang, J., and Sirringhaus, H. (2003) Inkjet printing of polymer thin-film transistor circuits. *Mater. Res. Soc. Bull.*, **28**, 829.
41. Speakman, S.S., Rozenberg, G.G., Clay, K.J., Milne, W.I., Ille, A., Gardner, I.A., Bresler, E., and Steinke, J.H.G. (2001) High performance organic semiconducting thin films: ink jet printed polythiophenes [rr-P3HT]. *Org. Electron.*, **2**, 65.
42. Plötnera, M., Wegener, T., Richter, S., Howitz, S., and Fischer, W.J. (2004) Investigation of ink-jet printing of poly-3-octylthiophene for organic field-effect transistors from different solutions. *Synth. Met.*, **147**, 299.
43. Arias, A.C., Daniel, J.H., Endicott, F., Lujan, R.A., and Krusor, B.S. (2006) Flexible (ink) jet printed backplanes for displays: materials development and integration. *Spring 2006. Meeting of the Materials Research Society*, 2006 April 17–21, San Francisco, CA, USA.
44. Kumar, A., Biebuyck, H.A., Abbott, N.L., and Whitesides, G.M. (1992) The use of self-assembled monolayers and a selective etch to generate patterned gold features. *J. Am. Chem. Soc.*, **114**, 9188.
45. Xia, Y. and Whitesides, G.M. (1998) Soft lithography. *Angew. Chem. Int. Ed.*, **37**, 550.
46. Rogers, J.A., Bao, Z., Makhija, A., and Braun, P. (1999) Printing process suitable for reel-to-reel production of high-performance organic transistors and circuits. *Adv. Mater.*, **11**, 741.
47. Rogers, J.A., Bao, Z., Baldwin, K., Dodabalapur, A., Crone, B., Raju, V.R., Kuck, V., Katz, H.E., Amundson, K., Ewing, J., and Drzaic, P. (2001) Paper-like electronic displays: Large-area rubber-stamped plastic sheets of electronics and microencapsulated electrophoretic inks. *Proc. Natl. Acad. Sci.*, **98**, 4835.
48. Leufgen, M., Lebib, A., Muck, T., Bass, U., Wagner, V., Borzenko, T., Schmidt, G., Geurts, J., and Molenkamp, L.W. (2004) Organic thin-film transistors fabricated by microcontact printing. *Appl. Phys. Lett.*, **84**, 1582.
49. Delamarche, E., Vichiconti, J., Hall, S.A., Geissler, M., Graham, W., Michel, B., and Nunes, R. (2003) Electroless deposition of Cu on glass and patterning with microcontact printing. *Langmuir*, **19**, 6567.
50. Zschiechang, U., Halik, M., and Klauk, H. (2008) Microcontact-printed self-assembled monolayers as ultrathin gate dielectrics in organic thin-film transistors and complementary circuits. *Langmuir*, **24**, 1665.
51. Zschieschang, U., Klauk, H., Halik, M., Schmid, G., and Dehm, C. (2003) Flexible organic circuits with printed gate electrodes. *Adv. Mater.*, **15**, 1147.
52. Wu, Y., Li, Y., Ong, B.S., Liu, P., Gardner, S., and Chian, B. (2005) High-performance organic thin-film transistors with solution-printed gold contacts. *Adv. Mater.*, **17**, 184.
53. Wu, Y., Li, Y., and Ong, B.S. (2006) Printed silver ohmic contacts for high-mobility organic thin-film transistors. *J. Am. Chem. Soc.*, **128**, 4202.
54. Wu, Y., Li, Y., Liu, P., Gardner, S., and Ong, B.S. (2006) Studies of gold nanoparticles as precursors to printed conductive features for thin-film transistors. *Chem. Mater.*, **18**, 4627.
55. Kagan, C.R., Breen, T.L., and Kosbar, L.L. (2001) Patterning organic-inorganic thin-film transistors using microcontact

printed templates. *Appl. Phys. Lett.*, **79**, 3536.
56. Park, S.K., Kim, Y.H., Han, J.I., Moon, D.G., and Kim, W.K. (2002) High-performance polymer TFTs printed on a plastic substrate. *IEEE Trans. Electron Devices*, **49**, 2008.
57. Takakuwa, A. and Azumi, R. (2008) Influence of solvents in micropatterning of semiconductors by microcontact printing and application to thin-film transistor devices. *Jpn. J. Appl. Phys.*, **47**, 1115.
58. Gray, C., Wang, J., Duthaler, G., Ritenour, A., and Drzaic, P.S. (2001) Screen printed organic thin film transistors (OTFTs) on a flexible substrate. *Process. SPIE*, **4466**, 89.
59. Bao, Z., Feng, Y., Dodabalapur, A., Raju, V.R., and Lovinger, A.J. (1997) High-performance plastic transistors fabricated by printing techniques. *Chem. Mater.*, **9**, 1299.
60. Lim, S.C., Seong, H., Yang, Y.S., Lee, M.Y., Nam, S.Y., and Ko, J.B. (2009) Organic thin-film transistor using high-resolution screen-printed electrodes. *Jpn. J. Appl. Phys.*, **48**, 081503.
61. Jo, J., Yu, J.-S., Lee, T.M., Kim, D.S., and Kim, K.Y. (2010) Roll-printed organic thin-film transistor using patterned poly(dimethylsiloxane) (PDMS) stamp. *J. Nanosci. Nanotechnol.*, **10**, 3595.
62. Yan, H., Chen, Z., Zheng, Y., Newman, C., Quinn, J., Dotz, F., Kastler, M., and Facchetti, A. (2009) A high-mobility electron-transporting polymer for printed transistors. *Nature*, **457**, 679.
63. Blanchet, G.B., Loo, Y.L., Rogers, J.A., Gao, G., and Fincher, C.R. (2003) Large area, high resolution, dry printing of conducting polymers for organic electronics. *Appl. Phys. Lett.*, **82**, 463.
64. Li, F.M., Dhagat, P., Jabbour, G.E., Haverinen, H.M., McCulloch, I., Heeney, M., and Nathan, A. (2008) Polymer TFT without surface pre-treatment on silicon nitride gate dielectric. *Appl. Phys. Lett.*, **93** (7), 073305.
65. Beecher, P., Servati, P., Rozhin, A., Colli, A., Scardaci, V., Pisana, S., Hasan, T., Flewitt, A.J., Robertson, J., Hsieh, G.W., Li, F.M., Nathan, A., Ferrari, A.C., and Milne, W.I. (2007) Ink-jet printing of carbon nanotube thin film transistors. *J. Appl. Phys.*, **102**, 043710.
66. Hsieh, G.W., Beecher, P., Li, F.M., Servati, P., Colli, A., Fasoli, A., Chu, D., Nathan, A., Ong, B., Robertson, J., Ferrari, A.C., and Milne, W.I. (2008) Formation of composite organic thin film transistors with nanotubes and nanowires. *Physica E Low Dimens. Syst. Nanostruct.*, **40**, 2406–2413.
67. (a) Dimatix Fujifilm (2007) *http://www.dimatix.com/* (accessed December 20); (b) Dimatix Materials Printer Manual, pp. 92–93.
68. Kaneda, M., Ishizuka, H., Sakai, Y., Fukai, J., Yasutake, S., and Takahara, A. (2007) Film formation from polymer solution using inkjet printing method. *AIChE J.*, **53**, 1100.
69. McArthur, C., Meitine, M., and Sazonov, A. (2003) Optimization of 75 °C amorphous silicon nitride for TFTs on plastics. *Mater. Res. Soc. Proc.*, **769**, 303.
70. Raja, M., Lloyd, G., Sedghi, N., Higgins, S.J., and Eccleston, B. (2002) Critical considerations in polymer thin-film transistor (TFT) dielectrics. *Mater. Res. Soc. Symp. Proc.*, **725**, P6.5.1–P6.5.6.
71. Lloyd, G., Raja, M., Sellers, I., Sedghi, N., Lucrezia, R.D., Higgins, S., and Eccleston, B. (2001) The properties of MOS structures using conjugated polymers as the semiconductor. *Microelectron. Eng.*, **59**, 323.
72. Iba, S., Sekitani, T., Kato, Y., Someya, T., Kawaguchi, H., Takamiya, M., Sakuri, T., and Takagi, S. (2005) Control of threshold voltage of organic field-effect transistors with double-gate structures. *Appl. Phys. Lett.*, **87**, 023509.
73. Bao, Z., Dodabalapur, A., and Lovinger, A.J. (1996) Soluble and processable regioregular poly(3-hexylthiophene) for thin film field-effect transistor applications with high mobility. *Appl. Phys. Lett.*, **69**, 4108.
74. Sirringhaus, H., Tessler, N., and Friend, R.H. (1998) Integrated optoelectronic devices based on conjugated polymers. *Science*, **280**, 1741.
75. Han, J.I., Kim, Y.H., Park, S.K., Moon, D.G., and Kim, W.K. (2004) Stability of

organic thin film transistors. *Mater. Res. Soc. Symp. Proc.*, **814**, I4.3.1.

76. Kawashima, N., Nomoto, K., Wada, M., and Kasahara, J. (2005) Organic-inorganic hybrid encapsulation for P3HT field-effect transistors. *Mater. Res. Soc. Symp. Proc.*, **817E**, I1.10.1. (pp. 37–42).

4
Gate Dielectrics by Plasma Enhanced Chemical Vapor Deposition (PECVD)

The demands on the gate dielectric layer in organic thin film transistors (OTFTs) are multiple, as summarized in Section 2.4.2. For instance, the material should form a pinhole free film with a high breakdown voltage and long term stability. The leakage current through the dielectric layer should be as low as possible to ensure low power consumption of the device. A high dielectric constant is desirable for low-voltage operation and a high current output. In the case of bottom-gate OTFTs, the dielectric layer presents a platform for the growth/formation of the organic semiconductor layer; thus, the surface properties of the gate dielectric play an important role in determining the molecular orientation, ordering, and microstructure of the organic semiconductor layer. These physical attributes are closely linked to the electrical properties of the semiconductor layer, which in turn influence the OTFT performance (e.g., field effect mobility, threshold voltage, on/off current ratio). Gate dielectric materials for OTFTs can be largely grouped into two categories: organic and inorganic. An overview of the two dielectric groups is presented in Section 4.1. The remainder of this chapter will focus on the development of inorganic gate dielectrics specific for OTFT applications.

4.1
Overview of Gate Dielectrics

4.1.1
Organic Dielectrics

The ultimate goal for OTFT research is to make devices with high field-effect mobility, high current on/off ratio, low threshold voltage, and subthreshold swing when operating the devices at low-voltage levels. The aforementioned transistor parameters depend critically on the properties of the gate dielectric. For example, dielectric surface properties affect the trapped charge densities at the interface between the semiconductor and the gate dielectric, thus they affect the field-effect mobility and the threshold voltage dramatically. The thickness of the gate dielectric, pinhole density, and dielectric constant are critical to the current on/off ratio, threshold voltage and subthreshold swing, and operating voltage. A variety of

Organic Thin Film Transistor Integration: A Hybrid Approach, First Edition. Flora M. Li, Arokia Nathan, Yiliang Wu, and Beng S. Ong.

organic materials have been used as gate dielectric in OTFTs, including poly(4-vinyl phenol) [1], poly(methyl methacrylate),2 polystyrene [2], poly(vinylchloride), poly (a-methylstyrene) [3], poly(vinyl alcohol) [4], poly(2-hydroxyethyl methacrylate) [5], poly(vinyl cinnamate) [6], polyimide [7, 8], poly(vinylidenfluoride) [9], cyano-ethylpullulan [10, 11], divinyltetramethyldisiloxane-bis(benzocyclobutene) [12], polypropylene [13], polypropylene-co-1-butene, and the amorphous fluoropolymer CYTOP, to name a few [13]. These organic gate dielectrics have been extensively reviewed previously [13, 14]. We focus here on two important topics that dramatically affect the aforementioned transistor parameters: the compatibility with semiconductors, and the production of high-capacitance dielectrics for operating transistors at low-voltage levels.

Our studies revealed that there was no universal dielectric for all organic semiconductors. Often requirements for the dielectric layer differ depending on the type of semiconductor, since the interfacial interactions of the semiconductor layer with the gate dielectric are specific. The selection of a gate dielectric and semiconductor pair with excellent interface properties remains one of the challenges for OTFT studies. One great example is the evolution of pentacene transistors. When bare silicon oxide was used as the gate dielectric, the device showed a low field effect mobility around 2×10^{-3} $cm^2\ V^{-1}\ s^{-1}$ [15]. Upon surface modification of the silicon oxide with a silane agent, such as octadecyltrichlorosilane, the mobility increased by several orders of magnitude to 1.0–1.7 $cm^2\ V^{-1}\ s^{-1}$ for the same pentacene semiconductor [16]. When a high-k dielectric Al_2O_3 was used and modified with phosphonic acid compounds, the device exhibited a further improvement of mobility by a factor of 2 [17]. The highest mobility, around 5.5 $cm^2\ V^{-1}\ s^{-1}$, was observed on a specific polymeric dielectric – poly(a-methylstyrene) [3]. Solution-processable polymeric semiconductors showed a similar dependence of device performance on the surface properties of the gate dielectric. For example, poly(3,3‴-didodecyl quarterthiophene) (PQT-12) showed very low field-effect mobility on a bare silicon oxide gate dielectric, but more than 2 orders of magnitude higher mobility upon modification of the silicon oxide with octyltrichlorosilane (OTS) [18]. Interestingly, this semiconductor showed medium mobility on a polymeric gate dielectric such as poly(4-vinylphenol). Modification of the polymeric gate dielectric with a polymeric silane interface material, poly(methyl silsesquioxane), resulted in a 50-fold enhancement of the mobility [19]. This dual-layer approach enables the production of high performance all-solution processed OTFTs on a flexible substrate [20]. The improvement in performance could be attributed to a reduction in surface charge trap densities and enhancement of the semiconductor morphology. Surface modification of the gate dielectric enhances not only the field-effect mobility, but also the threshold voltage. Modification of a silicon oxide gate dielectric with a poly(N-vinyl carbazole) (PVK) layer sharply reduced the pinch-off voltage of the device and improved the mobility [21]. The low pinch-off voltage is attributed to discharging from the PVK layer resulting in recombination of electrons and holes to trap holes in the channel. All the above studies clearly revealed that the transistor parameters can be tuned by carefully manipulating the surface characteristics of gate dielectrics. A proper combination of the semiconductor and the gate dielectric is essential for high performance devices.

In order to achieve a pinhole-free layer for low gate leakage from solution processable gate dielectrics, a thick film is often deposited. Due to the low dielectric constant of organic materials, the solution-processed gate dielectric usually offers a low capacitance, and thus a high operating voltage, low subthreshold swing, and so on. Therefore, another challenge remains: to achieve high capacitance organic gate dielectrics to enable the production of low-voltage OTFTs. Two general approaches have been described in the literature: a high-k gate dielectric and a thin dielectric layer. When a high-k gate dielectric is used, a thick film can be deposited that still has a high capacitance. Cai and coworkers used barium titanate and strontium titanate nanoparticle dielectrics, enabling the production of pentacene transistors operated at 10 V and PQT-12 transistors operated at 3 V with a high field effect mobility comparable to a silicon oxide gate dielectric and a reasonably good current on/off ratio [22]. Polymer/nanoparticle composites have also been applied as gate dielectrics for OTFT, using high-k inorganic nanoparticles such as Al_2O_3, ZrO_2, $BaTiO_3$, or barium strontium titanate to boost the overall dielectric constant [23–25]. Pentacene transistors with a high-k nanocomposite gate dielectric exhibited very low gate leakage, a small threshold voltage of 1.1 V, and a subthreshold swing of 100 mV dec^{-1} [23]. Alternatively, the capacitance can be increased by reducing the thickness of the gate dielectric. How to reduce pinholes in thin dielectrics is quite a challenge. Fine tuning the chemistry and the process were investigated. Bao and coworkers crosslinked a polyvinylphenol (PVP) gate dielectric through esterification reactions to achieve a thin gate dielectric with capacitance up to 400 nF cm^{-2}, which allows the production of low-voltage transistors with good mobility and a high on/off ratio of 10^3–10^5 [26]. This new gate dielectric gave a very low subthreshold swing of ~80 mV dec^{-1} with trimethyl-[2,5′5′, 2″, 5″, 2″]quarter-thiophen-5-yl-silane as the semiconductor [27]. Recently, Cheng and coworkers demonstrated a cross-linked CYTOP gate dielectric film as thin as 50 nm with low gate leakage, in contrast to the conventional 450–600 nm film [28]. This thin dielectric enabled production of top-gate OTFTs operating at a low voltage <5 V. Instead of using polymeric materials, self-assembled mono- or multilayers (SAMs) are used as ultrathin dielectric materials for OTFTs. For example, Halik and coworkers successfully used a SAM layer as the gate dielectric for OTFTs [29, 30], and the Northwestern group has developed self-assembled multilayers as reliable dielectric materials for low-voltage OTFTs [31, 32]. These approaches have been well summarized in a review paper [14].

Generally, an organic gate dielectric has a major impact on device performance since OTFTs are truly interface devices and the dielectric properties are dominant for some importance transistor parameters such as subthreshold swing and operating voltage. Although several classes of organic dielectric materials have been developed, a full understanding of the chemistry–morphology–function relationships and a well established film-deposition process require further investigation. Meanwhile, the challenges also indicate that gate dielectric materials provide new technological opportunities for high performance organic electronics.

4.1.2
Inorganic Dielectrics

This book explores the use of plasma enhanced chemical vapor deposited (PECVD) silicon nitride (SiN_x) and silicon oxide (SiO_x) films as gate dielectrics for OTFTs. PECVD dielectric materials have a number of appealing processing attributes, including relatively low deposition temperatures, large-area deposition capability, and high throughput. These attributes align with the technological drivers of organic electronics for low-cost flexible electronics application. Since the processing temperatures of most low-cost plastic substrates (e.g., poly(ethylene terephthalate) (PET), poly(ethylene naphthalate) (PEN), Kapton) are constrained to 150–200 °C or below [33], materials deposited at plastic-compatible temperatures are a requisite for realization of flexible circuits.

PECVD SiN_x is particularly attractive because it possesses excellent dielectric properties and good dielectric strength. More importantly, SiN_x has been a prevalent choice of gate and passivation dielectric materials for active matrix thin film transistor (TFT) backplanes for flat panel displays and imagers, demonstrating the reliability and technological maturity of this material [34–36]. PECVD SiO_x presents another gate dielectric option, offering the potential for low-temperature and large-area deposition. By comparison, low temperature SiN_x surpasses low temperature SiO_x in terms of electrical integrity with a lower leakage current, fewer pinholes, a higher breakdown field, higher stability, and less hysteresis [37]. Thus, the principal focus of our investigation is on the development of PECVD SiN_x for OTFTs. A preliminary study on the application of the PECVD SiO_x gate dielectric in OTFTs is also being conducted. In the past, researchers have reported limited OTFT performance with the SiN_x gate dielectric, which was attributed to surface roughness and unfriendly (or non-organic-friendly) interfaces of SiN_x. To enable pairing of SiN_x with organic semiconductors and to enhance the performance of related devices, our research approach is geared toward finding an organic-compatible SiN_x film composition (or recipe) and developing a complementary interface treatment strategy. These are, in fact, the objectives for the dielectric investigation in this chapter and the interface engineering study in the next chapter.

Experimental details of our gate dielectric investigation are presented in Section 4.2; these include an overview of the PECVD SiN_x deposition conditions and the analytical tools used for material characterization. The material properties of PECVD SiN_x thin films are examined in Section 4.3, which provides a basis for analyzing the impact of different SiN_x dielectrics on OTFT device performance. Section 4.4 investigates the electrical characteristics of PQT-12 OTFTs with different PECVD gate dielectrics. The investigation consists of four experimental components. The first experiment considers SiN_x films of varying stoichiometry deposited at a substrate temperature of 300 °C and evaluates the impact of film composition on the electrical performance of OTFTs; these results are reported in Section 4.4.1. Gate dielectrics deposited at a lower temperature are desirable for integration with plastic substrates for flexible electronic applications. 150 °C SiN_x gate dielectrics are assessed in Section 4.4.2, and the resulting device characteristics are compared with devices based on a 300 °C SiN_x gate dielectric. The concept of a stacked SiN_x gate dielectric is explored in Section 4.4.3. OTFTs with a PECVD SiO_x gate dielectric are examined in Section 4.4.4. Finally, preliminary OTFTs fabricated on flexible plastic substrates are demonstrated in Section 4.4.5.

4.2 Experimental Details and Characterization Methods

Amorphous hydrogenated silicon nitride films (abbreviated as a-SiN_x:H or SiN_x) are extensively used as dielectric or passivation materials in semiconductor device applications, including TFT, optoelectronic devices, metal-insulator-semiconductor (MIS) devices, and solar cells [38–43]. PECVD is a well-established technique for the deposition of SiN_x films [2, 35, 36, 44]. PECVD technology can accommodate low deposition temperatures and large area depositions, thus it presents exceptional potential for flexible and organic electronics fabrication. In TFT applications, PECVD SiN_x has demonstrated excellent matching as a gate dielectric for non-crystalline silicon TFTs (e.g., amorphous silicon, polycrystalline silicon, nanocrystalline silicon (nc-Si) TFTs) [2, 35, 36, 45]. For these silicon-based TFTs, a nitrogen-rich SiN_x gate dielectric (i.e., $[N]/[Si] > 1.33$) is preferable as it provides better interface properties with silicon thin films than silicon-rich SiN_x [46].

This chapter examines the application of PECVD SiN_x as the gate dielectric for OTFTs. In particular, SiN_x films of varying compositions were investigated to determine an optimal choice for organic semiconductor devices. The deposition conditions for SiN_x films used in our experiment are identified in Section 4.2.1. The analytical tools and characterization techniques used to study the material properties of various SiN_x films are discussed in Section 4.2.2.

4.2.1 Deposition Conditions of PECVD Silicon Nitride (SiN_x)

The SiN_x films were deposited by a multi-chamber 13.56 MHz PECVD cluster tool with load lock, manufactured by MVSystems, Inc., Golden, CO. Two deposition temperatures were considered in this experiment: 300 and 150 °C; the deposition conditions are summarized in Table 4.1. At a given substrate temperature, SiN_x films of varying compositions were obtained by adjusting the ammonia (NH_3) to silane (SiH_4) gas flow ratio from 5 to 20. All other deposition parameters (i.e., RF power and pressure) were fixed at values optimized for the standard recipe with a NH_3/SiH_4 gas flow ratio of 20; this standard recipe gives an optimal SiN_x for non-crystalline silicon-based device applications.

4.2.2 Thin Film Characterization Methods

A number of analytical tools were used to characterize the structural and surface properties of the PECVD SiN_x dielectric films. A brief description of these

Table 4.1 Deposition conditions for PECVD SiN_x films prepared at substrate temperatures of 300 and 150 °C. SiN_x films were deposited in a parallel-plate PECVD reactor operating at an excitation frequency of 13.56 MHz.

Substrate temperature (°C)	Pressure (m Torr)	RF power (W)	Duration (min)	Gas mixtures (sccm)	NH_3/SiH_4 gas flow ratio
300	400	2	30	NH_3: 25–100, SiH_4: 5 s	20, 15, 10, 5
150	1000	15	30	NH_3: 10–40, SiH_4: 2, H_2: 80	20, 15, 10, 5

tools is presented in this section. Fourier-transform infrared (FTIR) spectroscopy, ellipsometry, and X-ray photoelectron spectroscopy (XPS) were used to measure bulk properties of SiN_x films. Atomic force microscopy (AFM), contact angle, and XPS are used for interface characterization. For the purpose of OTFT integration, interfacial properties play a crucial role in establishing device performance. Since the organic semiconductor layer is deposited on the dielectric surface under the influence of the physical and chemical interactions between the organic and dielectric layers, the OTFT performance depends strongly on the semiconductor/dielectric interface properties.

4.2.2.1 Fourier Transform Infrared Spectroscopy (FTIR)

The chemical bonding structure of the SiN_x films was studied by FTIR spectroscopy, using a Shimadzu FTIR-8400S spectrometer in transmission mode at normal incidence. FTIR measurements were carried out to examine changes in the relative content of Si–H and N–H bonds in SiN_x films as the deposition conditions were varied.

4.2.2.2 Ellipsometry

Ellipsometry was used to determine the thickness and optical constants (refractive index n, coefficient of absorption k) of the PECVD SiN_x thin films. Measurements were made using a WVASE32 spectroscopic ellipsometer, by J.A. Woollam Co., Inc. The measured refractive index was used to estimate the nitrogen-to-silicon ratio, $x = [N]/[Si]$, of the SiN_x films; the procedure is described in Section 4.3.1.3.

4.2.2.3 X-Ray Photoelectron Spectroscopy (XPS)

XPS, also known as electron spectroscopy for chemical analysis (ESCA), is a surface sensitive chemical analysis technique that can be used to analyze the surface chemistry of a material. It is a quantitative spectroscopic technique that measures the elemental composition, empirical formula, chemical state, and electronic state of the elements that exist within a material's surface. In this chapter, XPS is used to study the surface chemical composition (i.e., elemental concentration) of SiN_x thin films with varying compositions. In Chapters 5 and 6, XPS measurements are used to characterize changes in surface chemical composition after various types of surface treatment. The depth profile of SiN_x samples was analyzed by *in situ* XPS ion beam sputtering with argon (Ar). The results are reported in Section 5.4.2, where the changes in elemental concentrations of oxygen (O_2) plasma treated SiN_x samples were analyzed as a function of depth (or etch time) by interleaving XPS analysis and argon ion sputtering at 3 keV. XPS measurements were obtained using a VG Scientific ESCALab 250, with a monochromatic Al KR X-ray source (1486.6 eV).

4.2.2.4 **Atomic Force Microscopy (AFM)**

The film surface morphology was studied using AFM with a Veeco Digital Instruments Dimension 3100 scanning probe microscope (SPM). The film surface was scanned using tapping mode. The AFM was primarily used to characterize the surface roughness of the gate dielectric layer under various deposition and treatment conditions.

4.2.2.5 **Contact Angle Analysis**

Contact angle analysis characterizes the wettability (or surface energy) of a surface by measuring the surface tension of a solvent droplet at its interface with a homogeneous surface. More specifically, the contact angle measures the attraction of molecules within the droplet to each other versus the attraction or repulsion those droplet molecules experience toward the surface molecules. The *contact angle* is defined as the angle between the tangent to the drop's profile and the tangent to the surface at the intersection of the vapor, the liquid, and the solid, as illustrated in Figure 4.1. A large contact angle means that the surface is hydrophobic and has a low surface energy (Figure 4.1a), whereas a small contact angle between solid surface and droplet indicates that the surface is hydrophilic and has a high surface energy (Figure 4.1b). Depending on the geometry and location of the surface to be analyzed, a number of different techniques are available for contact angle characterization. The most commonly used technique is the static or sessile drop method, which is the chosen technique for this research.

Contact angle measurement provides a sensitive method for the detection and identification of functional groups on a surface layer [47]. Hydrophobicity is a repulsive force between nonpolar molecules and water, and is often linked to side chains of alkyl (C–H) groups. Thus, organic/polymer surfaces are typically hydrophobic. On the other hand, the hydrophilicity of a surface is related to

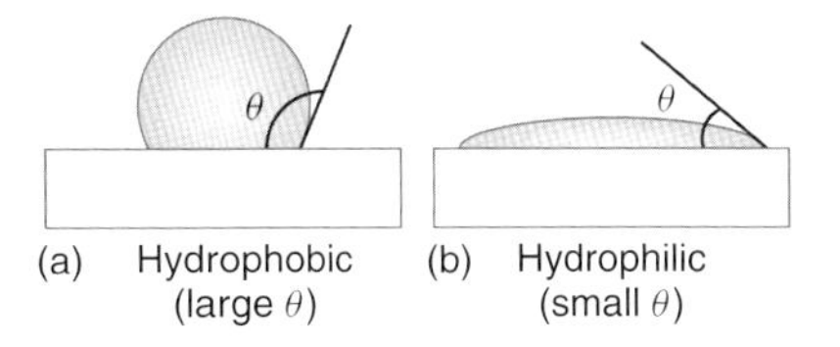

Figure 4.1 Contact angle (θ) of surfaces with different wettability: (a) water droplet on a hydrophobic surface showing large contact angle and (b) water droplet on a hydrophilic surface showing small contact angle.

its affinity to water and its ability to form H–bonds, where polar components typically lead to hydrophilic properties. Glass, SiO_2, and SiN_x surfaces are typically hydrophilic. Also, oxygen plasma treatment results in a hydrophilic surface with a decrease in water contact angle; this is due to the formation of a polar component on the exposed surface.

The contact angle provides a measure of surface energy or tension. Low surface energy is preferable for organic thin film deposition because less energy is required for molecules to assemble into an energetically-stable and well-ordered film. There is a general consensus that a dielectric with low surface energy (i.e., large contact angle) is favorable for the subsequent growth/deposition of the organic semiconductor layer. Therefore, researchers have experimented with a variety of interface treatment methods with the goal being to increase the contact angle (or hydrophobicity) of the surface. The contact angle measurements shown here have been used to study the surface wettability of various types of gate dielectric and surface treatments. This information, along with other surface characterization data, is used to explain and analyze electrical characteristics of the OTFTs.

4.3
Material Characterization of PECVD SiN_x Films

This section reports the material properties of various PECVD SiN_x films. The discussion is grouped into three categories: bulk/structural properties (Section 4.3.1), surface properties (Section 4.3.2), and electrical properties (Section 4.3.3). In terms of bulk or structural properties, the chemical bonding structure of the SiN_x films was studied by FTIR spectroscopy. The refractive index was measured by ellipsometry. Elastic recoil detection analysis (ERDA) provided another method to evaluate the chemical composition and relative atomic densities of the SiN_x films. The SiN_x film composition, denoted by the nitrogen-to-silicon ratio $x = [N]/[Si]$, was estimated from refractive index and ERDA data. Film thickness and surface profiles were measured using a Dektak 8 stylus profilometer (Veeco Instruments Inc.), after etching SiN_x to form step profiles.

Surface wettability was characterized by sessile drop contact angle measurements. Surface topography (including surface roughness) was characterized by AFM. Surface chemical composition was analyzed by XPS.

Electrical properties of SiN_x films were measured on MIS structures, using a Keithley 4200 semiconductor characterization system. Parameters of interest include the dielectric leakage current, the breakdown field, and the dielectric constant.

4.3.1
Bulk/Structural Characterization

4.3.1.1 FTIR Spectroscopy

The FTIR spectra of 300 and 150 °C SiN_x samples are presented in Figures 4.2 and 4.3, respectively. The main absorption peaks visible in the spectra are due to

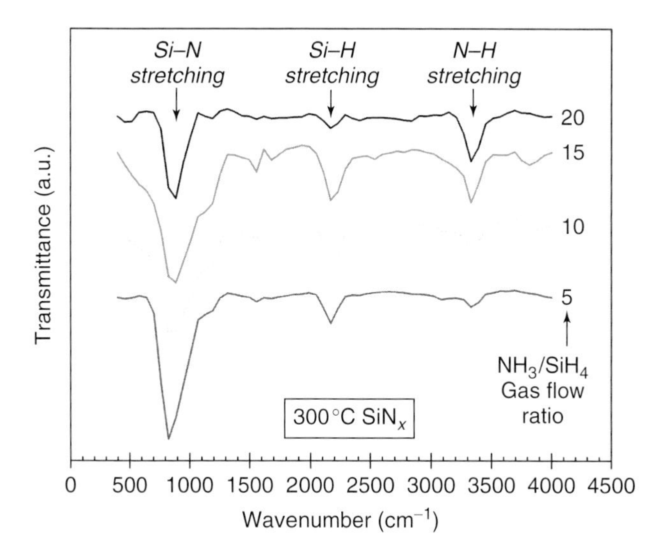

Figure 4.2 FTIR spectroscopy of the 300 °C PECVD SiN_x films.

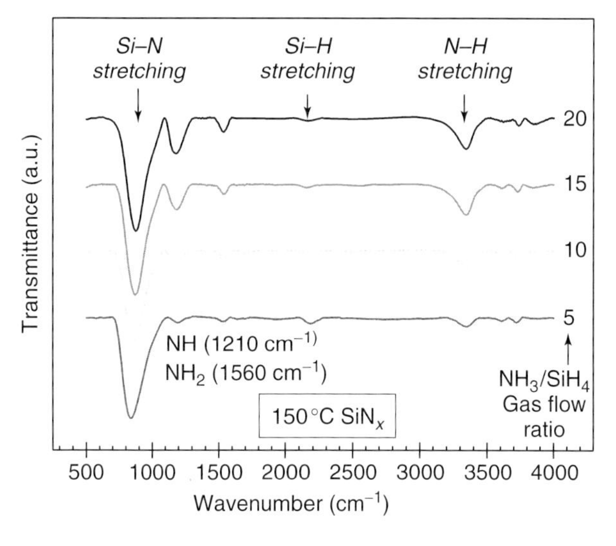

Figure 4.3 FTIR spectroscopy of the 150 °C PECVD SiN_x films.

Si–N bond stretching at 830–900 cm^{-1}, Si–H bond stretching at 2150–2180 cm^{-1}, and N–H bond stretching at 3340–3350 cm^{-1}. As shown in Figures 4.2 and 4.3, a decrease in the NH_3/SiH_4 gas flow ratio appears to strengthen the Si–H bond stretching mode, but weaken the N–H bond stretching mode, signifying a change in film composition as the gas ratio is varied.

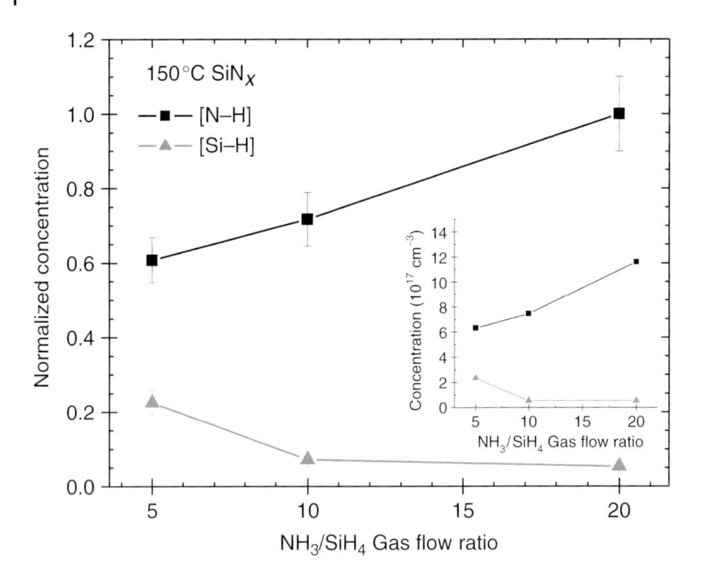

Figure 4.4 Normalized N–H and Si–H bond concentration as a function of NH_3/SiH_4 gas flow ratio for SiN_x films deposited at 150 °C. Data is normalized to [N–H] value for the sample deposited with NH_3/SiH_4 gas flow ratio of 20.

The N–H and Si–H bond densities were estimated following the method reported by Landford and Rand [48].[1)] Figure 4.4 shows the normalized N–H and Si–H bond concentrations as a function of the NH_3/SiH_4 gas flow ratio for films deposited at 150 °C. For higher values of the NH_3/SiH_4 gas flow ratio, there is an increase in N–H bond concentration, but a decrease in Si–H bond concentration. This suggests that with an increasing NH_3/SiH_4 gas flow ratio, the SiN_x films become more nitrogen-rich. Furthermore, Figure 4.4 shows that the N–H bond concentration is higher than the Si–H bond concentration; this is in agreement with the data shown in Figure 4.8, where 150 °C SiN_x samples are characterized by $[N]/[Si] > 1$.

The deposition recipes for 300 °C SiN_x films and 150 °C SiN_x films yield different concentrations of N–H and Si–H bonds, as shown in Figure 4.5. At a given value of the NH_3/SiH_4 gas flow ratio, 300 °C SiN_x is characterized by a higher Si–H bond concentration than 150 °C SiN_x, accompanied by a slight reduction in N–H bond concentration. This observation coincides with the larger Si–H peak observed in the FTIR spectra of 300 °C SiN_x in Figure 4.2 when compared to the FTIR spectra for 150 °C SiN_x in Figure 4.3. This observation also matches with the nitrogen-to-silicon ($x = [N]/[Si]$) value based on a refractive index in Figure 4.8. The consistent trends observed from the various characterization methods substantiate

1) The extraction method was developed for a certain set of calibration factors. Thus, the calculated absolute bond density values may exhibit a certain degree of error. Nonetheless, the relative change in bond densities can still provide useful information on film properties.

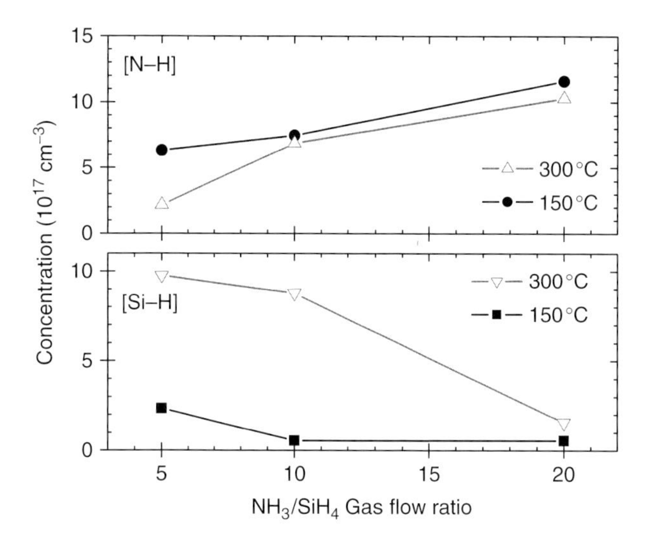

Figure 4.5 N–H and Si–H bond concentration as a function of NH_3/SiH_4 gas flow ratio for SiN_x films deposited at substrate temperatures of 150 and 300 °C.

the relative changes in nitrogen and silicon content as the deposition conditions (i.e., gas flow ratio, substrate temperature) are varied.

As the NH_3/SiH_4 gas flow ratio increases, the FTIR spectra in Figures 4.2 and 4.3 indicate a shift in the Si–N stretching band peak to a higher wavenumber. The changes in the Si–N stretching band wavenumber as a function of the NH_3/SiH_4 ratio for both 300 and 150 °C samples are plotted in Figure 4.6. The positive shift in the Si–N band wavenumber typically coincides with an increase in N–H bond density [44], which is consistent with our analysis here. The inset in Figure 4.6 plots the Si–N wavenumber, extracted from the FTIR spectra, against the [N]/[Si] calculated from the refractive index (see discussion in Section 4.3.1.3). This trend again corroborates with the convention that a higher Si–N wavenumber is associated with an increase in nitrogen content in the SiN_x film.

Stochiometric silicon nitride is represented by Si_3N_4, where $x = [N]/[Si] = 4/3 \approx 1.33$. Nitrogen-rich (N-rich) films are characterized by $x > 1.33$, whereas silicon-rich (Si-rich) films have $x < 1.33$. In our experiment, a high NH_3/SiH_4 gas flow ratio produces nitrogen-rich films and a low NH_3/SiH_4 gas flow ratio ($\sim$5) produces silicon-rich films. When a low NH_3/SiH_4 gas flow ratio is used, the SiH_4 species produce Si–Si and Si–H bonds in the growing film to form silicon-rich films [44].

4.3.1.2 **Refractive Index**

The influence of the NH_3/SiH_4 gas flow ratio on the refractive index (n) of the SiN_x films is shown in Figure 4.7. An increase in the refractive index is observed with a decreasing NH_3/SiH_4 gas flow ratio for both the 300 and 150 °C samples. A higher refractive index was measured for higher temperature nitrides at a given gas flow

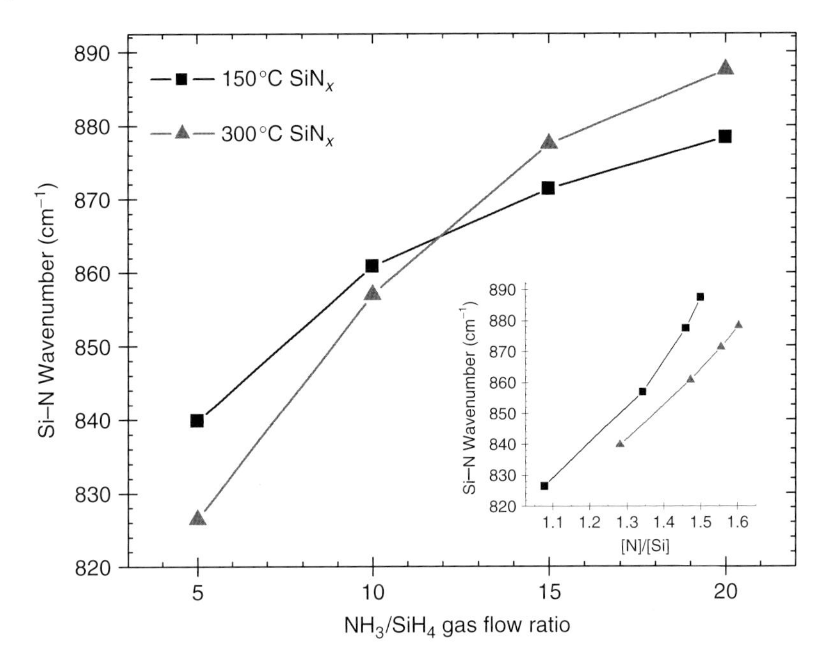

Figure 4.6 Wavenumber of the Si–N stretching band as a function of NH_3/SiH_4 gas flow ratio for PECVD SiN_x films. Inset is a plot of the changes in the Si–N wavenumber versus the [N]/[Si] ratio of the films.

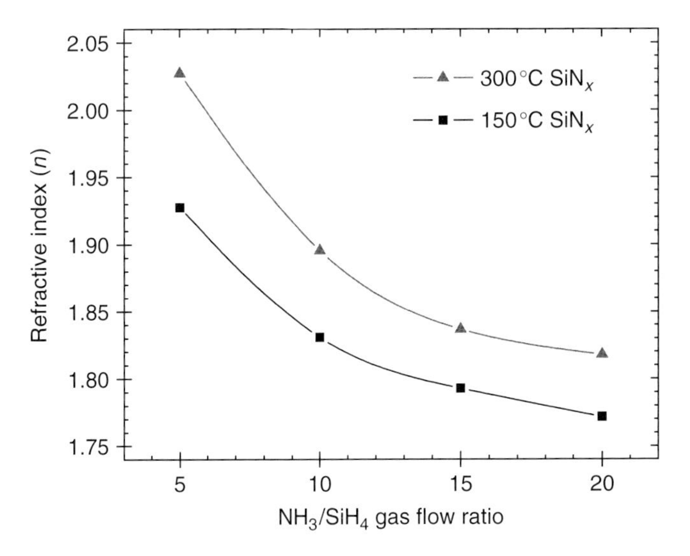

Figure 4.7 Refractive index at wavelength of 637 nm as a function of NH_3/SiH_4 gas flow ratio for 300 and 150 °C SiN_x samples.

ratio. The increase in the refractive index can be explained by the increase in the silicon (Si) content of the film. For example, it has been reported that a change in the refractive index from 2.3 to 3.6 occurs for samples with composition varying from silicon nitride to amorphous silicon; clearly, the increase in the refractive index is linked to an increase in Si concentration [44, 49]. Thus, as the NH_3/SiH_4 gas flow ratio decreases, more Si is being incorporated in the film, corresponding to an increase in the refractive index. Similarly, the higher refractive index in 300 °C films signifies richer Si content than in the 150 °C samples; this trend agrees well with the data in Figure 4.5 and is substantiated by the XPS results in Figure 4.12. Wang *et al.* attributed the increase in the refractive index to the densification of the film and the change in composition, which also corresponds to a higher Si content [44].

4.3.1.3 **[N]/[Si] Ratio**

The stoichiometry (x) of the SiN_x films is defined by the nitrogen to silicon ratio ([N]/[Si]). It is possible to estimate the [N]/[Si] ratio of SiN_x films from the refractive index (n) using the following formula [50, 51]:

$$\frac{[\mathrm{N}]}{[\mathrm{Si}]} = \frac{4}{3}\left[\frac{(3.3 - n)}{(n - 0.5)}\right] \tag{4.1}$$

Figure 4.8 plots the [N]/[Si] ratio, calculated from the refractive index (at the wavelength of 637 nm), as a function of the NH_3/SiH_4 gas flow ratio. As the NH_3/SiH_4 gas flow ratio changes from 5 to 20, the [N]/[Si] ratio of the SiN_x films increases, indicating an increase in nitrogen content relative to silicon. This trend is in agreement with the FTIR analysis in Section 4.2.2.1.

SiN_x films were also characterized using ERDA, which showed an increase in the [N]/[Si] ratio from 1 to 1.3 for 300 °C SiN_x when the NH_3/SiH_4 gas flow changes

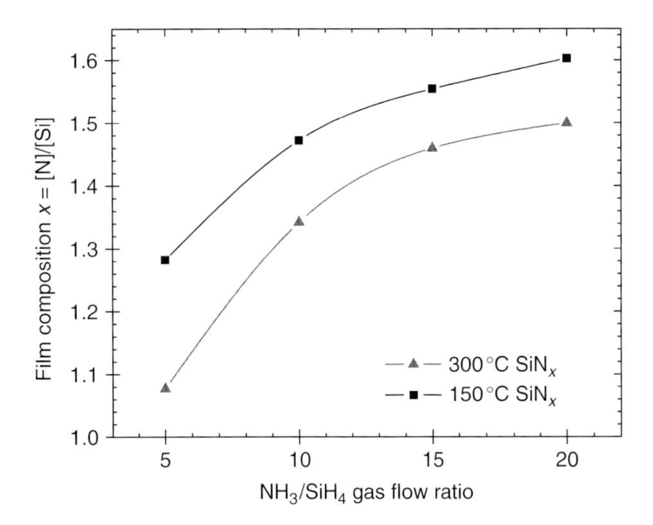

Figure 4.8 Nitrogen to silicon ratio ([N]/[Si]) of the 300 and 150 °C SiN_x films as a function of gas flow ratio.

from 5 to 20. This trend concurs with the [N]/[Si] data extracted from the refractive index. Composition information deduced from XPS measurements showed identical dependence of [N]/[Si] on the gas flow ratio. Collectively, FTIR, ERDA, XPS, and the refractive index measurements confirmed that, as the NH_3/SiH gas flow ratio increases, the SiN_x samples become more nitrogen rich, resulting in an increase in the [N]/[Si] ratio.

Comparing the effect of deposition temperature on [N]/[Si] in Figure 4.8, samples deposited at higher deposition temperatures displayed smaller [N]/[Si] at a given gas flow ratio. This change in composition is also reflected by an increase in Si–H bond concentration at higher deposition temperatures in Figure 4.5. A higher deposition temperature enhances the incorporation of Si in the film with respect to N, as evidenced by a decrease in [N]/[Si] to indicate a more silicon-rich film. Additionally, a higher temperature activates the release of non-bonded H and the breaking of weak H bonds so that the H content decreases, and the relative amount of N and Si increases, resulting in a denser film [44].

4.3.2 Surface Characterization

4.3.2.1 Contact Angle

The surface wettability is a key parameter of interest in organic electronic devices because it affects the molecular ordering of the organic film. In the case of bottom-gate OTFTs, ideally, the dielectric surface should be tailored to allow the formation of a well-ordered organic layer, so that the density of grain boundaries in the organic semiconductor and scattering at the interface are minimized. An untreated SiN_x surface is inherently hydrophilic. However, the majority of published results suggest that a hydrophobic surface is needed for the deposition of an organic semiconductor to achieve higher μ_{FE}. This mismatch can be resolved by proper surface treatment. Surface modification by alkyltrichlorosilane ($CH_3(CH_2)_{n-1}SiCl_3$) SAM is a popular choice for pre-treating hydroxylated dielectric surfaces, and has proven to significantly improve OTFT performance [52]. However, silane reagents often have poor adhesion on a bare SiN_x surface due to the lack of hydroxy groups. We address this limitation by pretreating the nitride samples with oxygen plasma prior to SAM deposition. It is believed that oxygen plasma exposure places hydroxy groups on the nitride surface to facilitate attachment of alkyltrichlorosilane molecules. Our experiments suggested that the duration of oxygen plasma exposure must be carefully chosen to yield high quality OTS SAM; more details are reported in Chapter 5. Oxygen plasma exposure has been used in a similar context for modifying other dielectric surfaces in OTFT fabrication, as reported in the literature [53, 54].

Contact angle measurements provide a means to evaluate the effectiveness of the surface modification recipe (i.e., oxygen plasma exposure followed by OTS SAM treatment). Figure 4.9 summarizes the contact angle data of SiN_x films before and after surface treatment. There are two key observations:

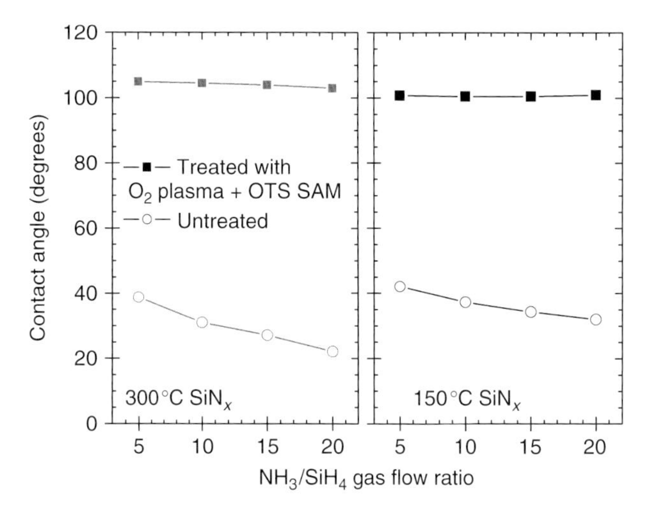

Figure 4.9 Water contact angle of SiN_x surface before (○) and after (■) surface treatment with oxygen plasma and OTS SAM, for four different SiN_x films (denoted by NH_3/SiH_4 gas flow ratios).

1) The relatively small contact angle for the untreated/bare nitride indicates a hydrophilic surface. In contrast, the rise in contact angle across all samples after surface treatment signifies effective surface modification to promote hydrophobicity of the nitride surface.
2) A dependence of the contact angle on the SiN_x film composition is evident for the untreated samples; the contact angle increases as the silicon content in the untreated nitride film increases. The dependence of the contact angle on film composition diminishes after surface treatment, suggesting that the dielectric surface property becomes largely dictated by the properties of the OTS SAM.

In-depth discussions on the impact of interface treatment on SiN_x surface properties and OTFT characteristics are presented in Section 3.8.

4.3.2.2 **Surface Morphology and Roughness**

Both the chemical and physical states of the gate dielectric surface affect OTFT performance. In terms of physical properties, surface roughness is particularly important because it hinders the movement of charge carriers. Rough dielectric surfaces can lead to the formation of voids or discontinuities in the overlying organic semiconductor film and can cause surface scattering. There have been a number of reports of reduction in mobility with increasing dielectric surface roughness. Chabinyc *et al.* showed that the field-effect mobility of OTFTs decreases nearly exponentially with the surface roughness of the gate dielectric [55]. Kim *et al.* evaluated changes in the mobility of solution-processed OTFTs with interface roughness. They reported that, below a critical roughness threshold, the mobility remained constant for low values of the interface roughness; however, for roughness

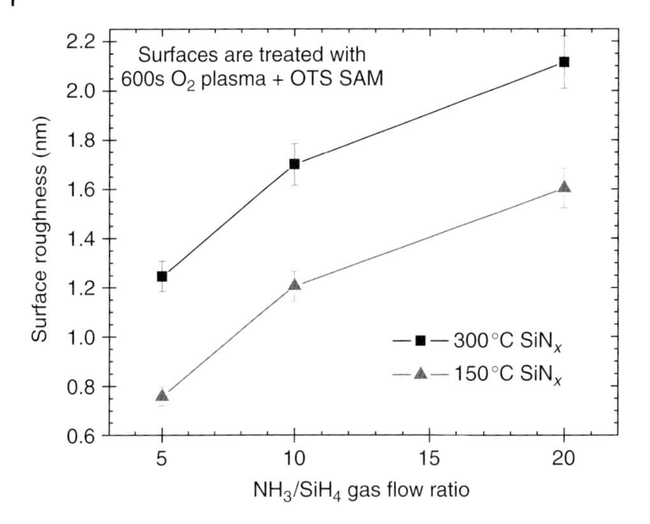

Figure 4.10 Surface roughness of 150 and 300 °C SiN_x films as a function of NH_3/SiH_4 gas flow ratios, after surface treatment with O_2 plasma and OTS SAM.

exceeding this threshold, a rapid drop in the mobility, by orders of magnitude, was observed [56].

AFM was used to characterize the surface roughness of the SiN_x samples. Figure 4.10 shows that as the NH_3/SiH_4 gas flow ratio decreases (i.e., as the film tends toward Si-rich), the surface roughness decreases. Another observation is that 150 °C SiN_x films have lower surface roughness than 300 °C SiN_x at a given gas flow ratio. These observations in surface roughness can offer valuable insights for analyzing/understanding OTFT characteristics; more discussions are presented in Sections 4.4.1 and 4.4.2.

4.3.2.3 Chemical Composition

XPS was used to study the elemental and chemical composition of the SiN_x surfaces. Surface composition is particularly important when analyzing the dielectric/semiconductor interface because the type and density of atoms present at the surface determines the type of bonding that takes place. Consequently, the type of interface bonding influences the quality, microstructure, and molecular ordering of the semiconductor layer, thus dictating the performance of OTFTs. XPS measurements were performed on various 300 and 150 °C SiN_x samples; their XPS spectra were used to calculate the elemental distribution on the sample's surface. Figure 4.11a plots the atomic distribution at the surface of three 300 °C SiN_x films. The displayed trend is consistent with the preceding discussions, where the sample labeled $x = 1.08$ has the highest atomic percentage in Si 2p, indicating a higher surface concentration of Si than the other two samples ($x = 1.34, 1.50$). This observation is in exact agreement with the definition of $x = [N]/[Si]$, where a smaller x means a larger Si content.

When using SiN_x as the gate dielectric in OTFTs, surface treatment steps are performed to prepare the dielectric surface for deposition of the organic semiconductor layer. (Detailed investigation of dielectric surface treatments is discussed in Chapter 5.) Therefore, XPS is used to study the chemical composition of the SiN_x surface *after* surface treatments to gain better insights on the desirable surface compositions for higher performance OTFTs. Figure 4.11b,c plots the atomic distribution of 300 °C SiN_x films after O_2 plasma treatment and O_2 plasma/OTS SAM treatment, respectively. A significant increase in the atomic percentage of O 1s after O_2 plasma treatment is observed by comparing Figure 4.11a–c. This is logical as O_2 plasma treatment is expected to attach oxygen atoms to the surface, forming an oxidized nitride surface. Figure 4.11c also shows a large increase in C 1s content, attributable to the presence of an OTS SAM. An interesting observation is that the Si 2p intensity remains highest for the $x = 1.08$ sample, even after O_2 plasma or O_2 plasma/OTS treatments. This Si-rich surface is believed to be a decisive factor for attaining high field-effect mobility for OTFTs on an SiN_x gate dielectric, which is explored or analyzed in-depth in Section 4.4.

Figure 4.12 compares the chemical composition of 300 and 150 °C SiN_x films, both deposited at the same NH_3/SiH_4 gas flow ratio. A higher Si concentration and lower N concentration (thus smaller [N]/[Si]) are observed in 300 °C SiN_x when compared to 150 °C SiN_x. This is in agreement with the data in Figure 4.8, where a smaller [N]/[Si] value is obtained for 300 °C SiN_x than for150 °C SiN_x, at a given NH_3/SiH_4 gas flow ratio. XPS of the O_2 plasma treated surfaces are displayed in Figure 4.12b; the increase in the O 1s signal signifies successful attachment of oxygen species on the SiN_x surface.

4.3.3 Electrical Characterization

Metal-insulator-semiconductor capacitance structures were fabricated using an $Al/SiN_x/Si/Al$ configuration to evaluate the leakage current, electrical breakdown field, and dielectric constant of the nitride films by means of current–voltage (I–V) and capacitance–voltage (C–V) measurements.

4.3.3.1 I–V Measurements

Figures 4.13 and 4.14 display the leakage current density as a function of the electric field for 300 and 150 °C SiN_x films, respectively. A leakage current density in the range 10–40 nA cm^{-2} for electric fields less than 2 MV cm^{-1} was observed across the different nitride films, comparable to results reported in Refs. [57, 58]. Electric breakdown was not observed for SiN_x films deposited with an NH_3/SiH_4 gas flow ratio of 20 and 10, for fields up to 6 MV cm^{-1}. However, a lower breakdown field was measured for SiN_x deposited at $NH_3/SiH_4 = 5$, suggesting a weakening of the dielectric strength as the NH_3/SiH_4 flow ratio decreases (i.e., for more silicon-rich films).

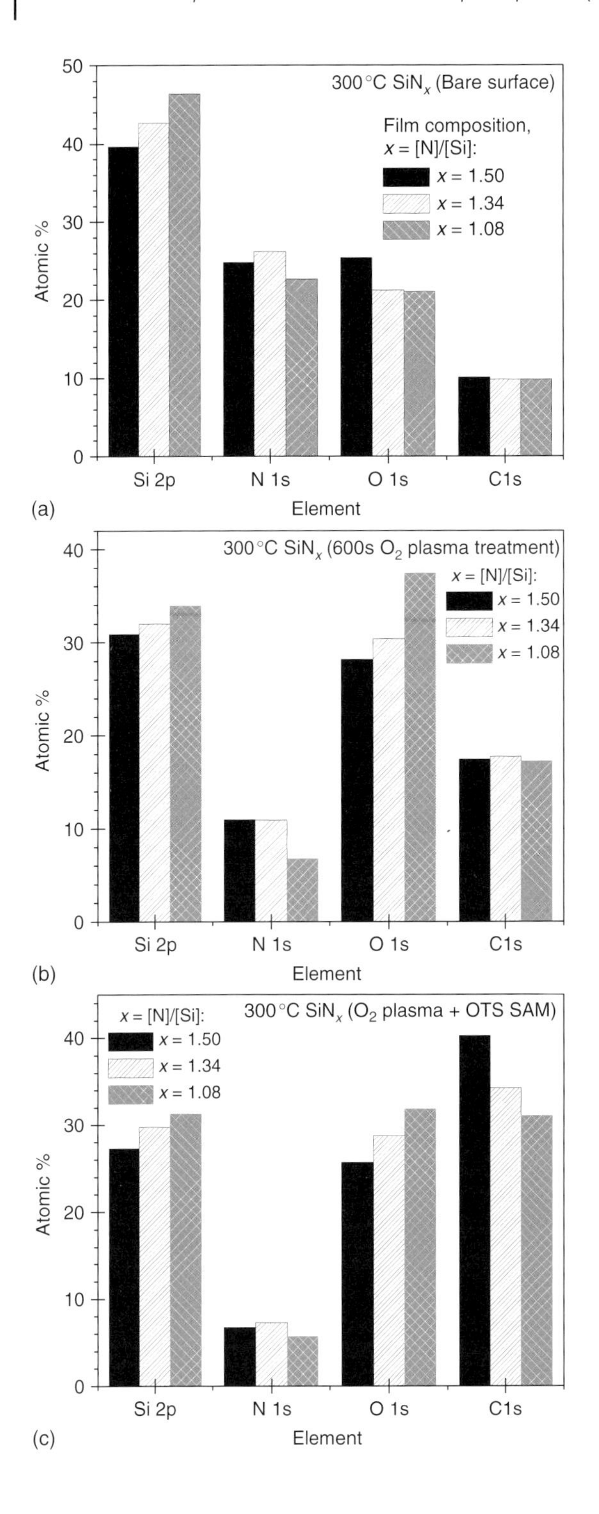
300 °C SiN$_x$ (Bare surface)
Film composition,
x = [N]/[Si]:
x = 1.50
x = 1.34
x = 1.08
Atomic %
Si 2p
N 1s
O 1s
C1s
Element
(a)
300 °C SiN$_x$ (600s O$_2$ plasma treatment)
x = [N]/[Si]:
x = 1.50
x = 1.34
x = 1.08
Atomic %
Si 2p
N 1s
O 1s
C1s
Element
(b)
x = [N]/[Si]:
300 °C SiN$_x$ (O$_2$ plasma + OTS SAM)
x = 1.50
x = 1.34
x = 1.08
Atomic %
Si 2p
N 1s
O 1s
C1s
Element
(c)

Figure 4.11 Chemical composition of three different 300 °C SiN_x films (x = 1.50, 1.34, 1.08), plotted as atomic distribution at the film surface as measured by XPS. Three surface conditions were considered for each of the 300 °C SiN_x films: (a) as-deposited (bare) surface, (b) O_2 plasma treated surface, and (c) O_2 plasma and OTS SAM treated surface.

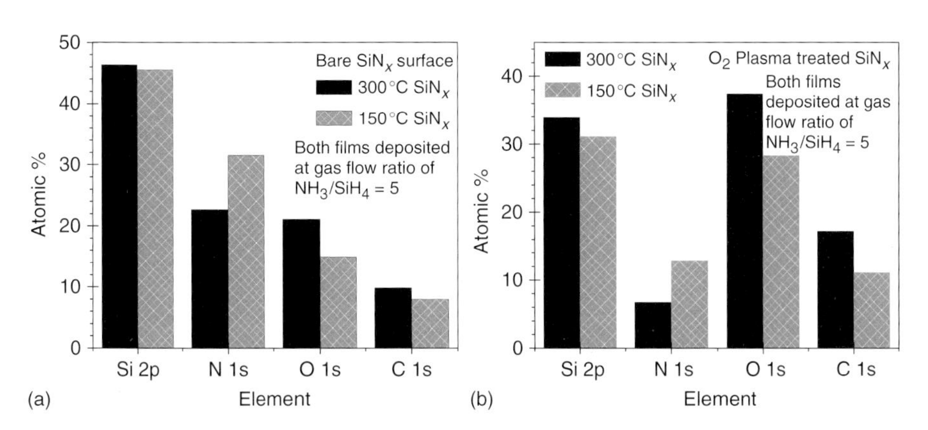

Figure 4.12 XPS measurement comparing the chemical composition of silicon-rich 300 and 150 °C SiN_x films deposited at a gas flow ratio of NH_3/SiH_4 = 5. Two surface conditions are considered: (a) as-deposited (bare) and (b) O_2 plasma treated surface.

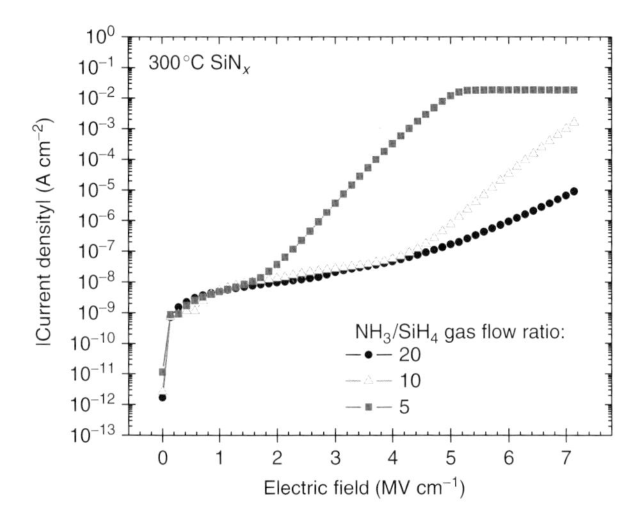

Figure 4.13 Leakage current density as a function of electric field for 300 °C SiN_x dielectrics deposited at various NH_3/SiH_4 gas flow ratios. The film thickness is approximately 300 nm. Measurements were performed using "a MIS capacitor" structure.

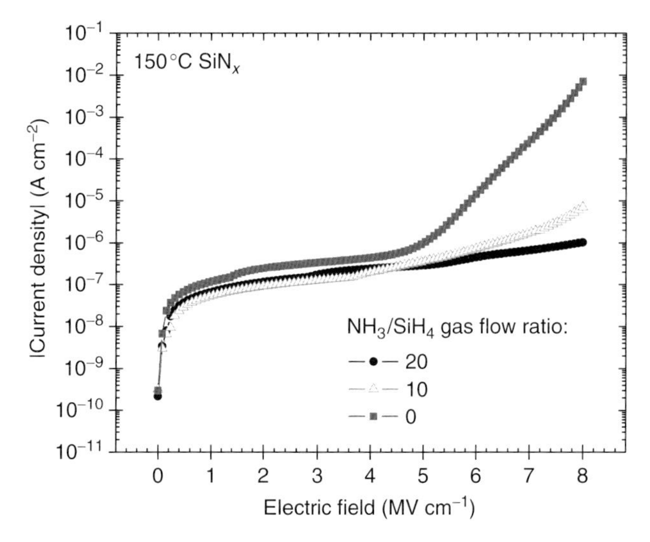

Figure 4.14 Leakage current density as a function of electric field for 150 °C SiN_x dielectrics deposited at various NH_3/SiH_4 gas flow ratios. The film thickness is approximately 300 nm. Measurements were performed using "a MIS capacitor" structure.

4.3.3.2 **C–V Measurements**

The relative dielectric constant (ε_r) of the various SiN_x films was obtained from capacitance measurements, using the formula:

$$C = \frac{\varepsilon_0 \varepsilon_r A}{d} \tag{4.2}$$

Figure 4.15 shows a decrease in the dielectric constant with a decreasing NH_3/SiH_4 gas flow ratio (i.e., increasing silicon content). $C-V$ measurements using MIS structures made with 300 and 150 °C SiN_x are displayed in Figures 4.16 and 4.17, respectively. Decrease in the gas flow ratio (i.e., more silicon-rich films) leads to a negative shift in flat band voltage (V_{FB}) and an increase in hysteresis. Such hysteresis (i.e., voltage shift between forward and reverse $C-V$ curves) is related to charge (electron) trapping within the SiN_x bulk. The data showed more hysteresis for films deposited at a lower NH_3/SiH_4 gas flow ratio (linked to a film with smaller [N]/[Si] ratio), suggesting an increase in charge trapping for the more silicon-rich SiN_x film. The observed hysteresis has been attributed to electron injection and trapping mechanisms in SiN_x, including: Fowler–Nordheim injection, trap-assisted injection, constant-energy tunneling from the silicon conduction band, direct tunneling from the silicon valence band, and hopping at the Fermi level [59]. These charge trapping mechanisms also lead to higher leakage currents (especially at high electric fields); this can be observed in the $I-V$ characteristics in Figures 4.13 and 4.14 for Si-rich SiN_x films.

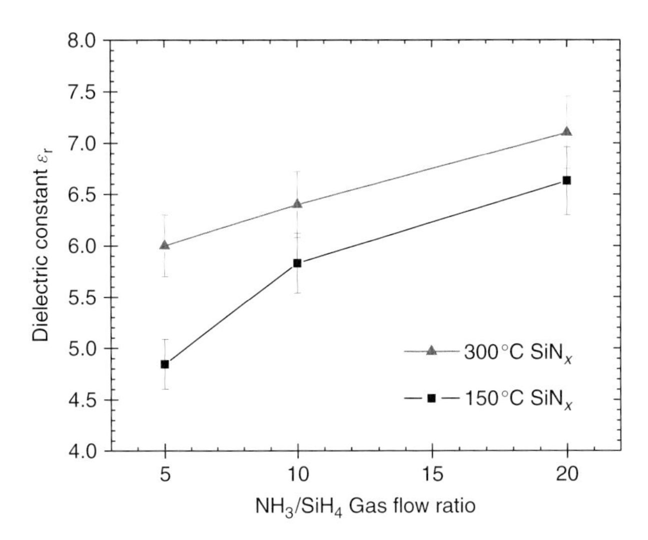

Figure 4.15 Dielectric constant (ε_r) for SiN_x films as a function of NH_3/SiH_4 gas flow ratio.

In conclusion, these electrical characteristics indicate that the composition of the SiN_x film, which depends on the deposition conditions, influences the electrical properties of the dielectric layer; this will in turn impact the OTFT device characteristics.

4.3.4 **Summary**

Amorphous hydrogenated silicon nitride films have been deposited by the PECVD technique, using ammonia (NH_3) and silane (SiH_4) as precursor gases. The gas flow ratio has been varied in order to study their effect on the properties of the films, which were characterized by FTIR, ellipsometry, ERDA, XPS, AFM, and contact angle.

As the NH_3/SiH_4 gas flow ratio was increased (by increasing the NH_3 partial pressure), the measurements indicated an increase in the nitrogen-to-silicon ratio ([N]/[Si]). Changes in the SiN_x film composition result in different surface (e.g., contact angle, surface roughness, surface composition) and electrical properties (e.g., dielectric constant, breakdown field, leakage current). When these SiN_x films are used as the gate dielectric in bottom-gate OTFTs, variation in surface states will influence the molecular ordering of the overlying organic semiconductor layer and the quality of the semiconductor/dielectric interface. These attributes have a strong bearing on the OTFT characteristics, including field-effect mobility, on/off current ratio, leakage current, threshold voltage, and subthreshold slope. Thus, proper control of the gate dielectric surface property is crucial for attaining higher

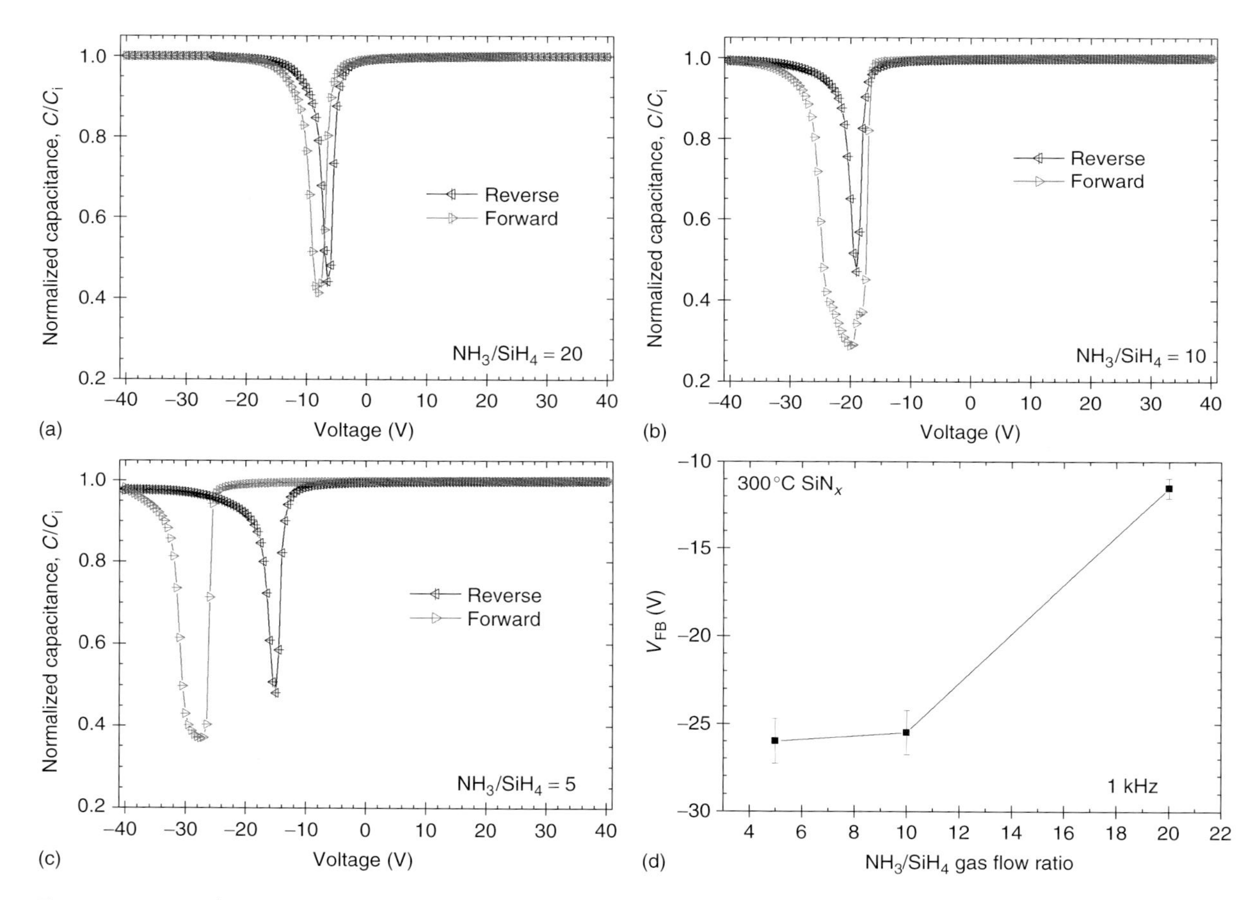

Figure 4.16 *C–V* characteristics of various 300 °C SiN_x dielectrics at 1 kHz, in both forward (negative to positive) and reverse (positive to negative) sweep directions. Flatband voltage (V_{FB}) is extracted from the forward curves and is plotted as a function of the NH_3/SiH_4 gas flow ratio.

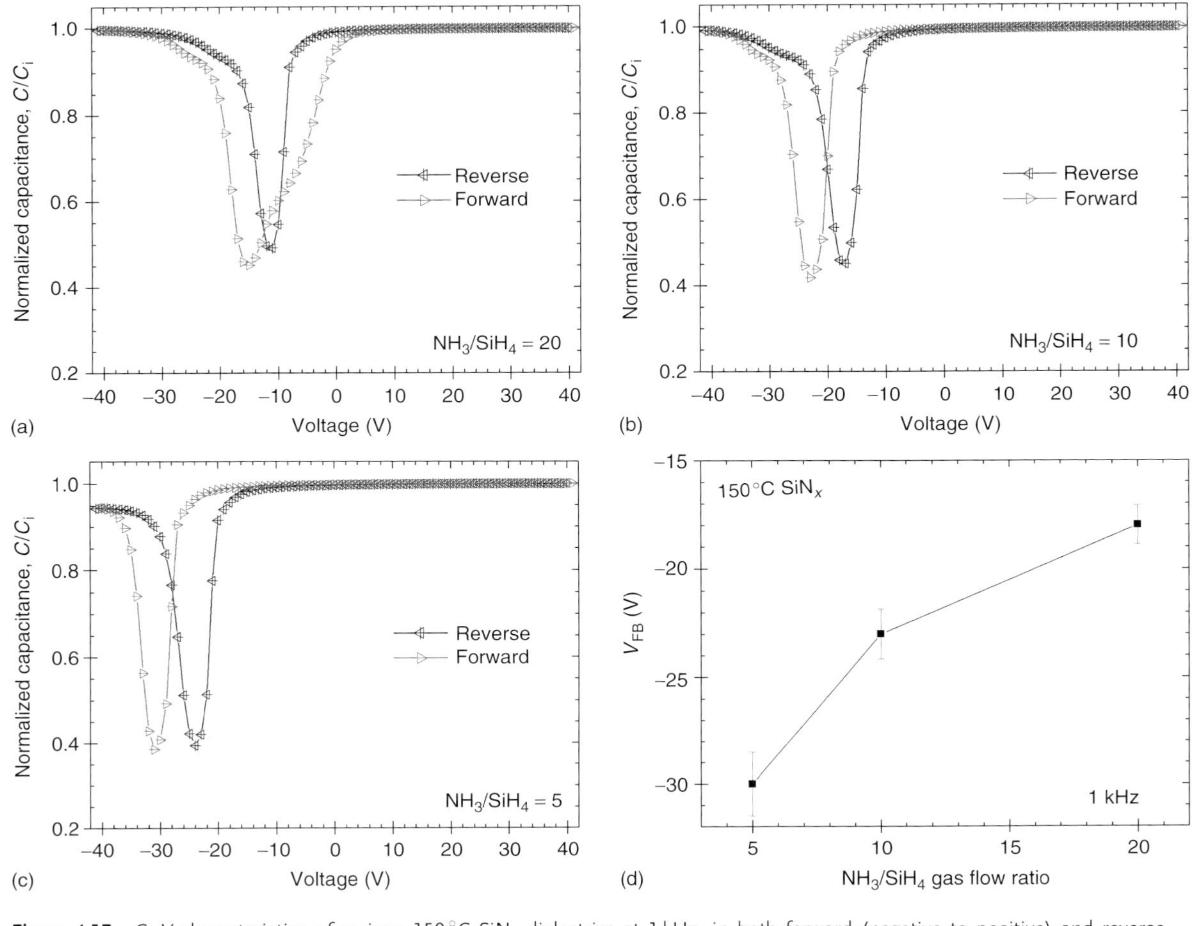

Figure 4.17 *C–V* characteristics of various 150 °C SiN_x dielectrics at 1 kHz, in both forward (negative to positive) and reverse (positive to negative) sweep directions. Flatband voltage (V_{FB}) is extracted from the forward curves and is plotted as a function of the NH_3/SiH_4 gas flow ratio.

performance OTFTs. In the next section, we examine the impact of changing the SiN_x gate dielectric properties on OTFT performance.

4.4
Electrical Characterization of OTFTs with PECVD Gate Dielectric

PECVD SiN_x and SiO_x thin films were evaluated as gate dielectrics for bottom-gate bottom-contact PQT-12 OTFTs. Devices were fabricated using the integration scheme outlined in Figure 6. Highly doped silicon wafers served as a common gate substrate. PECVD SiN_x or SiO_x (200–300 nm thick) formed the gate dielectric layer. Source/drain contacts, patterned by photolithography and a lift-off process, consisted of 60 nm of thermally evaporated Au on top of 5 nm of Cr adhesion layer. Au was used for the source/drain contacts because of its high work function (~5.1 eV), which theoretically should provide good matching to the highest occupied molecular orbital (HOMO) of the PQT-12 semiconductor, making hole injection from Au to PQT-12 more feasible. Solution processable PQT-12 was used as the organic semiconductor layer. PQT-12 semiconductor was synthesized as previously reported in Ref. [60] with a number average relative molecular mass (M_n) and polydispersity of 17 300 g mol^{-1} and 1.32. The PQT-12 films were spin-coated from a 0.3 wt% dispersion in 1,2-dichlorobenzene at a rate of 1000 rpm onto substrates [61]. The as-cast films were then dried in a vacuum oven at 80 °C and annealed at 140 °C, followed by a slow cooling to room temperature. Figure 4.18 illustrates the chemical structure of PQT-12. All processing and device characterization were carried out in the atmosphere.

Prior to deposition of the semiconductor layer, the dielectric surface was functionalized with a combination of oxygen (O_2) plasma treatment and an OTS ($CH_3(CH_2)_7SiCl_3$) SAM. O_2 plasma treatment was carried out using a Phantom II reactive ion etching (RIE) system by Trion Technology Inc., with chamber pressure 150 mTorr, O_2 gas flow rate of 30 sccm, and an exposure duration of 600 s. OTS SAM was formed by immersing the substrates in a 0.1 M solution of OTS in toluene for 20 min at 60 °C. The Au electrodes were functionalized with 1-octanethiol SAM by immersing the substrates in a 0.01 M solution in toluene for 30 min at room temperature to improve the contact properties. (In-depth discussions on dielectric and contact surface treatments are presented in Chapters 5 and 6, respectively.)

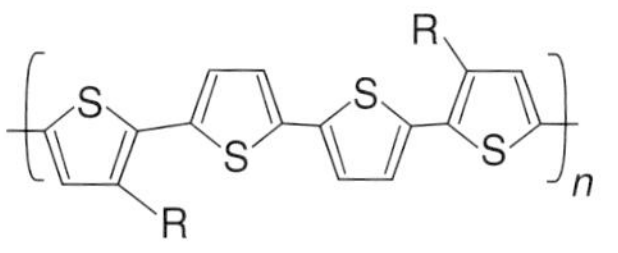

Figure 4.18 Chemical structure of poly(3,3‴-dialkylquarterthiophene) (PQT). In the case of PQT-12, alkyl side chain is R = n-C_{12}-H_{25}. (Adapted from [60].)

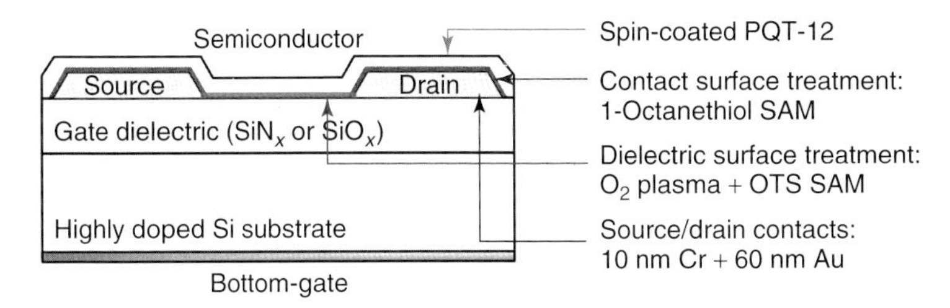

Figure 4.19 Schematic cross-section of the OTFT structure considered in this study.

A simplified schematic cross-sectional diagram of the OTFT structure is shown in Figure 4.19.

Various PECVD films were investigated as the gate dielectric layer, including 300 °C SiN_x, 150 °C SiN_x, and 200 °C SiO_x. The gate dielectric layers were deposited in a parallel-plate PECVD reactor operating at an excitation frequency of 13.56 MHz. The 300 °C SiN_x (300 nm thick) of various compositions were prepared by adjusting the ammonia (NH_3) to silane (SiH_4) gas flow rate ratio from 5 to 20. The 150 °C SiN_x (300 nm thick) was deposited using a mixture of silane, ammonia, and hydrogen as process gases. The 200 °C SiO_x (200 nm thick) was deposited with a mixture of silane and nitrous oxide gases. Table 4.1 summarizes the deposition conditions used for PECVD SiN_x films.

Electrical characterization of the OTFTs was carried out with a Keithley 4200-SCS parameter analyzer. Measurements were performed under ambient room conditions. From the measured transfer characteristic ($I_D - V_{GS}$), transconductance (g_m) was calculated and was used for device parameter extraction, as outlined in Section 2.2. Device parameters including effective field-effect mobility (μ_{FE}), on/off current ratio (I_{ON}/I_{OFF}), threshold voltage (V_T), and gate leakage current (I_G) are used to evaluate the performance of OTFTs with different PECVD gate dielectric platforms. The various groups of PECVD gate dielectrics examined include 300 °C SiN_x, 150 °C SiN_x, stacked SiN_x, 200 °C SiO_x, and low-temperature dielectrics on plastic substrates.

4.4.1
300 °C SiN_x Gate Dielectrics

The first experiment evaluates OTFTs fabricated with 300 °C SiN_x gate dielectric of varying film compositions. Figure 4.20a illustrates the transfer characteristics of a typical PQT-12 OTFT on a silicon-rich SiN_x ($x = 1.08$, 300 °C) gate dielectric. Parameter extraction from the saturation region ($V_{DS} = -40$ V) gives $\mu_{FE} = 0.092\,cm^2\,V^{-1}\,s^{-1}$, $I_{ON}/I_{OFF} = 10^7$, and $V_T = -3.8$ V. The output characteristics show well-defined saturation behavior, as illustrated in Figure 4.20b.

To examine the impact of SiN_x film composition on OTFT performance, three sets of PQT-12 OTFTs were fabricated on SiN_x gate dielectrics with different [N]/[Si] ratios (x = [N]/[Si] = 1.08, 1.34, 1.50) [62]. Each set of OTFTs was comprised

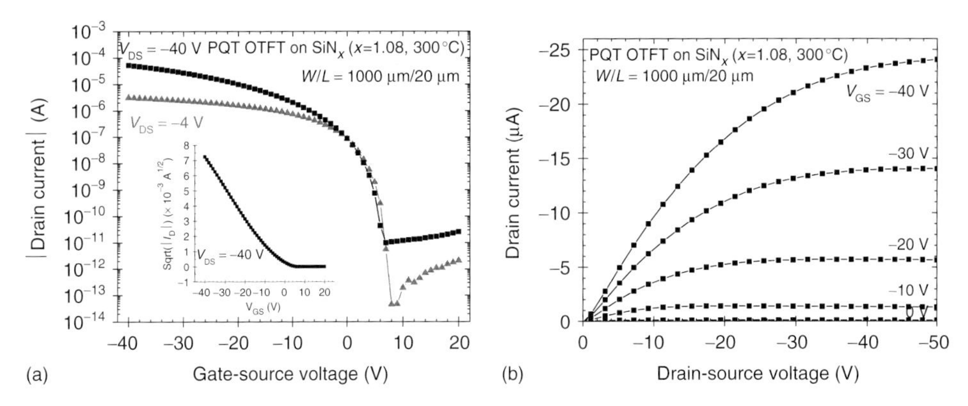

Figure 4.20 Electrical characteristics of PQT-12 OTFT on 300 °C SiN_x (x = 1.08) gate dielectric with W/L = 1000 µm/20 µm: (a) transfer characteristics ($\log|I_D| - V_{GS}$) and (b) output characteristics ($I_D - V_{DS}$). Inset of (a) shows a well-behaved linear plot of $\sqrt{I_D} - V_{GS}$ in the saturation regime.

of devices with different channel geometries (W/L ratio). Figure 4.21 compares the transfer characteristics, effective mobility, on/off current ratio, and threshold voltage as a function of the [N]/[Si] ratio of the SiN_x gate dielectrics. As the silicon content of the nitride dielectric increases (i.e., [N]/[Si] ratio decreases), I_{ON} increases, μ_{FE} increases, I_{ON}/I_{OFF} increases, and V_T shifts in the negative direction. A 125% (or 2.25 times) increase in μ_{FE} was observed using the silicon-rich SiN_x gate dielectric when compared to nitrogen-rich films. These results suggest that the silicon-rich SiN_x gate dielectric is a preferable choice for attaining OTFTs with improved μ_{FE}, I_{ON}/I_{OFF}, and current drive.

It is well recognized that a key criterion for attaining high mobility is to assemble an organic semiconductor film into a highly ordered molecular structure by a combination of clever molecular design and a properly prepared deposition surface [63–66]. The dependence of OTFT mobility on nitride composition observed here can be largely ascribed to differences in the surface properties (e.g., composition, topography, surface energy) of the various nitride dielectrics. A higher density of silicon atoms appears to increase the density of OTS grafted on the SiN_x surface, thus increasing the hydrophobicity of the dielectric surface (as seen in the contact angle measurements in Figure 4.9). It has been reported that polycrystallites of regioregular polythiophenes typically self-organize into a highly oriented structure when deposited on a hydrophobic surface, which is the desired molecular arrangement for improved transistor performance [63].

The AFM measurements in Section 4.3.2.2 revealed a decrease in surface roughness with a decrease in [N]/[Si] of SiN_x film (i.e., toward silicon rich). This characteristic favorably supports the observed improvement in mobility for Si-rich

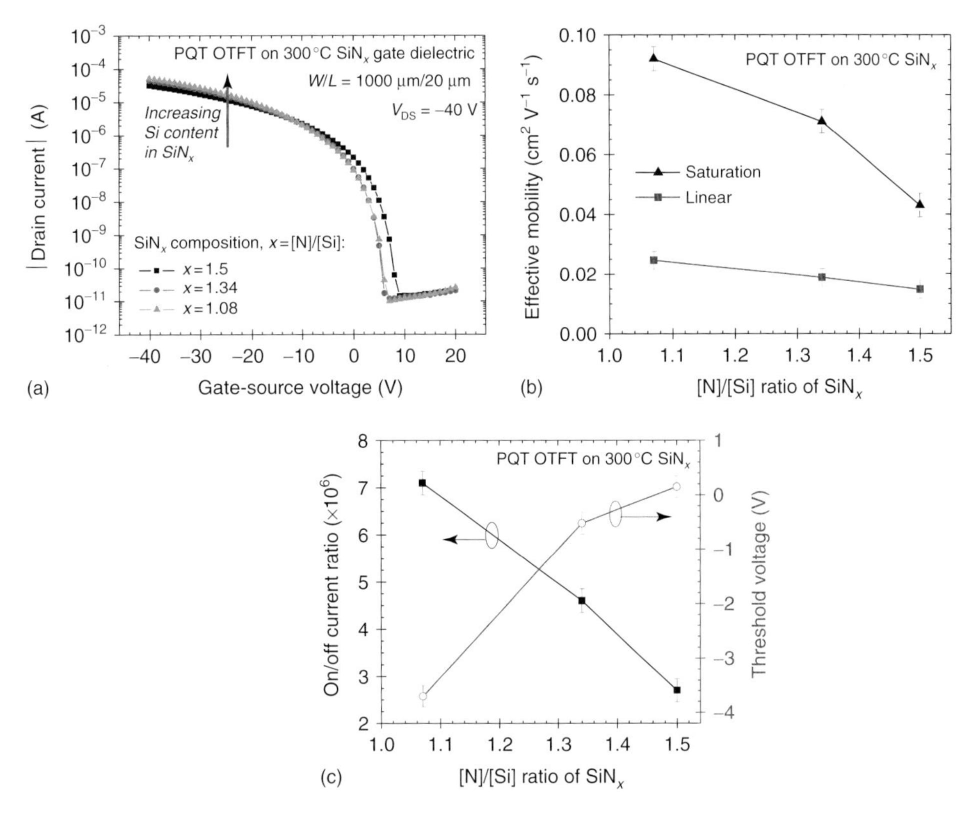

Figure 4.21 Comparison of PQT-12 OTFT characteristics on 300 °C SiN_x gate dielectric of varying film composition. (a) Transfer characteristics, (b) effective field-effect mobility, (c) on/off current ratio and threshold voltage as a function of the [N]/[Si] ratio of the SiN_x film.

SiN_x. This observation is consistent with the typical trend where an increase in dielectric surface roughness results in reduced mobility [55].

Considering the surface chemical composition of the various 300 °C SiN_x films plotted in Figure 4.11, there is a noticeable increase in the Si 2p peak as the [N]/[Si] ratio decreases. This result indicates an increase in silicon content at the surface, which agrees perfectly with our extracted values of the [N]/[Si] ratio. The Si 2p peak remained the largest for the Si-rich SiN_x ($x = 1.08$) sample relative to the other SiN_x films ($x = 1.50$, 1.34) even after O_2 plasma and/or OTS SAM treatments, as indicated in Figure 4.11b,c. It is hypothesized that the increased availability of Si atoms at the SiN_x dielectric surface is a criterion for enhanced OTFT performance. It has been suggested that chemical interaction between the gate dielectric, SAM, and organic semiconductor occurs via –Si–O–C– bonds at the interface [67]. Therefore, a higher concentration of Si atoms at the interface facilitates and promotes bonding/interaction with the SAM and the organic semiconductor layer, leading to improved device performance for OTFTs with a Si-rich SiN_x gate dielectric. The role of Si and O atoms at the dielectric surface in facilitating bonding to SAM/organic is further highlighted in Figure 4.11b,c, where the SiN_x sample with $x = 1.08$ displays the highest Si 2p and O 1s content compared to the other SiN_x samples after O_2 plasma and/or OTS treatments.

Based on the results reported in this experiment, it is concluded that the improvement in PQT-12 OTFT characteristics observed with a silicon-rich SiN_x gate dielectric (as opposed to N-rich SiN_x) can be attributed to various surface properties of silicon-rich SiN_x: (i) lowest surface energy (i.e., largest contact angle), (ii) lowest surface roughness, and (iii) highest [Si] and [O] content at the surface (after O_2 plasma and O_2 plasma/OTS treatment) when compared to the N-rich SiN_x samples. A summary of these key surface qualities of silicon-rich SiN_x, along with their implications, is presented in Table 4.2.

Interestingly, the improvement in OTFT observed with silicon-rich SiN_x is a new and unique discovery. This is because the opposite trend is typically seen in amorphous silicon or nc-Si TFTs, where nitrogen-rich SiN_x is more preferable in silicon-based TFTs [45, 46]. A possible rationalization to account for the different device behavior observed in OTFT and nc-Si TFT with the use of silicon-rich SiN_x gate dielectric is presented in Table 4.3.

Table 4.2 Unique surface properties of silicon-rich SiN_x (when compared to nitrogen-rich SiN_x) to account for the improved mobility in PQT-12 OTFTs. The qualities quoted are measured on O_2 plasma and OTS SAM treated Si-rich SiN_x surfaces.

Surface properties of Si-rich SiN_x	Mechanisms	Implications
Larger contact angle	Si-rich SiN_x has more Si–Si and Si–H bonds, which are *less polar* (or electronegative) than the N–H and Si–N bonds in N-rich SiN_x	Large contact angle → lower surface energy → more energetically favorable and stable
Lower surface roughness	More Si species are available for bonding with OTS via Si–O–C bonds, thus Si-rich SiN_x facilitates formation of a smoother OTS layer	Less interface scattering; less (physical) disruption of the overlying organic layer
More Si–O bonds	More Si–O bonds are available for bonding with polymer semiconductor via Si–O–C to generate an oxide-like surface composition	Favorable for deposition of OTS and polymer semiconductor

Table 4.3 Impact of silicon-rich SiN_x gate dielectric on OTFT and nc-Si TFT.

Impact of Si-rich SiN_x gate dielectric on	Electrical characteristics	Possible mechanisms and explanations
Organic TFT	Higher μ_{FE} and I_{ON}/I_{OFF}	• Device improvements with Si-rich SiN_x can be attributed to interfacial and microstructure effect! • Enhanced dielectric/semiconductor interface between SiN_x and PQT-12 • Si-rich SiN_x provides a more organic-friendly dielectric surface (larger contact angle, lower surface roughness) to facilitate well-ordered organic layer formation • Even if Si-rich SiN_x possesses more Si–Si dangling bonds, their effect might be masked by O_2 plasma and OTS SAM treatment
Nanocrystalline silicon (nc-Si) TFT	Lower μ_{FE} and I_{ON}/I_{OFF}	• Reduced device performance with Si-rich SiN_x can be linked to electrical and defect-related effect! • Poorer dielectric/semiconductor interface between Si-rich SiN_x and nc-Si • Higher interface trap density • More pronounced charge trapping effects with Si-rich SiN_x (Si–Si dangling bonds)

Our experimental results demonstrate enhancements in the static operation of OTFTs with a silicon-rich SiN_x gate dielectric. More detailed evaluation of the electrical stability and dynamic characteristics of OTFTs is needed to gain a more complete understanding of the overall impact of SiN_x composition on OTFT performance. An attempt was made to assess the stability of these OTFTs. However, it was difficult to isolate the effect of electrical stress from the effect of environmental stress on OTFT characteristics using our measurement set-up (under ambient conditions). Specially designed environmental chambers (with separate control on levels of humidity, air, oxygen, and vacuum) can facilitate thorough assessment of the impact of an Si-rich SiN_x gate dielectric on an OTFT's electrical stability.

4.4.2 150 °C SiN_x Gate Dielectrics

Due to constraints in the processing temperatures of most low-cost plastic substrates (e.g., PET, PEN, Kapton), the maximum fabrication temperature for transistors on these substrates is typically in the range 150–200 °C [33]; this

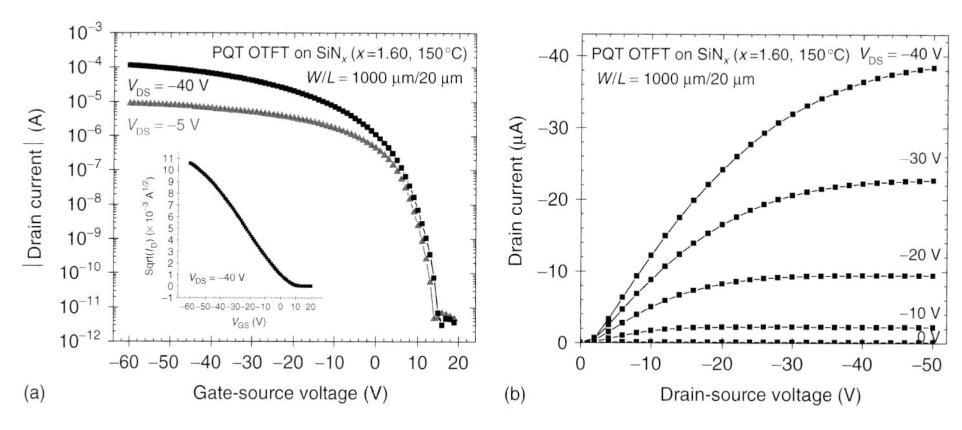

Figure 4.22 Electrical characteristics of PQT-12 OTFT on 150 °C PECVD SiN$_x$ (x = 1.60) gate dielectric with W/L = 1000 μm/20 μm: (a) transfer characteristics (log $|I_D| - V_{GS}$) and (b) output characteristics (I_D–V_{DS}). Inset of (a) shows a well-behaved linear plot of $\sqrt{I_D} - V_{GS}$ in the saturation regime.

temperature restriction also applies to gate dielectric deposition. PECVD SiN$_x$ is of particular interest for flexible electronics in view of its excellent dielectric properties, even at low temperature. In this section, the electrical performance of PQT-12 OTFTs on a 150 °C SiN$_x$ gate dielectric is evaluated.

Figure 4.22 depicts the transfer characteristics of a typical device on 150 °C SiN$_x$ (x = 1.60). The device parameters in the saturation region (V_{DS} = −40 V) are $\mu_{FE} = 0.06$ cm^2 V^{-1} s^{-1}, $I_{ON}/I_{OFF} = 1.3 \times 10^8$, $V_T = 5.74$ V, and I_G around 10^{-10} A. Well-behaved saturation characteristics can be observed in the output characteristics in Figure 4.22b; however, contact resistance is evident from the non-linear behavior at low drain voltages near the origin.

Figure 4.23 compares the effective mobility and on/off current ratio of PQT-12 OTFT with various 150 °C SiN$_x$ gate dielectrics, plotted as a function of gas flow ratio. The results showed an improvement in mobility with more Si-rich films (i.e., lower NH_3/SiH_4 gas flow ratio). This trend is consistent with outcomes for OTFTs with a 300 °C SiN$_x$ gate dielectric (see Figure 4.21). More importantly, Figure 4.23 demonstrates that PQT-12 OTFTs implemented on the 150 °C SiN$_x$ possessed higher mobility and on/off current ratio than its 300 °C SiN$_x$ counterpart! The improvement can be attributed to the larger contact angle and smaller surface roughness of the 150 °C SiN$_x$ films compared to 300 °C SiN$_x$, as illustrated in Figures 4.9 and 4.10, respectively. A larger contact angle implies lower surface energy, which is favorable for deposition of subsequent organic layers. A smoother dielectric surface provides a better platform for formation of a higher quality organic semiconductor layer and dielectric/semiconductor interface, and hence enhanced device characteristics. An attempt to compare the chemical composition of 300 and 150 °C SiN$_x$ film surfaces via XPS was made. Unfortunately, the resulting data

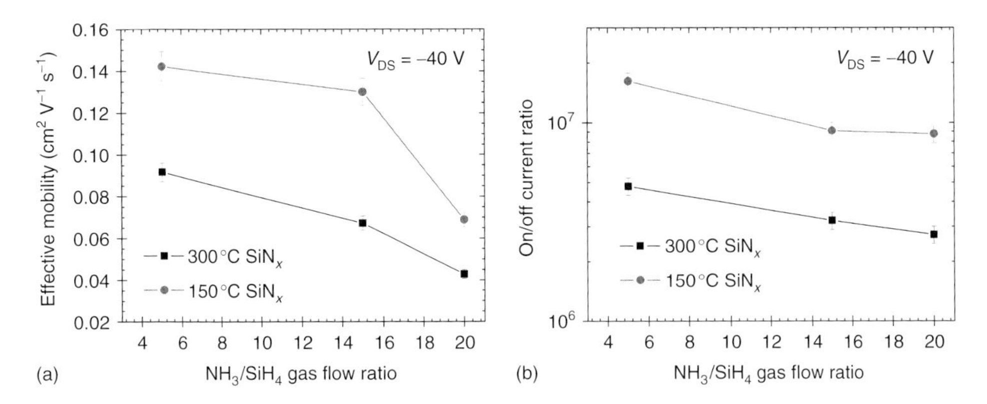

Figure 4.23 (a) Effective field-effect mobility and (b) on/off current ratio of PQT-12 OTFT ($W/L = 1000\ \mu m/20\ \mu m$) plotted as a function of NH_3/SiH_4 gas flow ratio for 150 and 300 °C SiN_x gate dielectric. Data correspond to measurements in the saturation ($V_{DS} = -40\,V$) region.

in Figure 4.12 were unable to provide sufficient justification to account for the observed OTFT trends.

Overall, these results showcase promising potential for implementing OTFT circuits with our low temperature 150 °C SiN_x for flexible electronic applications. Nonetheless, more extensive electrical characterization to examine the OTFTs' electrical stability, lifetime, and dynamic behavior are needed to gain more insight into the overall device performance.

4.4.3 Stacked SiN_x Gate Dielectrics

Electrical characterization of the SiN_x films in Section 4.3 demonstrates that N-rich SiN_x has lower leakage current, higher breakdown field, and less hysteresis than Si-rich SiN_x films. However, OTFT characterization in Sections 4.4.1 and 4.4.2 reveals that an Si-rich SiN_x gate dielectric is preferred for OTFTs because it leads to higher mobility, a higher on/off current ratio, and larger current driving capability. Our analysis showed that these device improvements are largely ascribed to a more organic-friendly interface with Si-rich SiN_x. In an attempt to combine the best qualities from both types of SiN_x films, stacked SiN_x gate dielectrics are considered. This design takes advantage of the excellent bulk properties of N-rich SiN_x and the desirable surface properties of Si-rich SiN_x. OTFTs featuring a stacked SiN_x gate dielectric are compared with OTFTs with a unilayer SiN_x gate dielectric (i.e., results from Section 4.4.1), to investigate possible improvements that can be achieved with the stacked dielectric structure.

Figure 4.24 depicts a cross-sectional diagram of an OTFT incorporating a stacked SiN_x gate dielectric. The stacked dielectric film consisted of a Si-rich top layer

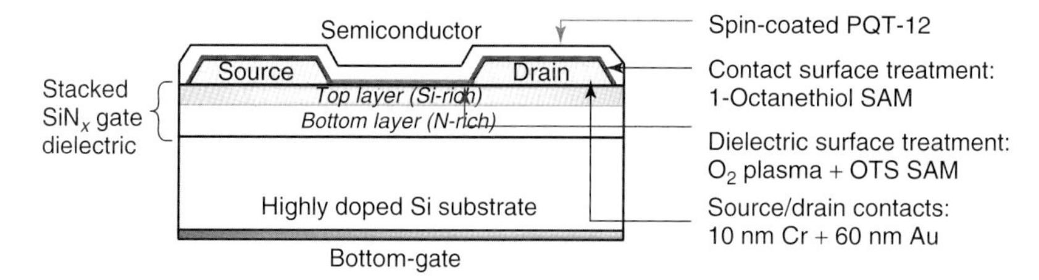

Figure 4.24 Schematic cross-section of the OTFT structure with stacked SiN_x gate dielectric.

Table 4.4 Description of stacked SiN_x gate dielectric samples.

Sample ID	Ratio of deposition time using N-rich recipe to Si-rich recipe	Estimated thickness (nm)		Estimated % thickness	
		N-rich	Si-rich	N-rich	Si-rich
SiN_x-stacked[2 : 1]	2 : 1	200	100	67	33
SiN_x-stacked[5 : 1]	5 : 1	250	50	83	17

Note: N-rich recipe refers to films deposited using NH_3/SiH_4 gas flow ratio of 20; Si-rich recipe refers to films deposited using NH_3/SiH_4 gas flow ratio of 5. SiN_x deposition was performed at substrate temperature of 300 °C, RF power of 2 W, pressure of 400 mTorr, and total duration of 30 min. The estimated thickness is based on the crude assumption that deposition rate is constant.

and an N-rich bottom layer. It was deposited in a single/continuous deposition run by adjusting the gas flow ratio at a selected time during the deposition. Two combinations of stacked SiN_x dielectric were tested for OTFTs, obtained by varying the "ratio of deposition time using an N-rich recipe to an Si-rich recipe." A description of these two stacked SiN_x samples is presented in Table 4.4. For this experiment, 300 °C SiN_x deposition recipes were used, where N-rich and Si-rich refer to recipes with gas flow ratio NH_3/SiH_4 of 20 and 5, respectively. Please refer to Table 4.1 for specific deposition parameters.

Figure 4.25 shows the FTIR spectrum of the two stacked SiN_x films. Both spectra display a larger N–H stretching peak than Si–H peak, implying that the bulk of the film is N–rich. Comparing the two stacked samples, Stacked[2 : 1] displayed a larger Si–H peak than Stacked[5 : 1]. This is logical as the Stacked[2 : 1] sample is designed to have a larger volume of Si-rich layer than the Stacked[5 : 1] sample. Material characterization confirmed that the properties of these stacked SiN_x films are intermediate between N-rich SiN_x and Si-rich SiN_x. As shown in Table 4.5, the dielectric constant, refractive index, and extracted [N]/[Si] value of stacked SiN_x films fall within the data interval bounded by N-rich SiN_x and Si-rich SiN_x.

The effective field-effect mobility of PQT-12 OTFTs with a stacked SiN_x gate dielectric is summarized in Figure 4.26. Higher mobility is attained with the SiN_x-Stacked[2 : 1] sample than the SiN_x-Stacked[5 : 1] sample. This suggests that

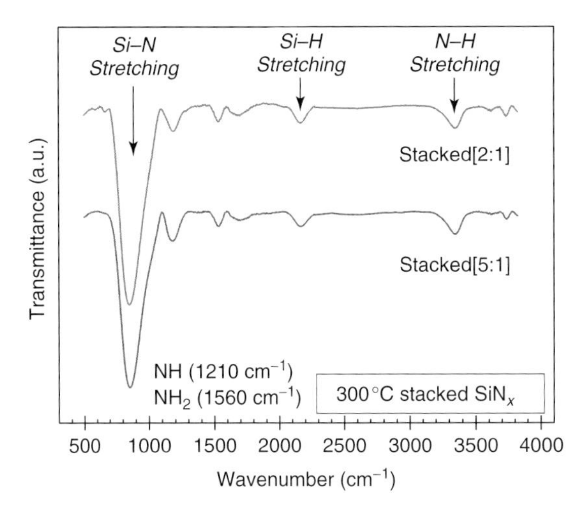

Figure 4.25 FTIR spectroscopy of the stacked SiN_x films, deposited by PECVD at a substrate temperature of 300 °C.

Table 4.5 Material characterization data for stacked SiN_x films, including refractive index, [N]/[Si] ratio, and dielectric constant. Data for Si-rich SiN_x and N-rich SiN_x are included for comparison.

SiN_x sample ID	Refractive index (n)	[N]/[Si]	Dielectric constant
Si-rich SiN_x	2.049	1.077	5.816
SiN_x-stacked[2 : 1]	2.039	1.093	5.862
SiN_x-stacked[5 : 1]	2.045	1.083	5.865
N-rich SiN_x	1.818	1.130	6.341

a thicker Si-rich top layer is preferred for enhanced mobility. When compared to the unilayer SiN_x, OTFTs with SiN_x-Stacked[2 : 1] demonstrated improved mobility over a N-rich gate dielectric. However, the Si-rich SiN_x gate dielectric provides superior mobility compared to stacked SiN_x gate dielectrics. These results clearly show that Si-rich SiN_x remains as the leading choice of gate dielectric for achieving high mobility OTFTs.

For ongoing research, more detailed device characterization and analysis should be performed to identify the potential strengths of this stacked structure. It is predicted that OTFTs fabricated on N-rich SiN_x or stacked SiN_x gate dielectrics can offer advantages in terms of reduced hysteresis, improved dielectric breakdown strength, enhanced AC/transient properties and improved electrical stability. Additional studies to understand the growth mechanisms of the stacked structure and the material properties of the transition region from N-rich to Si-rich would

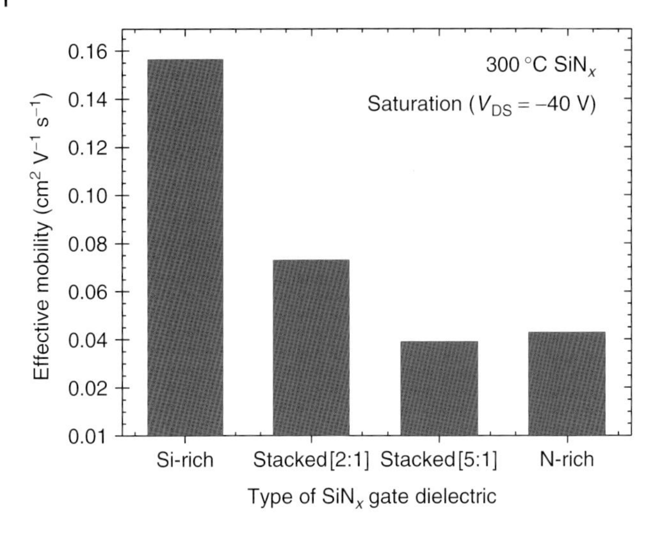

Figure 4.26 Effective field-effect mobility of PQT-12 OTFT with stacked SiN_x gate dielectrics. The mobility of OTFT with Si-rich SiN_x (x = 1.08) and N-rich SiN_x (x = 1.50) is also shown for comparison.

be beneficial. The deposition of stacked SiN_x was executed continuously by simply changing the gas flow ratio after a specific time duration. It is interesting to examine whether a momentary pause/break after deposition of the N-rich layer before restarting deposition for the Si-rich layer would alter the OTFT characteristics. An intermittent break (e.g., a few minutes) may help to purge the chamber and promote the growth of a higher quality Si-rich SiN_x at the interface. A better understanding of the growth process will allow more strategic design or optimization of the deposition scheme, to fulfill the intention of maximizing device performance by combining the benefits of N-rich SiN_x and Si-rich SiN_x in a stacked dielectric system.

4.4.4 200°C SiO_x Gate Dielectrics

PECVD SiO_x presents an alternative low-temperature gate dielectric option for large area flexible electronics integration. Most of the early research on OTFTs employed thermal silicon dioxide (SiO_2) on a highly-doped silicon wafer as a platform for material and device characterization; these experiments showed that excellent dielectric/semiconductor interface properties can be achieved between SiO_2 and the organic layer [52, 60, 64]. However, the use of thermal SiO_2 limits OTFT circuit integration because of the continuous gate structure (with the silicon wafer); more importantly, growth of thermal SiO_2 involves a high-temperature process that is not amenable to the low-temperature processing objectives of OTFTs. Therefore, it is of interest to explore PECVD SiO_x as the gate dielectric

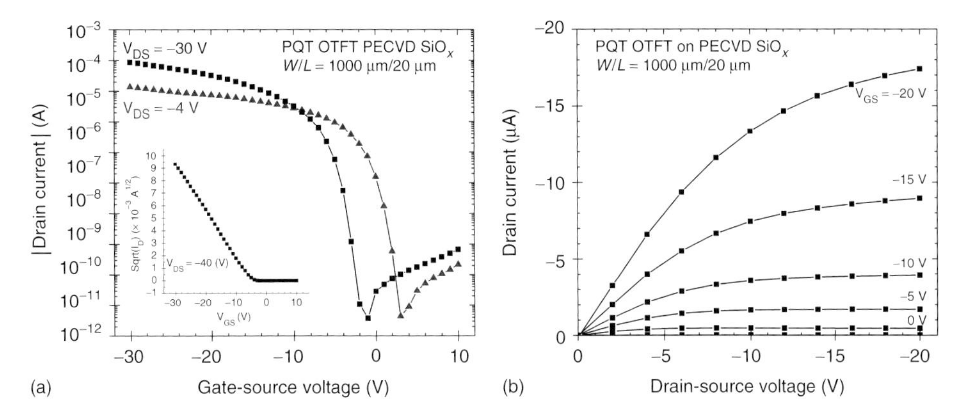

Figure 4.27 Electrical characteristics of PQT-12 OTFT on PECVD SiO_x gate dielectric with $W/L = 1000\ \mu m/20\ \mu m$: (a) transfer characteristics ($\log|I_D| - V_{GS}$) and (b) output characteristics ($I_D - V_{DS}$). Inset of (a) shows a linear plot of $\sqrt{I_D} - V_{GS}$ in the saturation regime.

for OTFTs. While being compatible with the processing temperatures of plastic substrates, low-temperature PECVD SiO_x can present a similar interface platform as thermal oxide to deliver high performance OTFTs. However, PECVD oxides are typically prone to poorer bulk properties than PECVD SiN_x [37], which may lead to high leakage currents, low breakdown fields, hysteresis effects, and V_T instability, in OTFTs. Thus, there is an ongoing research endeavor to enhance the quality of PECVD SiO_x thin film.

Transfer and output characteristics of a PQT-12 OTFT on SiO_x gate dielectric are shown in Figure 4.27. Parameter extraction from the saturation characteristics ($V_{DS} = -30$ V) gives $\mu_{FE} = 0.34\ cm^2\ V^{-1}\ s^{-1}$, $I_{ON}/I_{OFF} = 2.3 \times 10^8$, and $V_T = -5.04$ V. The output characteristics in Figure 4.27b demonstrate well-defined saturation behavior and reasonable contact properties [68]. It is worth highlighting that PECVD SiO_x produced relatively high mobility PQT-12 OTFT devices, in comparison to devices with PECVD SiN_x (as reported in previous sections). The enhanced mobility indicates that the PECVD SiO_x film, following appropriate interface treatment, delivers the needed surface platform for formation of a highly-ordered PQT-12 semiconductor layer, which is vital for attaining high field-effect mobility [69, 70]. Despite the improvement in mobility, OTFTs on the PECVD SiO_x gate dielectric exhibited high gate leakage current (from an initial value of 10^{-9} to 10^{-6} A) at high negative V_{GS} following successive electrical bias. PECVD SiO_x thin films deposited at low processing temperatures are typically susceptible to electrical integrity issues; the low mass density of SiO_x films often leads to considerable leakages [37]. Also, a bias-induced shift in V_T is apparent in the saturation and linear transfer characteristics in Figure 4.27a. This instability can be related to the higher density of fixed charge in the bulk SiO_x, charge trapping at the organic/oxide interface, and leakage paths in SiO_x.

In summary, PQT-12 OTFTs with a SiO_x gate dielectric offer superior mobility compared to devices on SiN_x. However, SiO_x is more susceptible to higher gate leakage, lower dielectric breakdown, and reduced electrical stability. Thus, for applications requiring low deposition temperatures, PECVD SiN_x is currently the preferred choice as it offers superior electrical integrity in terms of lower leakage current, fewer pinholes, a higher breakdown field, higher stability, and reduced hysteresis.

For future research, additional effort to improve the quality and integrity of PECVD SiO_x is needed to make SiO_x a feasible candidate for OTFT applications. In addition, it is of interest to explore a stacked SiN_x/SiO_x dielectric structure, with SiN_x serving as a bottom bulk layer and SiO_x as a top interface layer. This configuration takes advantage of the excellent bulk properties of SiN_x to ensure low gate leakage current, high dielectric breakdown strength, and good dielectric stability; this is combined with the preferable surface properties of SiO_x to form a quality dielectric/semiconductor interface for delivering high mobility OTFTs.

4.4.5 OTFTs on Plastic Substrates

PQT-12 OTFTs fabricated with low-temperature PECVD gate dielectrics have demonstrated very promising performance on rigid substrates, as presented in Sections 4.4.2 and 4.4.4 for 150 °C SiN_x and 200 °C SiO_x, respectively. This section examines the application of low-temperature PECVD dielectrics for demonstration of OTFTs on plastic substrates. Table 4.6 lists the various pairs of gate dielectrics and plastic substrates considered in this study, including a 150 °C SiN_x gate dielectric on

Table 4.6 Various plastic substrates and PECVD gate dielectric employed for PQT-12 OTFT fabrication in this study. The corresponding effective field effect mobility (μ_{FE}), on/off current ratio (I_{ON}/I_{OFF}), and threshold voltage (V_T), extracted in the saturation region, are shown.

ID	Substrate	PECVD gate dielectric	μ_{FE} (cm^2 V^{-1} s^{-1})	I_{ON}/I_{OFF}	V_T (V)	Remarks
#1	PET	150 °C SiN_x ($x = 1.60$)	0.0092	6.8×10^3	6.7	Higher μ_{FE} but lower I_{ON}/I_{OFF} compared to Kapton substrate
#2	Kapton	150 °C SiN_x ($x = 1.60$)	0.0051	1.6×10^6	5.0	Higher I_{ON}/I_{OFF} but lower μ_{FE} compared to PET substrate
#3	Kapton	180 °C SiO_x	0.0076	3.1×10^5	−3.3	Very low yield. High gate leakage current. Susceptible to dielectric breakdown

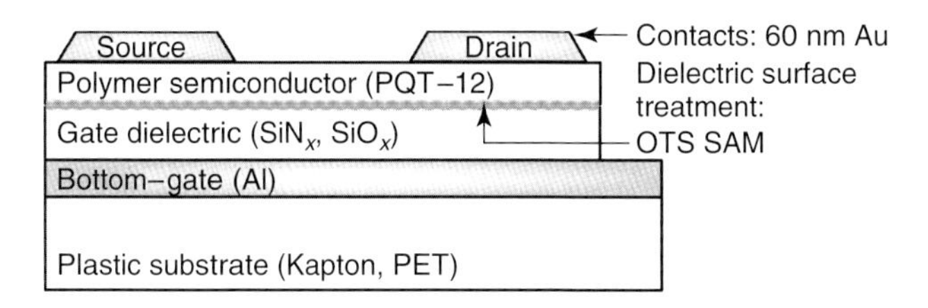

Figure 4.28 Schematic cross-section of the bottom-gate top-contact OTFT on a plastic substrate.

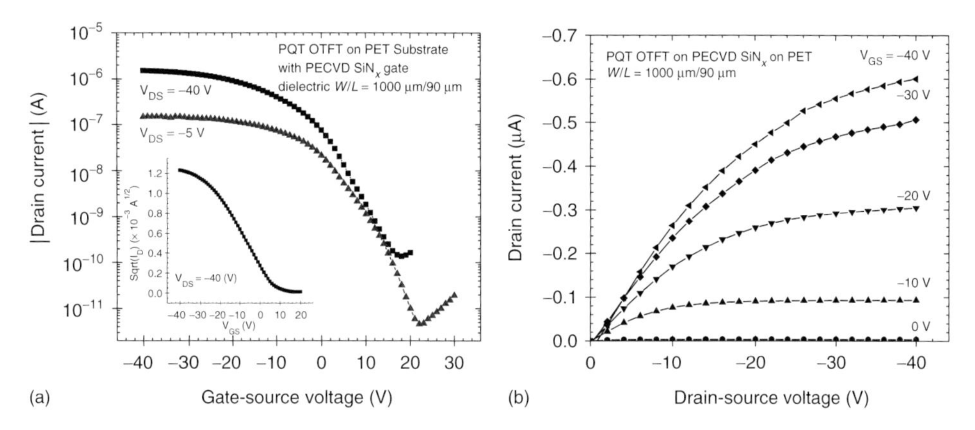

Figure 4.29 (a) Transfer and (b) output characteristics of PQT-12 OTFT on a PET substrate with PECVD SiN_x gate dielectric ($W/L = 1000$ μm/90 μm).

a PET substrate, 150 °C SiN_x on a Kapton (polyimide) substrate, and 180 °C SiO_x on a Kapton substrate.[2] The bottom-gate top-contact OTFT configuration, as illustrated in Figure 4.28, was used for this investigation. The source/drain Au contacts were defined using the shadow mask technique, to simplify the fabrication process and to avoid any process-induced degradation of devices on plastic substrates. The dielectric surface was pretreated with OTS SAM only; O_2 plasma treatment was omitted for preliminary tests to eliminate possible plasma-induced damage of the plastic substrate.

Figures 4.29–4.31 display the transfer and output characteristics of PQT-12 OTFTs on various plastic substrates and PECVD gate dielectrics. The effective field-effect mobility, on/off current ratio, and threshold voltage extracted from these measurements are summarized in Table 4.6. Each dielectric–substrate combination has its strengths and weaknesses.

2) Kapton® is a polyimide film developed by DuPont.

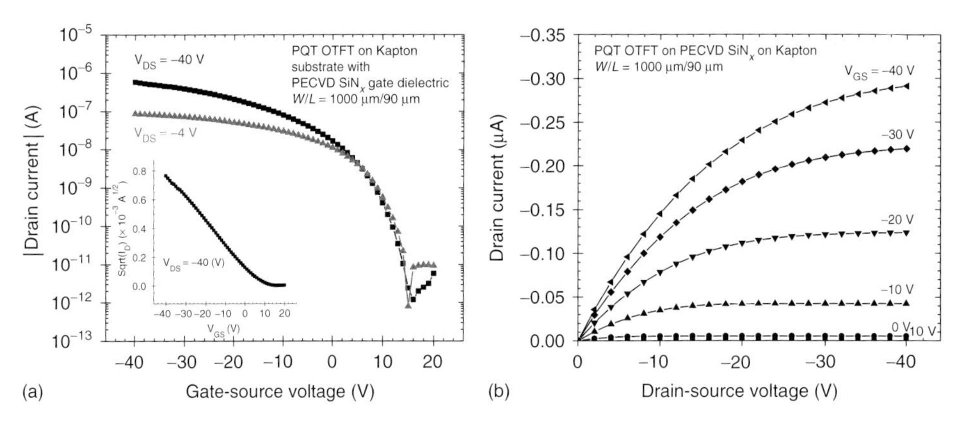

Figure 4.30 (a) Transfer and (b) output characteristics of PQT-12 OTFT on a Kapton substrate with PECVD SiN_x gate dielectric ($W/L = 1000$ μm/90 μm).

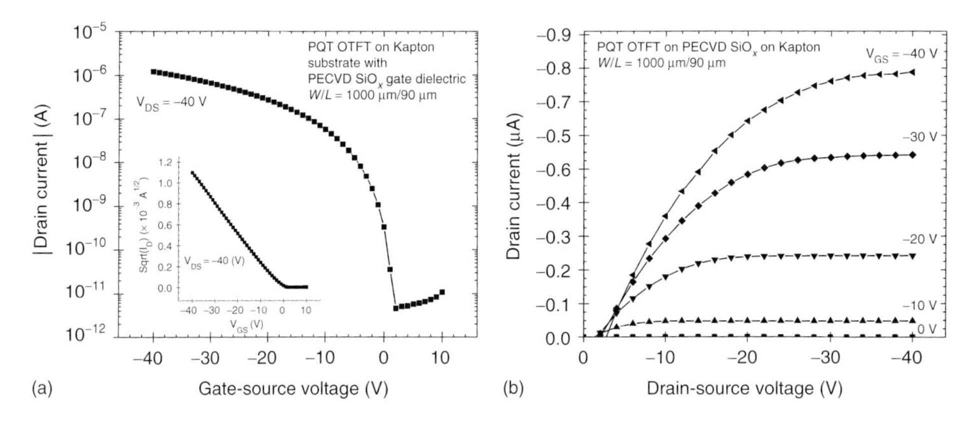

Figure 4.31 (a) Transfer and (b) output characteristics of PQT-12 OTFT on a Kapton substrate with PECVD SiO_x gate dielectric ($W/L = 1000$ μm/90 μm).

PQT-12 OTFTs with a PECVD SiO_x gate dielectric on Kapton exhibited higher mobility than devices with a SiN_x dielectric on Kapton. The higher mobility can be attributed to the favorable SiO_x/polymer interface. However, the device yield for the SiO_x on the Kapton substrate is very low. Moreover, the SiO_x/Kapton sample is highly prone to gate leakage current (10^{-9} A range) and dielectric breakdown (occurs around 40 V for 200 nm thick SiO_x). The offset of the I_D-V_{DS} curve from the origin ($V_{DS} = 0$ V) in Figure 4.31b is a result of a large gate leakage current in the device.

In contrast, OTFTs fabricated with a PECVD SiN_x dielectric on plastic substrates displayed much better dielectric strength and a lower dielectric leakage current compared to SiO_x, but at the expense of reduced mobility. Comparing SiN_x/Kapton with SiN_x/PET, devices on Kapton showed a higher on/off current ratio, better

Table 4.7 Surface roughness of 150 °C PECVD SiN_x on Si wafer and plastic substrate.

Substrate	Surface roughness (nm)
Silicon wafer	0.210
Kapton	3.37

saturation characteristics, and linear contact behavior (Figure 4.30); devices on PET showed higher mobility (Figure 4.29). Similar levels of gate leakage current were observed for the SiN_x/Kapton and SiN_x/PET samples.

Overall, there is a substantial reduction in device performance for OTFTs on plastic substrates when compared to OTFTs on rigid substrates. Considering PQT-12 OTFTs with a 150 °C SiN_x gate dielectric, μ_{FE} decreased by a factor of 10 when the substrate was changed from a rigid Si wafer to a flexible Kapton substrate. This reduction can be attributed to the higher surface roughness of SiN_x films on plastic substrates than on Si substrates, as indicated in Table 4.7. In general, plastic substrates display higher surface roughness; a passivation layer (e.g., benzocyclobutene (BCB)) can be used to smooth the substrate roughness. Chabinyc *et al.* showed that the field-effect mobility of OTFTs decreases nearly exponentially with the surface roughness of the gate dielectric [55]. Another possible explanation for the reduction in mobility is the omission of O_2 plasma treatment for the devices on a plastic substrate. As we will see in the next chapter, O_2 plasma treatment plays a critical role in achieving high mobility OTFTs with a SiN_x gate dielectric.

In summary, this experiment demonstrated the feasibility of fabricating OTFTs with a PECVD gate dielectric on plastic substrates. Preliminary devices showed functional devices, but the performance was inferior to devices on a rigid substrate. The surface roughness of plastic substrate is one of the key reasons for the limited mobility. It is recommended that ongoing research is focused on optimizing the properties of low-temperature gate dielectrics on plastic substrates to improve the dielectric properties (e.g., leakage, breakdown, stability), and on reducing the surface roughness of the resulting substrate via passivation techniques.

4.5 Summary and Contributions

PECVD SiN_x and SiO_x thin films were studied as gate dielectrics for PQT-12 OTFTs, with the objective being to develop materials for organic circuit integration on rigid and flexible substrates. The major observations from the investigations reported in this chapter are summarized in Table 4.8. Improvements in field-effect mobility and on/off current ratio were observed as the silicon content in the SiN_x gate dielectric increases, suggesting that silicon-rich SiN_x dielectrics are preferable over nitrogen-rich SiN_x. Two sets of OTFTs were evaluated, one with 300 °C SiN_x

Table 4.8 Summary of key observations on comparative study of OTFT with different PECVD gate dielectrics.

	Description	Key observations
Part A	300 °C SiN_x	SiN_x with smaller [N]/[Si] (i.e., Si-rich film) gives higher μ_{FE}, higher I_{ON}/I_{OFF}, smaller V_T
		As [N]/[Si] decreases, contact angle increases (slightly), surface roughness decreases, atomic percentage of Si at the surface increases; thus rendering an interface more favorable for OTFTs
Part B	150 °C SiN_x	Similar to 300 °C SiN_x, mobility increases as Si-content in SiN_x increases 150 °C SiN_x dielectric delivers higher mobility than 300 °C SiN_x
Part C	Stacked SiN_x gate dielectric	Performances with stacked dielectrics are inferior/poorer than unilayer Si-rich SiN_x dielectric
Part D	200 °C SiO_x	Achieved mobility as high as 0.5 $cm^2\,V^{-1}\,s^{-1}$ However, PECVD SiO_x has poorer dielectric integrity than PECVD SiN_x. SiO_x has lower breakdown fields, higher leakage currents, and lower yield/reproducibility More work on improving SiO_x properties is needed
Part E	PECVD dielectric on plastic substrates	Mobility is 1–2 orders lower on plastic substrates than on rigid substrate, accompanied by a reduction in device yield More work is needed to optimize the processing procedure for device fabrication on plastic substrates

gate dielectrics, and one with 150 °C SiN_x gate dielectrics. While the same film composition dependence was observed in both sets of OTFTs, OTFTs with 150 °C SiN_x gate dielectrics demonstrated higher mobility than devices with 300 °C SiN_x films. A strong correlation between the electrical characteristics of the OTFTs and the surface mechanics of the SiN_x films was observed. It was determined that the contact angle, surface roughness, and surface chemical composition of the SiN_x gate dielectric have an overriding impact on OTFT characteristics. Thus, it is concluded that the composition of the SiN_x film influences the dielectric surface properties, which in turn affects the quality of the dielectric/semiconductor interface and the molecular ordering of the overlying organic semiconductor layer; altogether, these factors have a strong bearing on the OTFT performance. These relationships are summarized in Figure 4.32.

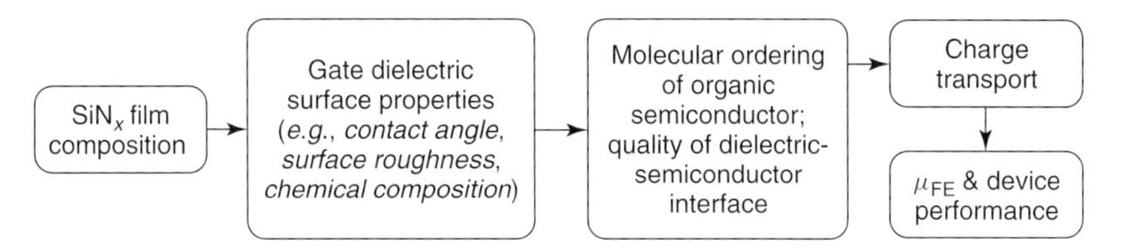

Figure 4.32 Flowchart illustrating the relationship between SiN_x film composition, interface properties, and OTFT performance.

PQT-12 OTFTs fabricated on PECVD SiO_x displayed superior mobilities compared to devices on SiN_x. This improvement is largely ascribed to organic-friendly surface properties of the SiO_x film. However, the electrical integrity and stability of low-temperature PECVD SiO_x are a concern. Extensive process and material optimization is necessary to make PECVD SiO_x useful for practical applications. Preliminary OTFTs on plastic substrates have been demonstrated; further process development is expected to generate enhanced device performance, taking us one step closer to the realization of flexible electronics. Overall, these results demonstrated the viability of using PECVD SiN_x and SiO_x as gate dielectrics for OTFT circuit integration, especially considering the vast and well-established infrastructure already in place for displays. The low temperature and large area deposition capabilities of PECVD SiN_x and SiO_x films present good compatibility and technological potential to facilitate integration of OTFT circuits on plastic substrates.

Further optimization of the PECVD gate dielectric composition and the dielectric/organic layer interface property using combinatorial techniques [71], may further improve OTFT performance and boost PECVD SiN_x and SiO_x as a serious gate dielectric alternative over the high-temperature thermal SiO_2, especially for applications on flexible plastic substrates. As we will see in the next chapter, by optimizing the interface treatment conditions, additional improvements (with mobility as high as $0.2\,cm^2\,V^{-1}\,s^{-1}$ with a 150 °C SiN_x gate dielectric) can be achieved.

As a continuation of this research to generate further device improvements and to develop additional understanding of device behavior, the following investigations are recommended:

- Expanding the range of deposition recipes of PECVD SiN_x, to evaluate if more silicon-rich films (e.g., films with $[N]/[Si] < 1$) can generate additional device improvements.
- AC/transient measurements and long term stress measurements to examine the impact of SiN_x gate dielectric composition on an OTFT's dynamic characteristics and stability.
- In-depth evaluation of all device parameters to draw a decisive conclusion on the impact of a Si-rich SiN_x gate dielectric on the overall OTFT performance.
- Further material development for PECVD SiO_x to improve dielectric integrity.
- Investigate stacked gate dielectric structure composed of SiO_x as top-layer and Si-rich SiN_x as bottom/bulk-layer. The SiO_x surface appears to render a favorable

dielectric/semiconductor interface to achieve high device mobility. SiN_x provides good dielectric integrity, high breakdown strength, and low leakage current.

References

1. Klauk, H., Halik, M., Zschieschang, U., Schmid, G., Radlik, W., and Weber, W. (2002) High-mobility polymer gate dielectric pentacene thin film transistors. *J. Appl. Phys.*, **92**, 5259.
2. Peng, X., Horowits, G., Fichou, D., and Garnier, F. (1990) All-organic thin-film transistors made of alpha-sexithienyl semiconducting and various polymeric insulating layers. *Appl. Phys. Lett.*, **57**, 2013.
3. Kelly, T.W., Muyres, D.V., Baude, P.F., Smith, T.P., and Jones, T.D. (2003) High performance organic thin film transistors. *Mater. Res. Soc. Symp. Proc.*, **771**, L6.5.
4. Schroeder, R., Majewski, L.A., and Grell, M. (2003) A study of the threshold voltage in pentacene organic field-effect transistors. *Appl. Phys. Lett.*, **83**, 3201.
5. Park, J., Park, S.Y., Shim, S., Kang, H., and Lee, H.H. (2004) A polymer gate dielectric for high-mobility polymer thin-film transistors and solvent effects. *Appl. Phys. Lett.*, **85**, 3283.
6. Stadlober, B., Zirkl, M., Beutl, M., Liesing, G., Bauer-Gegonea, S., and Bauer, S. (2005) High-mobility pentacene organic field-effect transistors with a high-dielectric-constant fluorinated polymer film gate dielectric. *Appl. Phys. Lett.*, **86**, 242902.
7. Bao, Z., Feng, Y., Dodabalapur, A., Raju, V.R., and Lovinger, A. (1997) High-performance plastic transistors fabricated by printing techniques. *Chem. Mater.*, **9**, 1299.
8. Kato, Y., Iba, S., Teramoto, R., Sekitani, T., and Someya, T. (2004) High mobility of pentacene field-effect transistors with polyimide gate dielectric layers. *Appl. Phys. Lett.*, **84**, 3789.
9. Naber, R.C.G., Tanase, C., Blom, P.W.M., Gelinck, G.H., Marsman, A.W., Touwslager, F.J., Setayesh, S., and de Leeuw, D.M. (2005) High-performance solution-processed polymer ferroelectric field-effect transistors. *Nat. Mater.*, **4**, 243.
10. Horowitz, G., Deloffre, F., Garnier, F., Hajlaoui, R., Hmyene, M., and Yassar, A. (1993) All-organic field-effect transistors made of π-conjugated oligomers and polymeric insulators. *Synth. Met.*, **54**, 435.
11. Taniguchi, M. and Kawai, T. (2004) Vertical electrochemical transistor based on poly(3-hexylthiophene) and cyanoethylpullulan. *Appl. Phys. Lett.*, **85**, 3298.
12. Chua, L.L., Ho, P.K.H., Sirringhaus, H., and Friend, R.H. (2004) High-stability ultrathin spin-on benzocyclobutene gate dielectric for polymer field-effect transistors. *Appl. Phys. Lett.*, **84**, 3400.
13. Veres, J., Ogier, S., Lloyd, G., and de Leeuw, D. (2004) Gate insulators in organic field-effect transistors. *Chem. Mater.*, **16**, 4543.
14. Facchetti, A., Yoon, M.H., and Marks, T.J. (2005) Gate dielectrics for organic field-effect transistors: new opportunities for organic electronics. *Adv. Mater.*, **17**, 1705.
15. Horowitz, G., Peng, X.-Z., Fichou, D., and Garnier, F. (1992) Role of the semiconductor/insulator interface in the characteristics of π-conjugated-oligomer-based thin-film transistors. *Synth. Met.*, **51**, 419.
16. Lin, Y.Y., Gundlach, D.J., Nelson, S.F., and Jackson, T.N. (1997) Pentacene-based organic thin-film transistors. *IEEE Trans. Electron Devices*, **44**, 1325.
17. Kelly, T.W., Boardman, L.D., Dunbar, T.D., Muyres, D.V., Pellerite, M.J., and Smith, T.P. (2003) High-performance OTFTs using surface-modified alumina dielectrics. *J. Phys. Chem. B*, **107**, 5877.
18. Wu, Y., Liu, P., Ong, B.S., Srikumar, T., Zhao, N., Botton, G., and Zhu, S. (2005) Controlled orientation of liquid-crystalline polythiophene semiconductors for high-performance organic

thin-film transistors. *Appl. Phys. Lett.*, **86**, 142102.
19. Wu, Y., Liu, P., and Ong, B.S. (2006) Organic thin-film transistors with poly(methyl silsesquioxane) modified dielectric interfaces. *Appl. Phys. Lett.*, **89**, 013505.
20. Liu, P., Wu, Y., Li, Y., Ong, B.S., and Zhu, S. (2006) Enabling gate dielectric design for all solution-processed, high-performance, flexible organic thin-film transistors. *J. Am. Chem. Soc.*, **128**, 4554.
21. Wang, Y., Liu, Y., Song, Y., Ye, S., Wu, W., Guo, Y., Di, C., Sun, Y., Yu, G., and Hu, W. (2008) Organic field-effect transistors with a low pinch-off voltage and a controllable threshold voltage. *Adv. Mater.*, **20**, 611.
22. Cai, Q.J., Gan, Y., Chan-Park, M.B., Yang, H.B., Lu, Z.S., Li, C.M., Guo, J., and Dong, Z.L. (2009) Solution-processable barium titanate and strontium titanate nanoparticle dielectrics for low-voltage organic thin-film transistors. *Chem. Mater.*, **21**, 3153.
23. Zirkl, M., Haase, A., Fian, A., Schön, H., Sommer, C., Jakopic, G., Leising, G., Stadlober, B., Graz, I., Gaar, N., Schwödiauer, R., Bauer-Gogonea, S., and Bauer, S. (2007) Low-voltage organic thin-film transistors with high-k nanocomposite gate dielectrics for flexible electronics and optothermal sensors. *Adv. Mater.*, **19**, 2241.
24. Jang, Y., Lee, W.H., Park, Y.D., Kwak, D., Cho, J.H., and Cho, K. (2009) High field-effect mobility pentacene thin-film transistors with nanoparticle polymer composite/polymer bilayer insulators. *Appl. Phys. Lett.*, **94**, 183301.
25. Schroeder, R., Majewski, L.A., and Grell, M. (2005) High-performance organic transistors using solution-processed nanoparticle-filled high-k polymer gate insulators. *Adv. Mater.*, **17**, 1535.
26. Roberts, M.E., Queralto, N., Mannsfeld, S.C.B., Reinecke, B.N., Knoll, W., and Bao, Z. (2009) Cross-linked polymer gate dielectric films for low-voltage organic transistors. *Chem. Mater.*, **21**, 2292.
27. Liu, Z., Oh, H.J., Roberts, M.E., Wei, P., Paul, B.C., Okajima, M., Nishi, Y., and Bao, Z., (2009) Solution-processed flexible organic transistors showing very-low subthreshold slope with a bilayer polymeric dielectric on plastic. *Appl. Phys. Lett.*, **94**, 203301.
28. Cheng, X., Caironi, M., Noh, Y.-Y., Wang, J., Newman, C., Yan, H., Facchetti, A., and Sirringhaus, H., (2010) Air stable cross-linked cytop ultrathin gate dielectric for high yield low-voltage top-gate organic field-effect transistors. *Chem. Mater.*, **22**, 1559.
29. Halik, M., Klauk, H., Zschieschang, U., Schmid, G., Ponomarenko, S., Kirchmeyer, S., and Weber, W. (2003) Relationship between molecular structure and electrical performance of oligothiophene organic thin film transistors. *Adv. Mater.*, **15**, 917.
30. Halik, M., Klauk, H., Zschieschang, U., Schmid, G., Dehm, C., Schütz, M., Maisch, S., Effenberger, F., Brunnbauer, M., and Stellacci, F. (2004) Low-voltage organic transistors with an amorphous molecular gate dielectric. *Nature*, **431**, 963.
31. Yoon, M.H., Facchetti, A., and Marks, T.J. (2005) σ-π molecular dielectric multilayers for low-voltage organic thin-film transistors. *Proc. Natl. Acad. Sci. U.S.A.*, **102**, 4678.
32. Ha, Y., Facchetti, A., and Marks, T.J. (2009) Push-pull π-electron phosphonic-acid-based self-assembled multilayer nanodielectrics fabricated in ambient for organic transistors. *Chem. Mater.*, **21**, 1173.
33. Macdonald, W. (2004) Engineered films for display technologies. *J. Mater. Chem.*, **14**, 4–10.
34. Stryakhilev, D., Sazonov, A., and Nathan, A. (2002) Amorphous silicon nitride deposited at 120 °C for OLED-TFT arrays on plastic substrates. *J. Vac. Sci. Technol. A*, **20**, 1087.
35. Wagner, S., Gleskova, H., Cheng, I.C., and Wu, M. (2003) Silicon for thin-film transistors. *Thin Solid Films*, **430**, 15.
36. McArthur, C., Meitine, M., and Sazonov, A. (2003) Optimization of 75 °C Amorphous Silicon Nitride for TFTs on Plastics. *Mater. Res. Soc. Symp. Proc.*, **769**, 303.

37. Meitine, M. and Sazonov, A. (2003) Low temperature PECVD silicon oxide for devices and circuits on flexible substrates. *Mater. Res. Soc. Symp. Proc.*, **769**, 165.
38. Cai, L., Rohatgi, A., Han, S., May, G., and Zou, M. (1998) Investigation of the properties of plasma-enhanced chemical vapor deposited silicon nitride and its effect on silicon surface passivation. *J. Appl. Phys.*, **83**, 5885.
39. Knipp, D., Street, R.A., Völkel, A., and Ho, J. (2003) Pentacene thin film transistors on inorganic dielectrics: Morphology, structural properties, and electronic transport. *J. Appl. Phys.*, **93**, 347.
40. Peláez, R., Castán, E., Dueñas, S., Barboll, J., Redondo, E., Mártil, I., and González-Díaz, G. (1999) Electrical characterization of electron cyclotron resonance deposited silicon nitride dual layer for enhanced Al/SiN_x:H/InP metal–insulator–semiconductor structures fabrication. *J. Appl. Phys.*, **86**, 6924.
41. Hugon, M.C., Delmotte, F., Agius, B., and Courant, J.L. (1997) Electrical properties of metal–insulator–semiconductor structures with silicon nitride dielectrics deposited by low temperature plasma enhanced chemical vapor deposition distributed electron cyclotron resonance. *J. Vac. Sci. Technol. A*, **15**, 3143.
42. Kessels, W.M.M., Hong, J., van Assche, F.J.H., Moschner, J.D., Lauinger, T., Soppe, W.J., Weeber, A.W., Schram, D.C., and van de Sanden, M.C.M. (2002) High-rate deposition of a-SiN_x:H for photovoltaic applications by the expanding thermal plasma. *J. Vac. Sci. Technol. A*, **20**, 1704.
43. Rohatgi, A. and Jeong, J.W. (2003) High-efficiency screen-printed silicon ribbon solar cells by effective defect passivation and rapid thermal processing. *Appl. Phys. Lett.*, **82**, 224.
44. Wang, L., Reehal, H.S., Martinez, F.L., San Andres, E., and del Prado, A. (2003) Characterization of nitrogen-rich silicon nitride films grown by the electron cyclotron resonance plasma technique. *Semicond. Sci. Technol.*, **18** (7), 663.
45. Esmaeili-Rad, M.R., Li, F., Sazonov, A., and Nathan, A. (2007) Stability of nanocrystalline silicon bottom-gate thin film transistors with silicon nitride gate dielectric. *J. Appl. Phys.*, **102**, 064512.
46. Martínez, F.L., San Andrés, E., del Prado, A., Mártil, I., Bravo, D., and López, F. (2001) Temperature effects on the electrical properties and structure of interfacial and bulk defects in Al/SiN_x:H/Si devices. *J. Appl. Phys.*, **90**, 1573.
47. Youn, B., Kim, D., and Jeon, S. (2005) Improvement of interfacial performance on silicone insulating and semiconducting interface by plasma treatment. 2005 IEEE Conference on Electrical Insulation and Dielectric Phenomena (CEIDP), pp. 75–78.
48. Landford, W.A. and Rand, M.J. (1978) The hydrogen content of plasma-deposited silicon nitride. *J. Appl. Phys.*, **49**, 2473.
49. Bulkin, P.V., Swart, P.L., and Lacquet, B.M. (1994) Electron cyclotron resonance plasma deposition of SiN_x for optical applications. *Thin Solid Films*, **241**, 247.
50. Mäckel, H. and Lüdemann, R. (2002) Detailed study of the composition of hydrogenated SiN_x layers for high-quality silicon surface passivation. *J. Appl. Phys.*, **92**, 2602.
51. Bustarret, E., Bensouda, M., Habrard, M.C., Bruyère, J.C., Poulin, S., and Gujrathi, S.C. (1988) Configurational statistics in a-$Si_xN_yH_z$ alloys: A quantitative bonding analysis. *Phys. Rev. B*, **38**, 8171.
52. Kosbar, L.L., Dimitrakopoulos, C.D., and Mascaro, D.J. (2001) The effect of surface preparation on the structure and electrical transport in an organic semiconductor. *Mater. Res. Soc. Symp. Proc.*, **665**, C10.6.1.
53. Song, C., Koo, B., Lee, S., and Kim, D. (2002) Characteristics of pentacene organic thin film transistors with gate insulator processed by organic molecules. *Jpn. J. Appl. Phys.*, **41**, 2730.
54. Sung, M.M., Kluth, G.J., and Maboudian, R. (1999) Formation of alkylsiloxane self-assembled monolayers on Si_3N_4. *J. Vac. Sci. Technol. A*, **17** (2), 540.
55. Chabinyc, M.L., Lujan, R., Endicott, F., Toney, M.F., McCulloch, I.,

and Heeney, M. (2007) Enhancement of field-effect mobility due to surface-mediated molecular ordering in regioregular polythiophene thin film transistors. *Appl. Phys. Lett.*, **90**, 233508.

56. Kim, D.H., Park, Y.D., Jang, Y., Yang, H., Kim, Y.H., Han, J.I., Moon, D.G., Park, S., Chang, T., Chang, C., Joo, M., Ryu, C.Y., and Cho, K. (2005) Effects of the surface roughness of plastic-compatible inorganic dielectrics on polymeric thin film transistors. *Adv. Funct. Mater.*, **15** (1), 77.

57. Park, K.J. and Parsons, G.N. (2004) Bulk and interface charge in low temperature silicon nitride for thin film transistors on plastic substrates. *J. Vac. Sci. Technol. A*, **22** (6), 2256.

58. Lau, W.S., Fonash, S.J., and Kanicki, J. (1989) Stability of electrical properties of nitrogen-rich, silicon-rich, and stoichiometric silicon nitride films. *J. Appl. Phys.*, **66**, 2765.

59. Powell, M.J. (1983) Charge trapping instabilities in amorphous silicon-silicon nitride thin-film transistors. *Appl. Phys. Lett.*, **43**, 597.

60. Ong, B.S., Wu, Y., Liu, P., and Gardner, S. (2004) High-performance semiconducting polythiophenes for organic thin-film transistors. *J. Am. Chem. Soc.*, **126**, 3378.

61. Ong, B.S., Wu, Y., Liu, P., and Gardner, S. (2005) Structurally ordered polythiophene nanoparticles for high-performance organic thin film transistors. *Adv. Mater.*, **17**, 1141.

62. Li, F.M., Wu, Y., Ong, B.S., and Nathan, A. (2007) Organic thin-film transistor integration using silicon nitride gate dielectric. *Appl. Phys. Lett.*, **90**, 133514.

63. Street, R.A. (2006) Conducting polymers: the benefit of order. *Nat. Mater.*, **5**, 171.

64. Bao, Z., Dodabalapur, A., and Lovinger, A.J. (1996) Soluble and processable regioregular poly(3-hexylthiophene) for thin film field-effect transistor applications with high mobility. *Appl. Phys. Lett.*, **69**, 4108.

65. Kline, R.J., McGehee, M.D., and Toney, M.F. (2006) Highly oriented crystals at the buried interface in polythiophene thin-film transistors. *Nat. Mater.*, **5**, 222.

66. Sirringhaus, H., Tessler, N., and Friend, R.H. (1998) Integrated optoelectronic devices based on conjugated polymers. *Science*, **280**, 1741.

67. Calhoun, M.F., Sanchez, J., Olaya, D., Gershenson, M.E., and Podzorov, V. (2008) Electronic functionalization of the surface of organic semiconductors with self-assembled monolayers. *Nat. Mater.*, **7**, 84–89.

68. Li, F.M., Wu, Y., Ong, B.S., Vygranenko, Y., and Nathan, A. (2008) Effects of film morphology and gate dielectric surface preparation on the electrical characteristics of organic-vapor-phase-deposited pentacene thin-film transistors in *Organic Thin-Film Electronics – Materials, Processes, and Applications*, eds. A.C. Arias, J.D. MacKenzie, A. Salleo, and N. Tessler, *Mater. Res. Soc. Symp. Proc.*, **1003E**, Warrendale PA, 2007, 1003E-O06-49.

69. Shtein, M., Mapel, J., Benziger, J.B., and Forrest, S.R. (2002) Study of PECVD silicon nitride and silicon oxide gate dielectrics for organic thin-film transistor circuit integration. *Appl. Phys. Lett.*, **81**, 268.

70. Salleo, A., Chabinyc, M.L., Yang, M.S., and Street, R.A. (2002) Polymer thin-film transistors with chemically modified dielectric interfaces. *Appl. Phys. Lett.*, **81**, 4383.

71. Lucas, L.A., DeLongchamp, D.M., Vogel, B.M., Lin, E.K., Fasolka, M.J., Fischer, D.A., McCulloch, I., Heeney, M., and Jabbour, G.E. (2007) Combinatorial screening of the effect of temperature on the microstructure and mobility of a high performance polythiophene semiconductor. *Appl. Phys. Lett.*, **90**, 012112.

5
Dielectric Interface Engineering

The overall organic thin film transistor (OTFT) performance is largely dictated by the properties and quality of the device material layers (e.g., semiconductor, dielectric, contact, etc.). Of nearly equal importance is the interfacial interaction and material compatibility between device layers. In particular, the operation of OTFTs is strongly dependent on the properties of two device interfaces:

1) The interface between the semiconductor and the gate dielectric, where charge transport takes place in the semiconductor layer, and
2) The interface between the semiconductor and the source/drain contacts, where charge injection occurs from the source/drain contacts into the semiconductor.

Optimization and modification of these interfaces for device improvement have been an active area of OTFT research [1–3]. Our focus is on engineering these two critical device interfaces to improve device performance. Interface treatment techniques for the dielectric/semiconductor interface are examined in this chapter and the contact/semiconductor interface is addressed in Chapter 6. The investigations are tailored to bottom-gate bottom-contact poly(3,3‴-dialkylquarterthiophene) (PQT-12) OTFTs on a SiN_x gate dielectric with Au contacts.

The purpose of this work was to improve the performance of PQT-12 OTFTs by systematically introducing and applying various surface treatments to the dielectric/semiconductor (SiN_x/PQT) interface. Four types of dielectric surface conditions were investigated: no treatment (i.e., bare SiN_x), oxygen (O_2) plasma treatment, octyltrichlorosilane (OTS, $CH_3(CH_2)_7SiCl_3$) self-assembled monolayer (SAM) treatment, and dual O_2 plasma/OTS treatment. The correlations between the surface properties (e.g., wettability, surface roughness, chemical composition) and the field-effect mobility of OTFTs were analyzed. This work demonstrates that, by suitably optimized dielectric surface treatment, the field-effect mobility of PQT-12 on a SiN_x gate dielectric can be enhanced nearly 30-fold over untreated devices, resulting in mobility as high as $0.2\,cm^2\,V^{-1}\,s^{-1}$. The studies reveal that the surface wettability and the roughness of the gate dielectric are decisive parameters that control device performance.

Organic Thin Film Transistor Integration: A Hybrid Approach, First Edition. Flora M. Li, Arokia Nathan, Yiliang Wu, and Beng S. Ong.

5.1 Background

High field-effect mobility is generally associated with a high degree of structural order of the organic semiconductor films. For bottom-gate OTFTs, the organic semiconductor film is deposited onto the dielectric layer under the influence of physical and chemical interactions between the semiconductor and the dielectric layer. Furthermore, since charge transport in the active channel typically occurs in the first few monolayers of the semiconductor closest to the interface [4, 5], it is expected that the surface state of the underlying dielectric layer will have a significant influence on the charge carrier mobility of the OTFTs. In turn, the quality of the dielectric/semiconductor interface and the molecular ordering of the organic semiconductor layer have a strong bearing on the OTFT characteristics, including the field-effect mobility (μ_{FE}), on/off current ratio (I_{ON}/I_{OFF}), subthreshold slope (S), threshold voltage (V_T), and leakage current (I_{leak}). Thus, proper control of the gate dielectric surface properties (both physical and chemical properties) is crucial for attaining higher performance OTFTs.

Ideally, the dielectric surface should provide an environment to allow formation of a well-stacked and highly-ordered organic film, so that the disorder phases in the organic film and scattering at the interface are minimized. To tailor the interface, two surface treatment approaches are used: SAMs [6, 7] and oxygen plasma treatment [8]. A background overview of these two techniques is presented next.

5.1.1 Self Assembled Monolayer (SAM)

SAMs are highly ordered, two-dimensional structures that form spontaneously on a variety of surfaces, and serve as an excellent surface modification system. Depending on the application, different interface properties can be achieved by tuning the interfacial properties through varying the rigidity, length, and terminal functional group of the molecule. This in turn influences the stacking, alignment, packing, conformation, uniformity, polarity, and charge density of the surface [1]. In general, molecules that form SAMs contain a head group, a body, and an end/terminal group. A simple illustration of the three main parts of a SAM system is given in Figure 5.1a. The most common and expedient method to deposit SAMs is by immersing a substrate into a dilute solution of SAM precursor molecules dissolved in an organic solvent [9]. The process is self-limiting and the resulting film is a dense organization of molecules (see Figure 5.1b). The orientation of the individual molecules and structure are dependent on the substrate material's structure and cleanliness. SAMs are useful as passivating layers, for the controlled modification of surface properties, and for the surface functionalization in molecular growth processes.

In the context of OTFTs, SAMs are used to chemically modify the gate dielectric surface prior to deposition of the organic semiconductor layer. The formation of

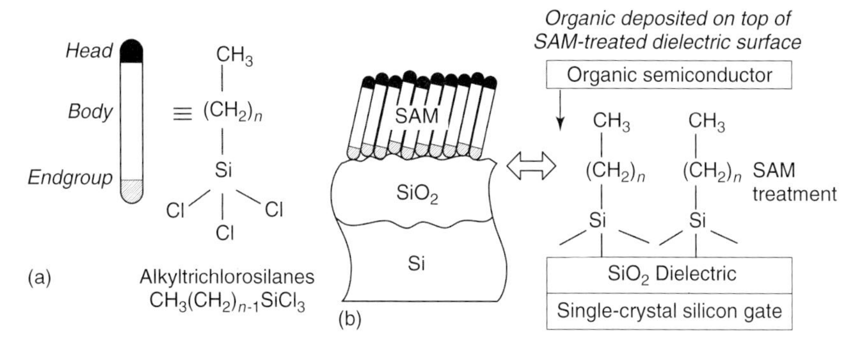

Figure 5.1 (a) Simplified illustration of three main parts of a SAM system and (b) the formation of an alkylsiloxane SAM by adsorption of alkyltrichlorosilanes from solution on a Si–SiO_2 substrate (adapted from [9]).

organic thin films using the self-assembly technique refers to the spontaneous formation of an ordered monolayer of organic molecules on a surface. Alkylsilane SAMs on oxide surfaces have been widely used as molecular platforms in the fabrication of OTFTs, especially to control the orientation, morphology, and grain size of the organic semiconductor films, as well as to aid charge transport. Within the alkyltrichlorosilane ($CH_3(CH_2)_{n-1}SiCl_3$) family (OTS, $n = 8$) and octadecyl-trichlorosilane (ODTS, $n = 18$) are frequently studied for pre-treating the SiO_2 surfaces. Hexamethyldisilazane (HMDS) is another popular surface modifier for SiO_2. Metal surfaces can also be treated with SAMs to enhance the characteristics of bottom-contact OTFTs. SAMs of alkanethiols ($CH_3(CH_2)_{n-1}SH$), such as 2-mercapto-5-nitro-benzimidazole (MNB, $C_7H_5N_3O_2S$), have been applied on metal contact surfaces, and have been shown to improve the quality of the overlying organic semiconductor film, reduce contact resistance, and enhance field-effect mobility in OTFTs [7–11].

The formation of organic SAMs typically reduces the surface energy of gate dielectric or metal electrodes, giving a more favorable environment for the deposition of organic semiconductor molecules. Numerous papers have reported enhancement in semiconductor film quality and in OTFT performance after ODTS SAM treatment of the SiO_2 dielectric surface [6, 7]. For instance, Song *et al.* studied the effect of various treatment techniques on the performance of bottom-gate pentacene OTFT with SiO_2 gate dielectric on a Si substrate [11]. The treatments considered include: annealing, iodine doping, MNB SAM, HMDS SAM, and ODTS SAM. The most effective treatments were MNB, HMDS, and ODTS SAMs, which generated two to three order increases in mobility compared to the as-deposited samples. The SiO_2 gate dielectric treated with ODTS SAM delivered the highest mobility, the highest on/off current ratio, the lowest leakage current, and close to zero threshold voltage. ODTS SAM forms a hydrophobic surface and reduces the surface tension of SiO_2, thus promoting good molecular ordering of the subsequently-deposited organic semiconductor layer and prohibiting penetration of moisture into the organic layer (i.e., ODTS can act as a moisture inhibitor). The investigations reported here use OTS SAM; previous experiments have found that OTS SAM yields higher mobility than ODTS SAM for PQT-12 OTFTs on an SiO_2 gate dielectric.

Studies have shown that, depending on the properties of the substrate surface that result from SAM treatment, the organic polymer semiconductor molecules may adopt two different orientations: perpendicular (edge-on) or parallel (face-on) to the dielectric surface (Figure 5.2) [12]. Increased field-effect mobility has often been reported when the polymer nanocrystals are oriented perpendicularly (i.e., edge-on) with respect to the dielectric substrate compared to those with parallel (face-on) orientation.

Additional evidence on the influence of substrate properties on the molecular orientation and grain morphology of the organic semiconductor is found in studies on vacuum deposited pentacene. It was shown that pentacene molecules stand almost perpendicular to the substrate when deposited onto flat, inert (or hydrophobic) substrates (e.g., polymeric dielectric). In contrast, if the substrate is more reactive (or hydrophilic), pentacene molecules tend to lie flat on the substrate due to increased pentacene–substrate interaction [1]. The enhancement in device mobility on SAM-modified surfaces was attributed to the increased grain size of the semiconductor, high molecular surface mobility and reduced interactions with a hydrophobic substrate surface.

The general consensus among the initial research in OTFTs is that an inert hydrophobic substrate surface is most desirable for the formation of high quality, well-ordered organic semiconductor layers to deliver higher device performance.

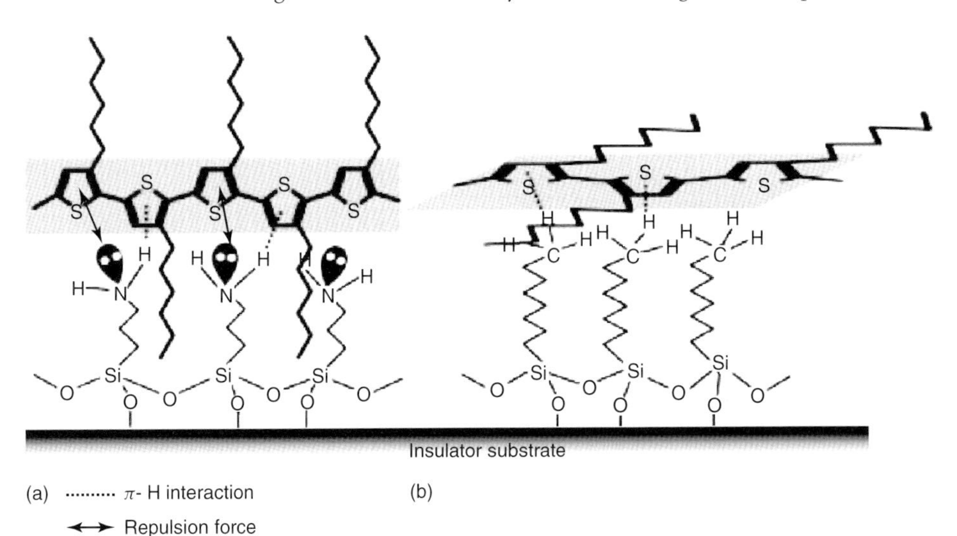

Figure 5.2 Schematics of different P3HT chain orientations, (a) edge-on and (b) face-on, according to interfacial characteristics (adapted from [12]).

Table 5.1 The impact of surface conditions on the molecular orientation of an organic semiconductor (e.g., pentacene and P3HT).

Semiconductor	Edge-on orientation	Face-on orientation	References
Pentacene	Inert (non-polar) surfaces → Higher μ	Reactive (polar) surface → Lower μ	[13, 14]
P3HT	Surfaces with polar endgroups → Higher μ	Surfaces without polar endgroups → Lower μ	[12]

Interestingly, a more exhaustive review of the scientific literature revealed that the preferable conditions of the dielectric surface to give high carrier mobility vary with the choice of material system; two examples are given in Table 5.1. In the case of pentacene, it was reported that inert surfaces (e.g., non-polar) are favorable for higher mobility [13, 14]. In contrast, a polar surface was observed to attain higher mobility for poly(3-hexylthiophene) (P3HT) devices, as reported by Kim *et al.* [12]. But in either case, higher mobility was observed when the organic semiconductor molecules adapted an edge-on or perpendicular orientation with respect to the substrate surface.

A potential concern of using SAM treatment is that it involves a wet deposition process, which requires immersing OTFT substrates in a SAM solution. In some instances, the immersion process may remove the contact metals, and the solvent may attack the plastic substrates (which are employed for flexible electronics). To circumvent these wet processing issues, surface treatment based on a dry process was explored. We investigated oxygen plasma treatment as an alternative approach for dielectric surface modification. This technique is discussed next.

5.1.2 Oxygen Plasma Treatment

The effects of oxygen (O_2) plasma treatment on the performance of bottom-gate pentacene OTFT with a thermal SiO_2 gate dielectric were studied by Lee *et al.* [8]. Prior to pentacene deposition, the substrates were treated with O_2 plasma using reactive ion etching (RIE) equipment. Compared to the as-deposited OTFTs, the electrical characteristics of OTFTs with an O_2 plasma treated gate dielectric are superior, with a 10-fold increase in field-effect mobility, an order of magnitude increase in I_D, higher on/off current ratio, sharper subthreshold slope, and reduction in device resistance (including channel resistance and contact resistance) [8]. It was concluded that O_2 plasma treatment facilitated the growth of larger pentacene grains and enhanced ordering of the pentacene molecules, resulting in a decrease in the total device resistance and a lowering of the contact resistance [8]. These improvements can be ascribed to a plasma cleaning effect, where removal of surface defects and contaminants by ashing typically occurs in the initial phase of plasma exposure. The enhancement of grain size and molecular ordering can also be attributed to the modification of surface potential energy. A lower surface potential energy barrier produces a larger diffusion length of pentacene molecules on the dielectric surface, thus larger grains are formed and the crystallinity of the pentacene film is improved. Plasma treatment also improved device uniformity across the wafer [8]. However, excessive exposure can cause a reduction in the field-effect mobility, threshold voltage, and off-current [8]. In summary, O_2 plasma treatment with a suitably controlled exposure time can produce OTFTs with enhanced, as well as uniform, performance across the wafer.

Wang *et al.* have shown that O_2 plasma treatment of an organic polymer gate dielectric can be used to systematically modify the threshold voltage (V_T) of pentacene OTFTs at the process level [15]. They reported that O_2 plasma treatment resulted in a positive V_T shift in pentacene OTFTs, but this was accompanied by an increase in field effect mobility. The shift in V_T was attributed to process-induced trap states, more specifically, an increase in fixed charges and mobile charges, at the semiconductor/dielectric interface.

5.1.2.1 Basics of Plasma Processing (Etching)

To understand the effect of O_2 plasma treatment on surface properties, an overview of plasma processing technology is provided here. By definition, plasma is a partially ionized gas consisting of positively and negatively charged radicals and neutral species. Sometimes considered as the fourth state of matter, it is highly conductive due to the presence of charged particles and is magnetically controllable. Plasma technology is widely employed in the microelectronics industry in many dry etching or thin-film deposition systems. The ability of plasmas to transfer the lithographically defined patterns into underlying layers reliably and controllably, and etch anisotropically without requiring crystallographic orientation at low or intermediate pressures makes them very attractive for most dry etching mechanisms.

The basic principle of dry etching by plasma involves ionizing a gas to create highly reactive positively and negatively charged species, which react chemically with the material to be etched; the reactions form volatile compounds that can be removed [16]. The Trion Phantom II RIE/ICP hybrid system was used in this project for oxygen plasma treatment of the gate dielectric [17]. During RIE processing, process gases flow through the chamber, and RF power is applied between the two electrodes to generate a plasma (where highly reactive radicals, electrons, photons, and ions are generated through ionization of the feed gas by electron impact dissociation). The chemically reactive species diffuse to the surface of the wafer and are adsorbed on the wafer surface. The chemical species undergo surface diffusion and react with the material to be etched forming volatile products. These volatile products are desorbed from the surface and transported away from the chamber. Ion bombardment of the wafer also takes place through the generation of a DC bias on the lower electrode. The DC bias is a potential drop between the plasma and the wafer that accelerates ions toward the wafer to produce directional etching, as illustrated in Figure 5.3. A reactive ion etch mechanism is also called *ion-beam*

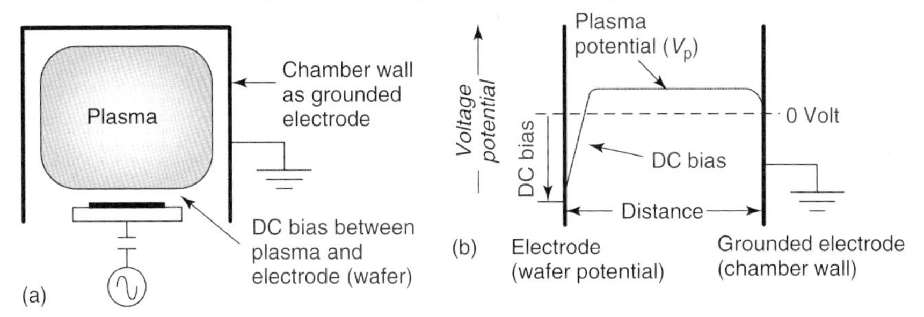

Figure 5.3 (a) Diagram of an RIE system. (b) Potential distribution as a function of the distance between the electrodes. Wafer is placed on a smaller electrode, which is biased negatively relative to the plasma. Ions accelerate from the plasma to the wafer due to a DC bias electric field.

assisted radical etching. Depending on the etch mechanism, plasma etch profiles can vary from isotropic to anisotropic (i.e., directional). The anisotropy level is controlled by the plasma conditions. The rate of etching or the rate of reaction in plasma processing is sensitive to various system parameters, including DC bias, RF power, pressure, and gas flow rate.

RIE power is specified by an externally applied DC bias parameter. Under the influence of the DC bias, either positive or negative ions (depending on whether the DC bias is negative or positive) are accelerated toward the electrode where the substrate is mounted. The directed motion imparted to the ions by the DC bias, as well as the chemical reactivity of the ions, results in the selective, anisotropic chemical etching of the substrate. The magnitude of the DC bias at the electrode significantly affects, for example, the degree of the anisotropic etching (the smaller the magnitude of the DC bias, the lower the degree of anisotropic etching) and thus the shape of the etch profile. In addition, the magnitude of the DC bias significantly affects the etch selectivity between a substrate layer to be etched and material layers not to be etched, for example, the masking material or other layers of substrate material. A DC bias just sufficient to impose directionality on the reactive species will draw the reactive species to the electrode at a speed sufficient to cause sputter removal of a relatively small amount of masking material or substrate material which is not to be removed. However, an increase in the magnitude of the DC bias significantly above that required to impose directionality will result in the sputter removal of an undesirably large amount of material which is not to be removed. Consequently, proper control of DC bias can help to achieve a desired etch profile, for example, an essentially vertical, inclined, or curved etch profile, to avoid undesirably large reductions in etch selectivity, and to avoid unintentional removal of material [18].

The plasma density becomes low in an RIE system when it is operated at low pressures, reducing the etch rates. Higher plasma densities can be generated with an inductively-coupled plasma (ICP) system, which heats the plasma inductively by the electric fields generated by a coil wrapped around the discharge chamber without coupling it to the substrate [16]. Apart from the high plasma density, ICP provides lower ion energy and the capability to operate at lower chamber pressures of less than 10 mTorr. Moreover, ion fluxes and ion energy can be controlled separately, thus increasing flexibility in optimizing etch responses. The higher plasma density afforded by the ICP system improves etch rate, anisotropicity, profile selectivity, and uniformity, places less stringent requirements on operating pressures, and reduces radiation damage from RIE [19]. Thus, in a hybrid ICP–RIE system, ICP is employed to generate a high concentration of reactive species which increases the etch rate, whereas a separate RF bias (i.e., RIE mode) is applied to the substrate to create directional electric fields that accelerate reactive species toward the substrate to achieve more anisotropic etch profiles. As the ICP–RIE mode provides high density without increasing the DC bias self-voltage, it is an intrinsically low damage process.

O_2 plasmas are commonly employed in wafer cleaning procedures. Oxygen radicals react with organic contaminants on the wafer surface, forming volatile organic compounds (e.g., H_2O, CO, or CO_2) that are desorbed from the surface and removed from the plasma. These reactions are believed to contribute to changes in OTFT performance upon O_2 plasma treatment. For the present research, O_2 plasma treatment of the SiN_x surfaces was performed using an ICP/RIE system. It was found that the choice of ICP or RIE mode for O_2 plasma generation affects the device performance; more discussion will be presented in Section 5.4.

5.2 Experimental Details

Bottom-gate bottom-contact PQT-12 OTFTs with a plasma enhanced chemical vapor deposited (PECVD) SiN_x gate dielectric were used for this dielectric interface engineering study. The OTFTs were fabricated on highly-doped silicon substrates serving as the gate, with thermally-evaporated Cr–Au source/drain contacts, and solution-processed PQT-12 as the active organic semiconductor layer. The fabrication sequence was outlined in Figure 3.8. Two dielectric surface treatment techniques were investigated: O_2 plasma and OTS SAM. O_2 plasma exposure was carried out using the Phantom II RIE system, a plasma etch system designed by Trion Technology Inc. OTS SAM was formed by immersing the substrates in a 0.1 M solution of OTS in toluene for 20 min at 60 °C, followed by rinsing with toluene and isopropanol. These treatments were done on the SiN_x gate dielectric, prior to deposition of the PQT-12 layer. A simplified cross-section structure of the OTFT is illustrated in Figure 5.4.

The study of dielectric surface engineering is divided into two parts, as summarized in Table 5.2. The first experiment, reported in Section 5.3, examines the impact of individual dielectric surface treatment techniques, as well as their combination, on OTFT performance. The combination of O_2 plasma and OTS treatment was considered, as it is believed that O_2 plasma treatment can hydroxylate or oxidize

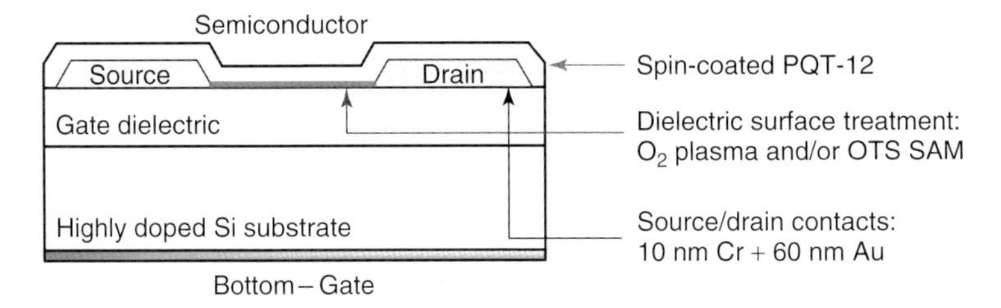

Figure 5.4 Schematic cross-section of the OTFT structure considered in this dielectric-surface treatment study.

Table 5.2 Dielectric/semiconductor interface engineering experiments.

	Description	**Detail**
Part A (Section 5.3)	Comparison of transistors with and without surface treatment	Examine device improvement resulting from O_2 plasma, OTS SAM, and their combinations
Part B (Section 5.4)	Effect of O_2 plasma exposure conditions	Evaluate impact of exposure duration and power on OTFT characteristics

the SiN_x surface, to assist the growth of SAM through a silanization reaction [4]. The second experiment, reported in Section 5.4, investigates the effect of O_2 plasma exposure conditions (i.e., duration and power) on device performance, to determine an optimal O_2 plasma exposure recipe for our devices. Electrical characterization of the OTFTs was carried out with a Keithley 4200-SCS parameter analyzer, under ambient conditions. For each dielectric surface treatment condition, at least five OTFTs with different W/L were fabricated and tested, and the average value of the device parameter is reported below. The parameter extraction procedure is presented in Section 2.2. Surface properties were characterized using contact angle measurements for surface wettability, atomic force microscopy (AFM) measurements for surface roughness, and X-ray photoelectron spectroscopy (XPS) measurements for chemical composition. The relationships between electrical properties and surface properties are analyzed to establish a better understanding of device behavior. Please refer to Section 4.1.2 for an overview of the various material characterization techniques.

5.3
Impact of Dielectric Surface Treatments

The first experiment evaluates the impact of different dielectric surface treatments on OTFT performance. Four types of device were prepared as outlined in Table 5.3.

Table 5.3 Dielectric surface treatment experiment: sample descriptions.

Sample ID	Description of the dielectric surface treatment
None or untreated	As-deposited SiN_x dielectric surface, without any treatment
O_2 plasma	Substrate exposed to O_2 plasma (60 s, 100 m Torr, 34 W. Refer to Recipe (I) in Table 5.6)
OTS	Substrate treated with OTS SAM treatment (0.1 M OTS in Toluene for 20 min at 60 °C)
O_2 plasma/OTS	Substrate first exposed to O_2 plasma (60 s), followed by OTS SAM

The electrical characteristics of the various OTFTs were compared to evaluate the effectiveness of each treatment recipe. The dielectric surface properties were characterized to investigate how these surface characteristics correlate with the observed device behavior.

5.3.1 Electrical Characterization

The transfer and output characteristics of PQT-12 OTFTs on a 300 °C SiN_x ($x = 1.08$) gate dielectric with different dielectric surface treatments are presented in Figure 5.5. As seen from the output characteristics, for a given gate voltage, the drain current increases by more than one order of magnitude from the untreated sample to the O_2 plasma/OTS treated sample. Using the extraction method outlined in Section 2.2, the changes in various device parameters, extracted from the OTFT characteristics in Figure 5.5, are summarized in Figure 5.6 and Table 5.4. The data show an obvious improvement in field-effect mobility and on/off current ratio after surface treatment (as shown in Figure 5.6). This was accompanied by a reduction in the inverse subthreshold slope (as shown in Table 5.4), suggesting improved dielectric/semiconductor interface properties as a result of the treatments. V_T shifts to increasingly positive voltages after surface treatment, and is possibily related to electron trapping at the interface.

When compared to untreated devices, remarkable improvement in mobility is observed after dielectric surface treatments. In particular, the combinatorial treatment using O_2 plasma and OTS SAM generates the largest enhancement, where the mobility increased 27-fold (2614%) (Figure 5.6b). Figure 5.7 plots the improvement in mobility of the treated device relative to the untreated samples (i.e., calculated using "mobility of treated device"/"mobility of untreated device"). To understand and account for these device variations, interface properties were examined to seek scientific insights.

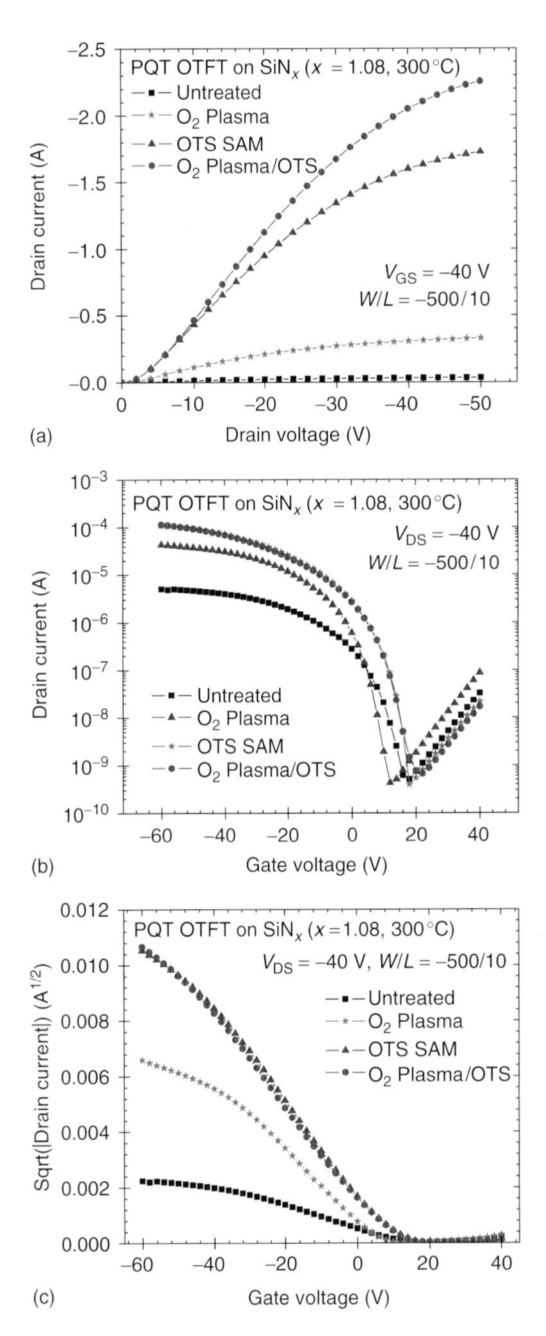

Figure 5.5 (a) Output and (b,c) transfer characteristics of PQT-12 OTFTs on 300 °C PECVD SiN$_x$ (x = 1.08) with different dielectric surface treatments. The transfer characteristics are shown on a semi-logarithmic scale in (b) and as the square root of the drain current in (c).

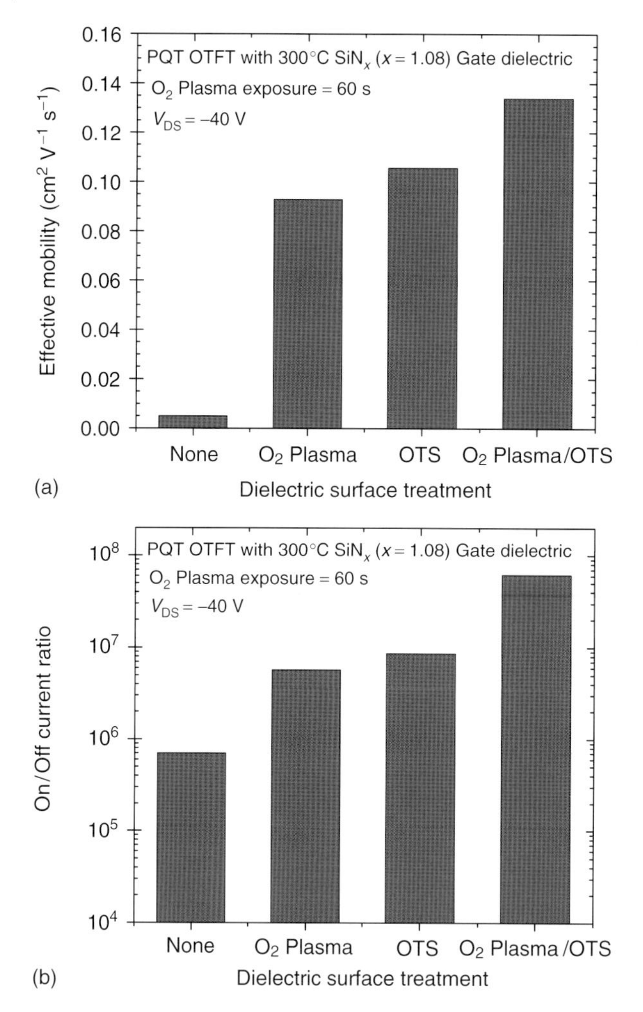

Figure 5.6 (a) Effective mobility and (b) on/off current ratio of PQT-12 OTFT with SiN$_x$ gate dielectrics under various dielectric surface treatment conditions. Measurements were collected in the saturation region. Each data point corresponds to an average value from three to six devices.

5.3.2
Interface Characterization

The electrical characterization data indicated that the dielectric surface treatments have a very strong influence on the OTFT performance. The dielectric surface properties (e.g., contact angle, roughness, and composition) of the samples were analyzed, in an attempt to establish correlations with the observed electrical characteristics. The results showed that dielectric surfaces with a large contact

Table 5.4 Changes in threshold voltage V_T, switch-on voltage V_{SO}, and inverse subthreshold slope S of OTFTs with different dielectric surface treatments. Data are extracted from PQT-12 OTFTs on 300 °C SiN_x ($x = 1.08$) gate dielectric, with $W/L = 500\,\mu m/10\,\mu m$, in the saturation region ($V_{DS} = -40\,V$) (see Figure 5.5).

Treatment	V_T (V)	V_{SO} (V)	S (V dec^{-1})
Untreated	3.54	18	4.28
O_2 plasma	5.04	12	3.08
OTS	9.46	18	2.84
O_2 plasma/OTS	9.19	20	3.81

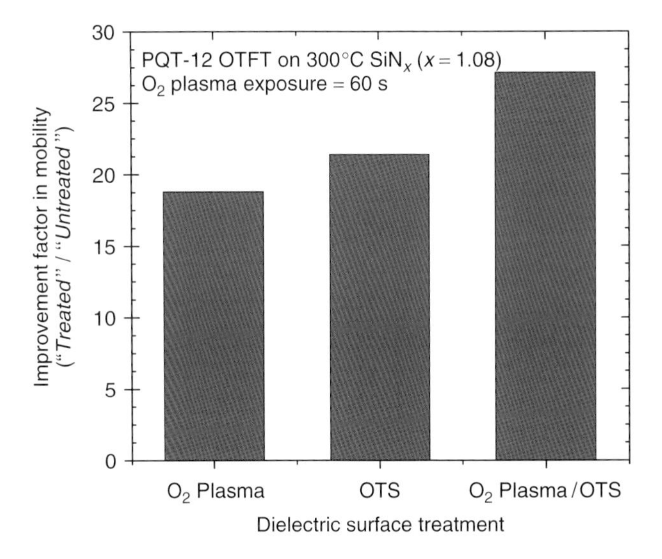

Figure 5.7 Factor of improvement in effective mobility, relative to untreated devices, for PQT-12 OTFT after various types of dielectric surface treatments. Measurements were collected in the saturation region.

angle and low surface roughness are desirable for OTFTs. However, in some scenarios, they may behave as competing processes where the relative impact of one attribute may override the other.

5.3.2.1 **Contact Angle**

Figure 5.8 displays photographs from the water contact angle measurement of an SiN_x surface upon various surface treatment conditions; the measurement data is plotted in Figure 5.9. Compared to an untreated SiN_x surface, an O_2 plasma treated sample (Figure 5.8b) displayed a smaller contact angle, indicating hydrophilic

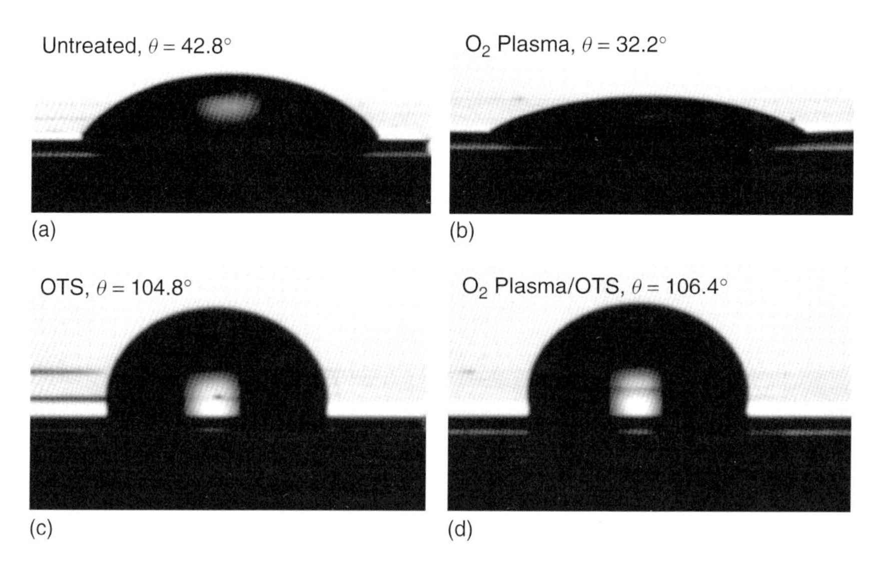

Figure 5.8 (a–d) Photographs from contact angle measurement of an SiN_x surface after various dielectric surface treatments.

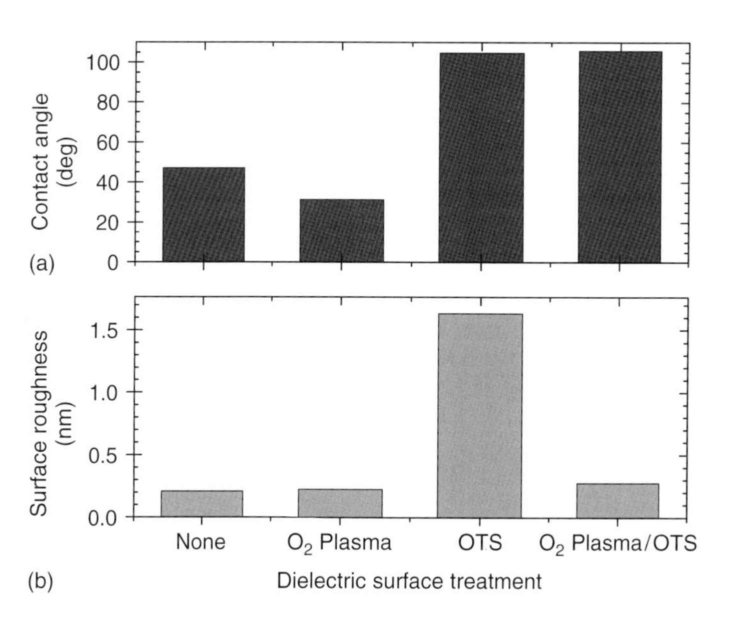

Figure 5.9 (a) Water contact angle measurement and (b) mean surface roughness of SiN_x surfaces after various dielectric surface treatments.

properties. In contrast, samples with OTS treatment (Figure 5.8c,d) showed larger contact angle values, indicating hydrophobic surface states. The untreated SiN_x is relatively hydrophilic, as indicated by its small contact angle value. Hydrophobicity is linked to the polarity (and thus the difference in electronegativity) of the bonds/molecules. Si–O, Si–N, and N–O bonds are polar, leading to a hydrophilic surface for (bare) SiO_2 and SiN_x. Si–C and C–C bonds are weakly polar (or non-polar), thus OTS and polymer surfaces are hydrophobic.

In general, most inorganic oxide and nitride surfaces (e.g., SiO_2 and SiN_x) show hydrophilic properties, whereas most organic materials (e.g., polythiophene) show hydrophobic properties. This mismatch in surface wettability/energy is believed to hinder the formation of well-ordered organic semiconductor molecules on inorganic dielectric surfaces. To overcome this mismatch, OTS SAMs are used to modify the dielectric surface from hydrophilic to hydrophobic; this modification can facilitate the formation of an organic semiconductor layer with enhanced crystallinity and ordering, thus resulting in increased carrier mobility [6].

On the other hand, O_2 plasma treatment leads to a reduction in the contact angle, indicating that the surface becomes more polar (or hydrophilic). O_2 plasma treatment can have multiple effects on surface properties. Very short O_2 plasma exposure duration (or initial stages of the exposure) typically causes removal of organic contamination from the dielectric surface; contaminants may originate from wet cleaning procedures or from the environment [15]. Removal of organic compounds reduces the contact angle. Furthermore, O_2 plasma can generate polar groups (e.g., O–H, Si–O, N–O) on the surface, thus the sample acquires a hydrophilic character after O_2 plasma treatment [16].

Most literature reports state that hydrophobic surfaces (i.e., large contact angle) are favorable for the formation of well-ordered organic semiconductor film, and have a positive impact on mobility [7]. This tendency is reflected in our experiment where OTS treatment displays the large contact angle and generates high device mobility, relative to the untreated sample. An interesting and striking observation in our experiment is that there is an 18-fold improvement in mobility in O_2 plasma treatment (despite its hydrophilic surface) when compared to an untreated OTFT, as shown in Figure 5.7. These results suggest that contact angle alone cannot give an accurate prediction of device behavior. In fact, interplay between contact angle, surface roughness, chemical composition, and other factors has a collective influence on the microstructure of the overlying organic semiconductor layer and on the interface quality. These factors are discussed next. In addition, deviation from the general hydrophobic-mobility trend is possible, depending on the material system, as discussed in Section 5.1.1 (Table 5.1).

5.3.2.2 **Surface Roughness**

Surface roughness affects the performance of thin film transistors (TFTs). It can influence the quality of the semiconductor layer, affect charge transport in the channel, and cause interface scattering. Recent work by Chabinyc *et al.* showed that the field-effect mobility of OTFTs decreases nearly exponentially with the surface roughness of the gate dielectric [20]. The surface roughness data are

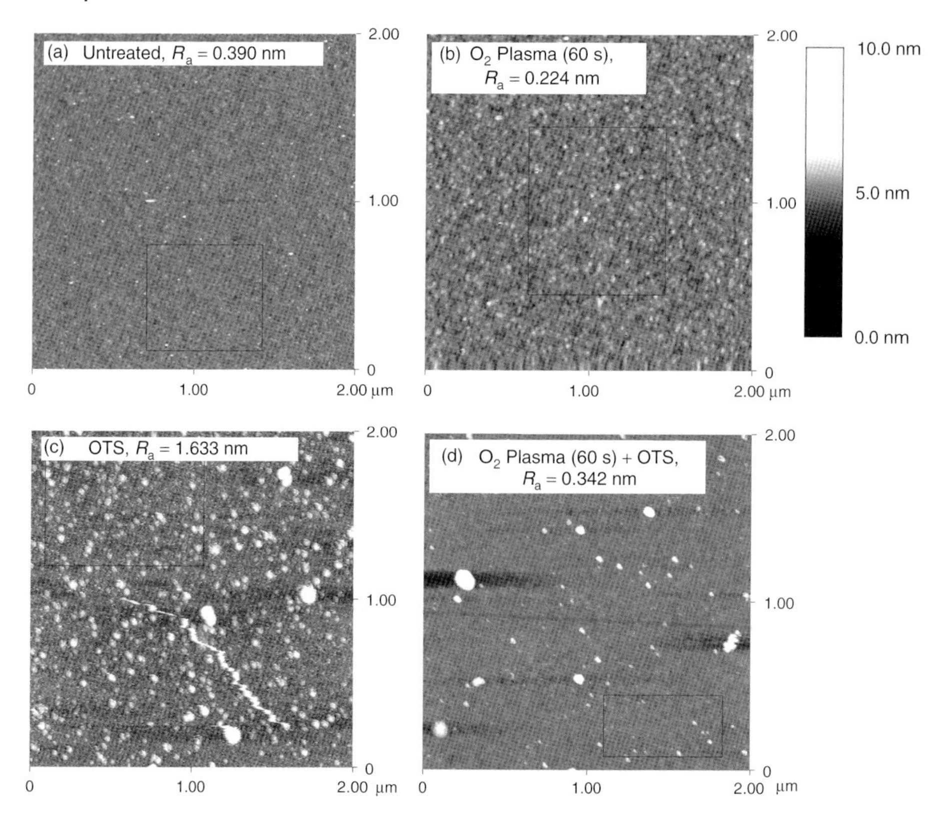

Figure 5.10 AFM images showing the surface topography of an SiN_x surface after various dielectric surface treatments. The mean surface roughness (R_a) is indicated.

summarized in Figure 5.9, and the AFM surface topography photographs in Figure 5.10. Interestingly, compared to the untreated SiN_x sample, O_2 plasma treatment (Figure 5.10b) led to a reduction in surface roughness; the smoother surface is believed to be a key reason for the improved mobility observed in O_2 plasma treated OTFTs. In contrast, OTS treatment (Figure 5.10c) resulted in an increase in surface roughness. The higher surface roughness of the OTS-treated SiN_x is speculated to be due to a more sporadic formation of OTS SAM on the bare SiN_x surface. On the other hand, combinatorial treatment with O_2 plasma and OTS SAM (Figure 5.10d) generated a much smoother surface, suggesting that pre-exposure of SiN_x to O_2 plasma is critical for the formation of a smooth high quality OTS SAM on SiN_x. It is hypothesized that O_2 plasma exposure places hydroxy groups on the surface, which are needed to facilitate attachment of OTS molecules on SiN_x. Overall, it is concluded that the combination of O_2 plasma and OTS treatment is the most preferable dielectric surface treatment approach; it delivers the highest TFT mobility, along with the largest contact angle and relatively low surface roughness for the dielectric surface.

Altogether, these results showed that a variety of surface parameters must be taken into account to explain OTFT behavior. In most cases, higher mobility is achieved with an increasing contact angle and decreasing surface roughness [20]. However, we conclude here that it is the combinatorial or collective effect of these parameters (along with many others) that dictate the resulting device performance. Table 5.5 compares the changes in field-effect mobility, contact angle, and surface roughness across samples with different dielectric surface treatments. The champion device (with highest mobility) is attained with the combinatorial O_2 plasma/OTS treatment, where the dielectric surface is simultaneously characterized by a large contact angle and low surface roughness. The moderate mobility observed in the O_2 plasma sample can be attributed to the low surface roughness, despite its low contact angle (or high surface energy). On the other hand, the high mobility observed in the OTS sample is attributed to its low surface energy (i.e., high contact angle), despite its higher surface roughness. Therefore, depending on the particular sample or material system, the effect of surface roughness may prevail over that of surface energy, and vice versa. In addition, it is important to consider other surface properties that might influence OTFT characteristics; surface chemical composition is examined next.

5.3.2.3 Chemical Composition

It is interesting to observe in Table 5.5 that both the "untreated" and "O_2 plasma" samples are characterized by a low contact angle (hydrophilic surface) and low surface roughness, but the O_2 plasma treated device has much higher mobility than the untreated device. The difference in device behavior can be related to the elemental chemical composition of the interface. The atomic composition at the SiN_x surface was studied by XPS, and the results are shown in Figure 5.11.

Table 5.5 Comparison of effective field-effect mobility, contact angle, and surface roughness for samples with different dielectric surface treatment. Data shown are obtained by averaging results from a number of samples.

Treatment	μ_{FE} ($cm^2 V^{-1} s^{-1}$)	Contact angle, θ (°)	Surface roughness (nm)	Remark
Untreated	0.0049	47.2	0.21	Low θ + low roughness → low μ
O_2 plasma	0.093	31.35	0.224	Low θ + **low roughness** → moderate μ
OTS	0.106	104.77	1.633	**High** θ + high roughness → moderate μ
O_2 plasma/OTS	0.134	105.73	0.276	**High** θ + **low roughness** → **high** μ

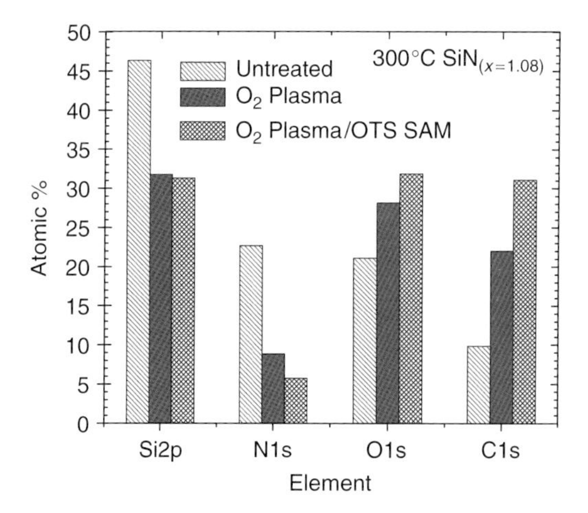

Figure 5.11 Atomic composition from XPS measurements of SiN_x surfaces with different surface treatments.

A key distinction between the untreated sample and the treated samples is that the former surface is dominated by Si and N, whereas the latter surfaces are dominated by Si, O, and C. As such, despite "untreated" and "O_2 plasma" samples having similar surface roughness and contact angles, the O_2 plasma surface has richer Si, O, C content than the untreated surface. The presence of Si–O or Si–C bonds is believed to improve the formation of the subsequent organic layer; thus, resulting in an improved dielectric/semiconductor interface.

5.3.3 Analysis

Compared to results published in the literature [4], the relatively large improvement in mobility with O_2 plasma treatment (as shown in Figures 5.6 and 5.7) is quite remarkable because it has been reported that a hydrophobic dielectric surface is necessary for achieving high mobility OTFTs. A hydrophobic surface, characterized by low mean free surface energy, provides a favorable platform for organic semiconductor molecules to assemble into a well-ordered film [13]. As noted in Figure 5.9, OTS treatment leads to a hydrophobic surface (large contact angle), whereas O_2 plasma treatment produces a hydrophilic surface (small contact angle). Therefore, the observation that O_2 plasma gives mobility nearly comparable to the OTS treatment is a unique discovery/breakthrough. Although the exact reason for this somewhat unconventional behavior is not completely clear, a few possible explanations are proposed here to account for the various observations in this experiment.

- **O_2 plasma vs. OTS**: The O_2 plasma treated device has comparable mobility to the OTS SAM treated device. The observation that the former has considerably smaller surface roughness than the latter provides an acceptable justification for the high mobility in O_2 plasma devices (see Figure 5.7). On the other hand, the small contact angle of the O_2 plasma treated surface (i.e., hydrophilic, high surface energy) does not corroborate with this mobility trend.

- **OTS vs. O_2 plasma/OTS**: There might be imperfections in the OTS SAM device when formed on a *bare* SiN_x surface. Since a hydroxylated surface is typically needed for the formation of OTS, bare SiN_x might not provide the most favorable environment for high quality OTS SAM formation. By exposing the SiN_x surface to O_2 plasma prior to OTS treatment, the quality of the OTS SAM is improved. As observed in Figure 5.7, the surface roughness for OTS on an O_2 plasma treated SiN_x is considerably lower than OTS on a bare SiN_x, suggesting the positive contribution or effectiveness of O_2 plasma treatment on the formation of high quality OTS SAM. The higher surface roughness for OTS on bare SiN_x likely compromises the mobility in OTS devices when compared to O_2 plasma/OTS devices. In terms of surface energy, the contact angle measurements showed very little difference between OTS surfaces and O_2 plasma/OTS surfaces; this suggests the samples have comparable surface energy (or wettability).
- **Untreated vs. O_2 plasma**: Both the untreated SiN_x surface and the O_2 plasma treated surface are characterized by low surface roughness and a low contact angle (i.e., hydrophilic surface). However, the O_2 plasma treated device showed significantly greater mobility than the untreated device. Therefore, surface roughness and surface energy alone cannot completely capture the interface properties to explain/understand OTFT behavior. Additional insight can be gained from studying the chemical composition of the surface. The XPS data in Figure 5.11 show that the treated samples are characterized by a higher O 1s concentration and a significantly lower Si 2p concentration compared to the untreated sample. It is believed that an oxide-like surface is beneficial to the assembly of the organic semiconductor layer. O_2 plasma treatment produces an oxynitride surface, where more Si–O bonds are present compared to the bare SiN_x surface. The dominance of Si–O bonds on the O_2 plasma treated surface enhances the dielectric/semiconductor interface properties compared to bare SiN_x (which is characterized by mainly Si–N bonds). This key distinction in surface composition accounts for the enhanced device mobility observed in the O_2 plasma treated OTFT compared to untreated OTFT. In addition, O_2 plasma treatment may improve the charge injection properties at the contacts to contribute to enhanced device performance. Recent studies reported a reduction in the hole injection barrier at the Au/pentacene interface by O_2 plasma treatment, leading to an increase in linear mobility for pentacene OTFTs [21].
- **Fluorinated O_2 plasma treated surface**: The XPS data for the O_2 plasma treated surface showed a very peculiar feature: a very weak presence of fluorine. It is speculated that fluorine species were introduced onto the surface during O_2 plasma exposure by the RIE process. There is a possibility that the presence of fluorine or the combinatorial effect of fluorine and oxygen on the SiN_x surface is particularly favorable for organic semiconductors. However, more controlled experiments must be conducted to justify this. Nonetheless, this

unique phenomenon of "fluorinated surface treatment" warrants more research attention to exploit its potential for practical use.

The above analyses confirmed that the OTFT device characteristics are influenced by surface roughness, surface energy, and surface composition. From this investigation, it is concluded that the ideal dielectric surface for bottom-gate OTFTs is characterized by low surface roughness, low surface energy (i.e., high contact angle), and an Si–O-like surface composition; this combination leads to higher mobility for PQT-12 OTFTs on SiN_x gate dielectrics. Grazing incidence X-ray diffraction (GIXRD) can be performed on the devices to compare the molecular microstructure of PQT-12 on the various treated surfaces, to understand how the molecular ordering of PQT-12 varies on the different surfaces.

5.4 Impact of Oxygen Plasma Exposure Conditions

The objective of this experiment was to systematically evaluate the effect of O_2 plasma exposure conditions on PQT-12 OTFT performance, and to determine the optimal exposure conditions for achieving high mobility devices. The first experiment examines the impact of exposure *duration*; the second considers the impact of exposure *power*. Bottom-gate bottom-contact transistors with a Si-rich 150 °C SiN_x ($x = 1.28$) gate dielectric and pre-patterned source/drain contacts served as the device platform for this experiment. Prior to deposition of PQT-12, all dielectric surfaces were pre-treated with O_2 plasma and OTS SAM and the Au contacts were treated with 1-octanethiol SAM. Three O_2 plasma exposure recipes were used and are listed in Table 5.6. The effects of O_2 plasma treatment were analyzed by examining the variation of performance parameters such as field-effect mobility, on/off current ratio, subthreshold slope, threshold voltage, and on-set voltage with respect to the plasma exposure time. The physical (e.g., surface roughness) and chemical properties (e.g., contact angle, chemical composition) of the O_2 plasma treated SiN_x surface were analyzed for elucidation of the observed TFT behavior.

Table 5.6 Various O_2 plasma recipes used in this experiment, with chamber pressure at 150 m Torr, and O_2 gas flow rate at 30 sccm. DC bias directly controls the RIE power.

O_2 plasma recipe	DC bias (V)	ICP power (W)	RIE power (W)	Exposure duration (s)
Recipe (I): RIE only	−300	0	34	0–900
Recipe (II): ICP only	0	300	0	200, 600
Recipe (III): RIE + ICP	−150	300	9	600

5.4.1 Electrical Characterization

5.4.1.1 Impact of Exposure Duration

The study on O_2 plasma exposure duration was executed using Recipe (I) listed in Table 5.6. The transfer and output characteristics of PQT-12 OTFTs (with $W/L = 500\,\mu m/10\,\mu m$) for different O_2 plasma exposure durations are shown in Figure 5.12. Table 5.7 summarizes the device parameters extracted from the OTFT characteristics in Figure 5.12. The data revealed an improvement in field-effect mobility and on/off current ratio as the O_2 exposure duration increased from 0 to 80 s; this is accompanied by a reduction in the inverse subthreshold slope, suggesting improved dielectric/semiconductor interface properties as a result of the treatments. However, for exposure duration from 80 to 900 s, a gradual reduction in mobility and on/off current ratio is observed. There is a clear shift in threshold voltage V_T and switch-on voltage V_{SO} toward higher positive voltages as the O_2 plasma exposure duration increases. This shift is likely related to O_2 plasma induced electron trapping at the interface [22]. Additional analysis is presented in Section 5.4.3

A set of five samples (with various W/L) were tested for each exposure duration. The average effective mobility and on/off current ratio for each set of devices, as a function of O_2 plasma duration, are plotted in Figure 5.13. The plots show a distinct peak near 80 s, where the maximum field effect mobility and on/off current ratio were observed. At an exposure duration of 80 s, an effective mobility of $0.216\,cm^2\,V^{-1}\,s^{-1}$ was measured, which represents a 6.3-fold (or 530%) increase in mobility compared to devices without O_2 plasma exposure (i.e., exposure time = 0 s). (Note: all devices were treated with OTS SAM.) For lengthened exposure ($t > 80$ s), μ_{FE} gradually decreased and plateaued at $0.05\,cm^2\,V^{-1}\,s^{-1}$. Figure 5.13b shows that shorter exposure duration also favors a higher on/off current ratio.

Table 5.7 OTFT device parameters as a function of O_2 plasma exposure duration. Data are extracted from PQT-12 OTFTs on 150 °C SiN_x ($x = 1.28$) gate dielectric, with $W/L = 500\,\mu m/10\,\mu m$, in the saturation region ($V_{DS} = -60\,V$) (refer to Figure 5.12).

Duration (s)	μ_{FE} ($cm^2\,V^{-1}\,s^{-1}$)	V_T (V)	V_{SO} (V)	I_{ON}/I_{OFF}	S ($V\,dec^{-1}$)
0	0.0476	−2.33	8	4.06×10^4	3.067
40	0.1481	−2.13	10	2.70×10^7	1.587
80	0.1746	1.33	12	2.15×10^9	0.830
160	0.1307	10.90	22	1.04×10^8	1.290
300	0.0912	19.59	34	4.17×10^7	1.691
600	0.0416	21.25	36	2.04×10^7	1.586
900	0.0247	24.18	40	3.76×10^7	3.081

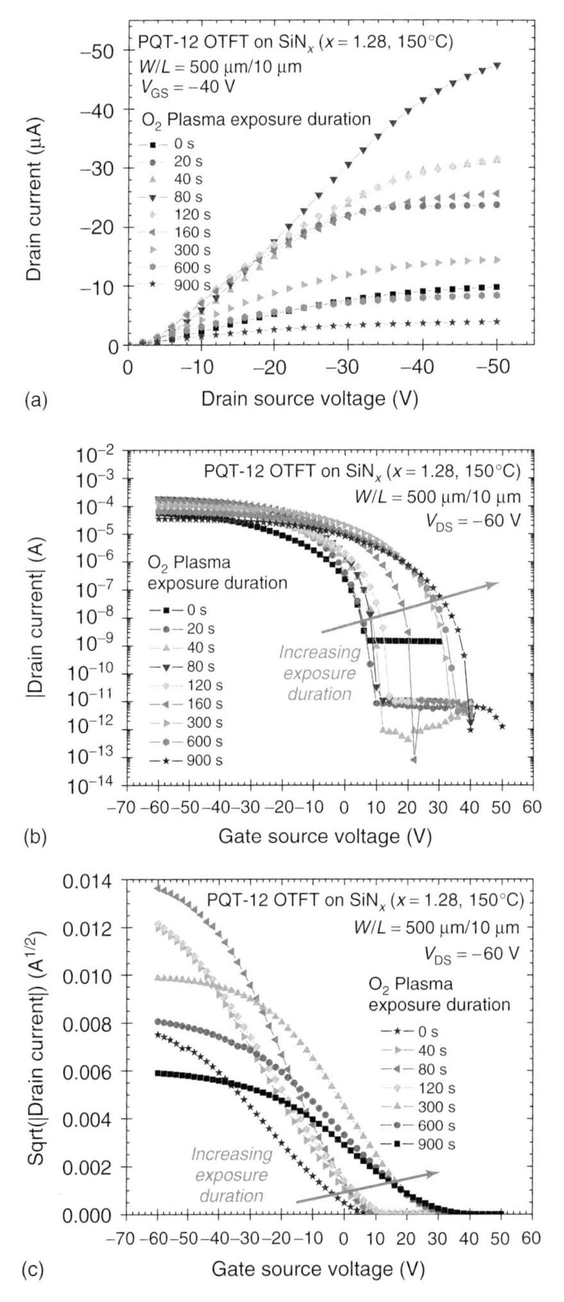

Figure 5.12 (a) Output characteristics and (b,c) transfer of PQT-12 OTFTs on 150 °C PECVD SiN$_x$ (x = 1.28) gate dielectric treated with different O$_2$ plasma exposure duration. The transfer characteristics are shown on a semi-logarithmic scale in (b) and as the square root of the drain current in (c). All samples were treated with O$_2$ plasma, OTS SAM, and 1-octanethiol SAM.

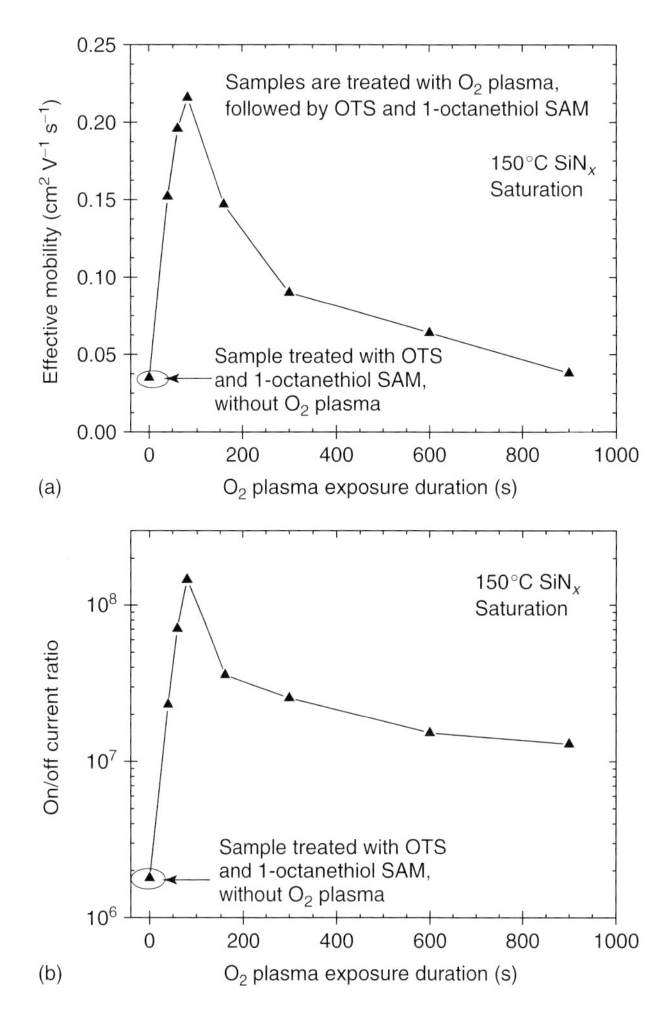

Figure 5.13 (a) Effective mobility and (b) on/off current ratio as a function of O_2 plasma exposure duration of PQT-12 OTFT with 150 °C SiN_x ($x = 1.28$) gate dielectric. Data were collected in the saturation region. All samples were treated with O_2 plasma, OTS, and 1-octanethiol. Here, each data point corresponds to an average value from three to six devices.

In the following/subsequent discussion, $t = 80$ s is referred to as the "*turn-around*" point. Potential interface mechanisms responsible for the observed OTFT behavior are analyzed in Section 5.4.2.

5.4.1.2 **Impact of Exposure Power**

The influence of O_2 plasma exposure power on the effective mobility of PQT-12 OTFTs is illustrated in Figure 5.14. An increase in effective mobility is observed as the RIE power increases (from 0 to 34 W). It is hypothesized that as the RIE power increases, there is an increase in the ion energy of reactive ions impinging

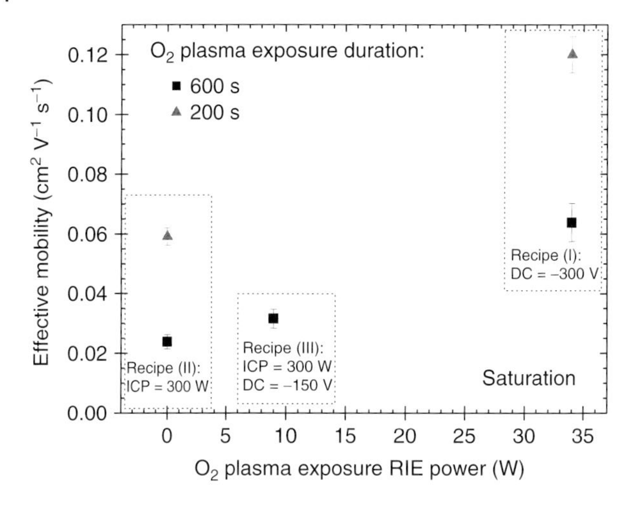

Figure 5.14 Effective mobility as a function of O_2 plasma exposure power of PQT-12 OTFT with SiN_x gate dielectric biased in the saturation region. All samples were treated with O_2 plasma and OTS SAM.

on the wafer; as a result, more oxygen species become attached/adsorbed on the substrate surface. This explanation can be verified by XPS measurements (see Section 5.4.2.3). The enhanced attachment of oxygen species on the surface is believed to facilitate and enhance subsequent formation of the OTS SAM and PQT-12 layer on the SiN_x gate dielectric.

The results also reveal that O_2 plasma generated by the RIE mode of operation is more favorable than that generated by the ICP mode for improving device mobility. The RIE mode creates a stronger drift field that directs the ions onto the substrate. This increases the reactivity of the SiN_x surface, enhancing the bonding with OTS SAM on the O_2 plasma treated SiN_x surface. In addition, Figure 5.14 reveals that 200 s exposure leads to higher mobility than 600 s exposure, for the two recipes plotted here. This substantiates the previous observation where shorter exposure duration led to higher mobility.

5.4.2
Interface Characterization

Contact angle, AFM, and XPS were used to study the interface properties of O_2 plasma treated SiN_x. The interface characterization data are correlated with electrical data to account for the mechanisms underlying the observed OTFT characteristics.

5.4.2.1 Contact Angle

Figure 5.15 shows the changes in the water contact angle (θ) value of the SiN_x surface as a function of O_2 plasma exposure duration. Two sets of samples are

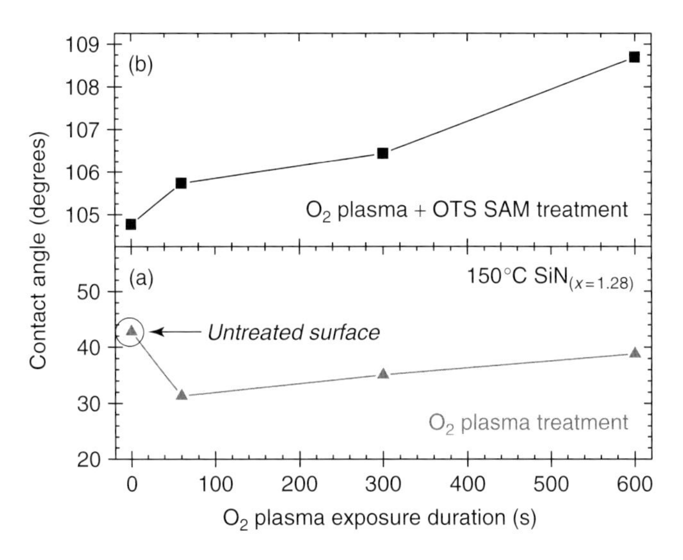

Figure 5.15 Water contact angle of 150 °C SiN_x ($x = 1.28$) surface treated with (a) O_2 plasma only and (b) O_2 plasma/OTS SAM, as a function of O_2 plasma exposure duration.

considered: the first set considers devices treated with O_2 plasma only, and the second set considers devices treated with O_2 plasma and OTS SAM. In the case of samples with the combined treatment, the surfaces are hydrophobic ($\theta > 107^\circ$) and the contact angle is weakly dependent on O_2 plasma exposure duration. As the exposure duration increases from 0 to 900 s, the change in contact angle is less than 5°. No conclusive/definite correlation can be drawn between contact angle data and field-effect mobility for these O_2 plasma/OTS samples.

On the other hand, for samples treated with O_2 plasma only, there is a distinct dependence of contact angle on exposure duration. Figure 5.15 shows that the contact angle is smallest at an exposure time around 80 s, which corresponds to the turnaround point observed in the OTFT's mobility vs. O_2 plasma exposure time curve in Figure 5.13a. However, a logical correlation between the two sets of data is currently unclear. Intuitively, mobility should depend on the dynamics of the dielectric surface after O_2 plasma/OTS SAM treatment. However, the contact angle of the O_2 plasma/OTS surface showed very little dependence on O_2 plasma exposure duration, and, hence, cannot provide sufficient justification for the observed mobility behavior. Similarly, the contact angle data displayed no obvious dependence on O_2 plasma exposure power for the O_2 plasma/OTS surfaces. Therefore, these contact angle observations are unable to provide a clear explanation of the observed changes in mobility under different O_2 plasma exposure conditions. Further analysis via surface roughness and surface chemistry measurements may offer insight on the underlying mechanism for the "turnaround" behavior observed in the OTFT electrical characteristics.

5.4.2.2 **Surface Roughness**

Figure 5.16 plots the surface roughness of SiN_x as a function of O_2 plasma exposure duration, for two sets of samples: (i) O_2 plasma treatment only and (ii) O_2 plasma/OTS SAM combined treatment. Overall, bare SiN_x and O_2-plasma treated SiN_x displayed very smooth surfaces, with measured surface roughness in the range 0.192–0.224 nm. There is a slight decrease in surface roughness with prolonged O_2 plasma exposure. In contrast, we observed a larger variation in surface roughness for samples treated with O_2 plasma/OTS; more importantly, a "turnaround" effect is apparent in the data. Here, a *minimum* surface roughness is observed near an O_2 plasma exposure duration of 80 s, which corresponds precisely to the point of *maximum* mobility in Figure 5.14a. Generally, as surface roughness decreases, OTFT mobility increases [20]. This behavior is evident by correlating the roughness data of the O_2 plasma/OTS sample in Figure 5.16 with the mobility data in Figure 5.14.

Figure 5.17 shows the surface roughness of SiN_x samples treated with O_2 plasma/OTS SAM as a function of O_2 plasma exposure power. With increasing RIE power, the surface roughness increases. Since higher energy ions are accelerated toward the substrate surface at higher RIE power, this potentially creates a rougher surface. The results also indicated that longer exposure duration (at 600 s) generates a rougher surface (than at 200 s), consistent with the observations in Figure 5.16. Interestingly, the roughness data follow an opposite trend to mobility. While higher RIE power enhances mobility (Figure 5.14), it is linked to higher surface roughness (Figure 5.17). This behavior diverges from the typical trend reported in the literature, stating that higher surface roughness degrades mobility. Such

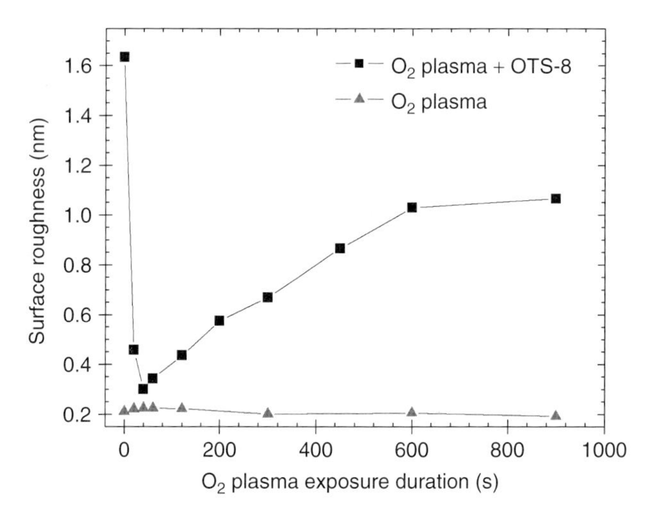

Figure 5.16 Surface roughness as a function of O_2 plasma exposure duration for SiN_x surface treated with O_2 plasma and OTS SAM. Measurements performed using tapping-mode AFM. These samples are treated with O_2 plasma Recipe (I) in Table 5.6.

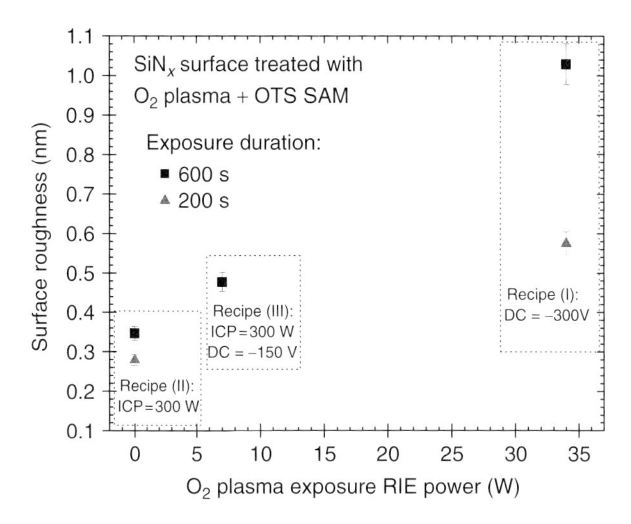

Figure 5.17 Surface roughness as a function of O_2 plasma exposure power for SiN_x surface treated with O_2 plasma and OTS SAM. Measurements performed using tapping-mode AFM.

divergence can be attributed to other surface mechanisms that might impose a more dominant impact on device characteristics than surface roughness. The chemical composition of the surfaces is examined in the next subsection to seek insight on this matter.

In summary, AFM data reveal an increase in surface roughness with O_2 plasma exposure time and power. During exposure, oxygen radicals may preferentially remove weak Si–Si bonds and form Si–O bonds at the surface [23], thus influencing the roughness and chemical reactivity of the surface. Surface roughness plays an important role in the molecular ordering and quality of the overlying organic semiconductor layers; the increased surface roughness may limit the quality of OTFTs.

5.4.2.3 **Chemical Composition**

Figure 5.18a compares the XPS survey spectrum of the untreated SiN_x surface and the O_2 plasma treated (80, 600 s) SiN_x surfaces. Upon O_2 plasma exposure, the SiN_x dielectric surface experienced an increase in oxygen (O 1s) and carbon (C 1s) peaks, accompanied by a decrease in nitrogen (N 1s) and silicon (Si 2p) peaks. Figure 5.18b indicates that the O 1s atomic percentage is highest for an exposure duration of 80 s; this is well-matched to the point of maximum effective mobility in Figure 5.14a.

The XPS data revealed that the untreated (as-deposited) SiN_x surface is dominated by Si–N bonds, while the O_2 plasma treated SiN_x surface becomes more dominated by Si–O bonds. Thorough inspection of the Si 2p spectrum indicates that the Si−O peak becomes larger than the Si−N peak after O_2 plasma treatment. This means that the SiN_x surface is oxidized. The SiN_x surface acquires an oxide-like

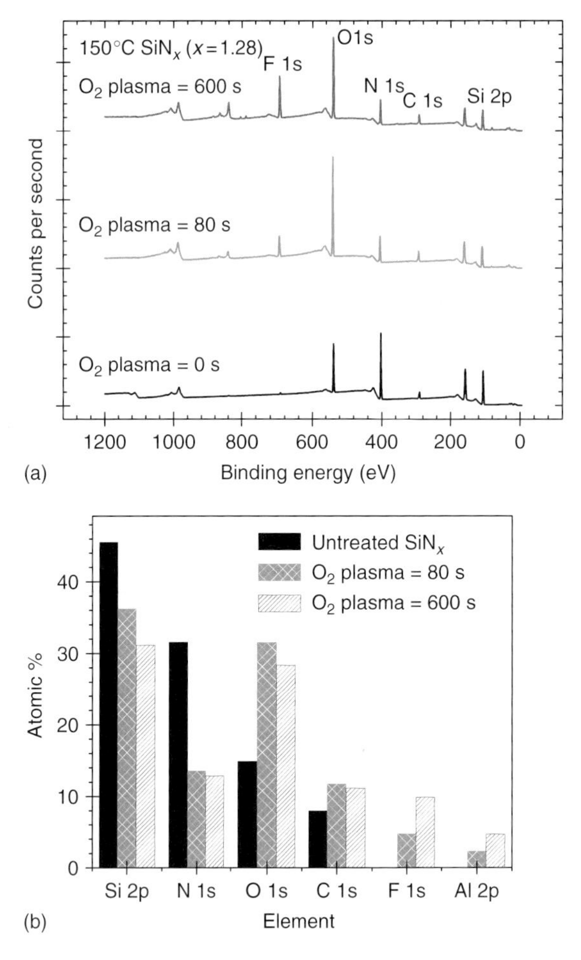

Figure 5.18 (a) XPS spectrum in survey scan mode of an untreated SiN$_x$ surface and O_2 plasma treated SiN$_x$ surfaces after 80 and 600 s exposure. (b) Chemical composition (represented by atomic percentage) of the SiN$_x$ top surface extracted from (a). O_2 plasma exposure was performed using Recipe (I) in Table 5.6.

or oxynitride-like character upon O_2 plasma exposure, which is highly favorable for deposition of OTS SAM and organic semiconductor layers. This oxide-like interface, along with the reduced surface roughness (Figure 5.16), of the O_2 plasma (80 s) treated SiN$_x$ favorably supports the observation that O_2 plasma treatment creates a desirable platform for OTFTs, thus yielding improved effective mobility (Figure 5.14). Additional analysis of the XPS data is presented next. Collectively, the lower surface roughness and the higher Si-O content observed at 80 s O_2 plasma exposure of the SiN$_x$ surface provide a solid explanation for the maximum mobility at this turnaround point.

Detection of Fluorine and Alumina on O_2 Plasma Treated SiN_x Surface Peculiarly, the XPS spectra of the O_2 plasma treated SiN_x surface in Figure 5.18 disclosed the presence of fluorine (F 1s) and aluminum (Al 2p). The intensities of the F 1s and Al 2p peaks are relatively weak compared to other components, but increase with O_2 plasma exposure duration. The unexpected presence of these elements is speculated to be a consequence of contamination from the plasma chamber/equipment. A Trion Phantom II RIE/ICP hybrid system was used to execute O_2 plasma treatment. This equipment is routinely used to run a variety of semiconductor process recipes (e.g., etching, cleaning) which may involve CF_4 and SF_6 as process gases. Residual fluorine compounds in the chamber may react with the sample surface during O_2 plasma exposure; this accounts for the detection of fluorine on the O_2 plasma treated SiN_x surface.

Regarding sources of aluminum contamination, the RIE chamber is equipped with anodized Al (i.e., alumina) as electrodes, and the sample/wafer is placed on the bottom electrode. Due to the nature of the RIE/plasma processes, physical sputtering is inevitable; this causes the anodization to deteriorate over time, exposing the underlying aluminum electrode to plasma. As a result, unintentional sputtering of aluminum may occur, which is manifested as aluminum contamination on the sample's surface. XPS measurements were repeated on three batches of samples over the course of two years; traces of F and Al were still detected despite diligent attempts to decontaminate the chamber via physical and chemical cleaning methods.

An interesting speculation is that the presence of F and Al actually functionalizes the interface in a manner that contributes positively to OTFT performance. Analysis of the F 1s region suggests the possibility of Si–F and C–F bonding or AlF_3 compounds, while the Al 2p spectrum displayed characteristics of Al_2O_3. SAMs of fluorinated silanes have recently been investigated for surface modification of the dielectric surfaces in OTFTs [24, 25]. Examples of fluorinated SAMs include (tridecafluoro-1,1,2,2-tetrahydrooctyl)trichlorosilane (FTS) and (tridecafluoro-1,1,2,2-tetrahydrooctyl)triethoxysilane. Kobayashi *et al.* reported that perfluoroalkylsilane self-assembled monolayers (F-SAM) with fluorine groups generated the highest mobility for p-type pentacene OTFTs compared to devices treated with alkylsilane SAMs (CH_3-SAM) or aminoalkylsilane SAM (NH_2-SAM) [24]. Calculation of the molecular dipole indicated that when the molecules are uniformly aligned, F-SAMs generate a local electric field that enhances the accumulation of holes in the channel, when compared to the CH_3-SAMs. This agrees well with the enhanced p-type carrier conduction in pentacene and the positive V_T shift observed in devices treated with F-SAMs [24].

Podzorov and coworkers reported larger improvement in rubrene transistor properties when using SAM with fluorine end-groups [25]. They observed that organosilane SAMs with larger fluorine content induced a higher surface conductivity owing to the larger electron-withdrawing ability of fluorinated compounds (when compared to non-fluorinated compounds). Such SAM-induced conductivity is due to a ground-state charge transfer at the interface, and the degree of charge transfer (or conductivity) of SAM-functionalized samples depends on the electron-withdrawing ability of organosilane molecules [25].

Moreover, plasma fluorinated surfaces have been shown to lower the surface energy, due to formation of C–F bonds [26]. Fluorinated surfaces also displayed improved stability, where the presence of the C–F functionalities hinders the adsorption of O_2, N_2, CH_4, and CO gases. These characteristics are favorable in the case of OTFTs, where a reduction in surface energy and inhibition of O_2 adsorption of the fluorinated surface can improve the quality of the organic semiconductor layer and enhance the device stability, respectively.

It is hypothesized that our fluorinated interface provoked similar improvement in PQT-12 OTFTs. It is speculated that the substantial improvement seen in our O_2 plasma treated samples is related to the unique presence of F and Al at the interface. This speculation is substantiated by the fact that the degree of improvement achieved by our O_2 plasma treatment on SiN_x outperformed other O_2 plasma treated OTFT devices reported in the literature (specifically, for devices without SAM treatment) [8]. Our O_2 plasma treatment resulted in an 18-fold improvement in mobility relative to the untreated device (Figure 5.7). In comparison, Lee *et al.* reported a 9.3-fold improvement in mobility for their O_2 plasma treated pentacene OTFTs on a SiO_2 gate dielectric [8]. Further experiments are needed to systematically study the impact of F and Al on interface properties and on OTFT characteristics, in order to develop a better understanding of the role of a fluorine/alumina-functionalized dielectric interface.

Effect of O_2 Plasma Exposure Power As reported in Section 5.4.1.2, a higher RIE power of O_2 plasma exposure gives higher effective mobility of the OTFT device. Contact angle and surface roughness measurements were unable to account for this exposure power dependence. Changes in the chemical composition of the O_2 plasma treated SiN_x surface as a function of RIE power are plotted in Figure 5.19. The results revealed that higher RIE power generated more oxygen species on the

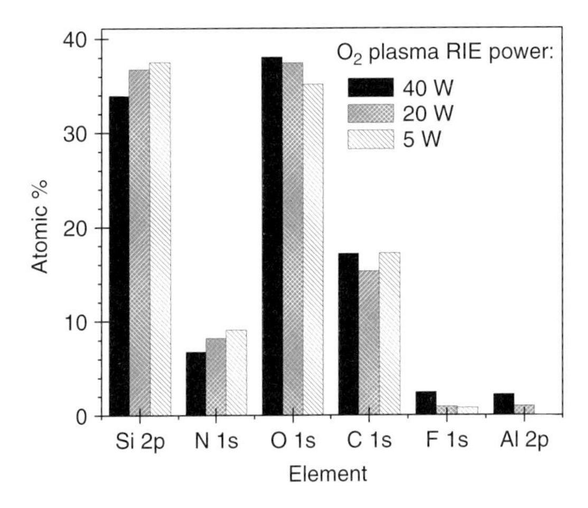

Figure 5.19 Changes in chemical composition (represented by atomic percentage) of an O_2 plasma treated SiN_x surface at various exposure powers.

surface. It appears that higher exposure power rendered a more oxidized surface, which is believed to be favorable for the subsequent binding of organosilane SAM or organic semiconductor layers. As the bias power increases, there is an increase in the energy of the reactive ions impinging on the wafer. This increase can partly be confirmed from the experiment, where the oxygen content increased with RIE bias power. However, more in-depth studies are needed to develop a better understanding of the underlying science/mechanism.

5.4.2.4 XPS Depth Profile Analysis

Figure 5.20 compares the corresponding Si 2p, N 1s, and O 1s XPS spectra as a function of the total sputtering time for three SiN_x samples treated for different O_2 plasma durations. For this XPS depth analysis, the estimated sputtering rate using Ar ions is 1 nm min^{-1}. For untreated SiN_x, XPS depth profile measurements

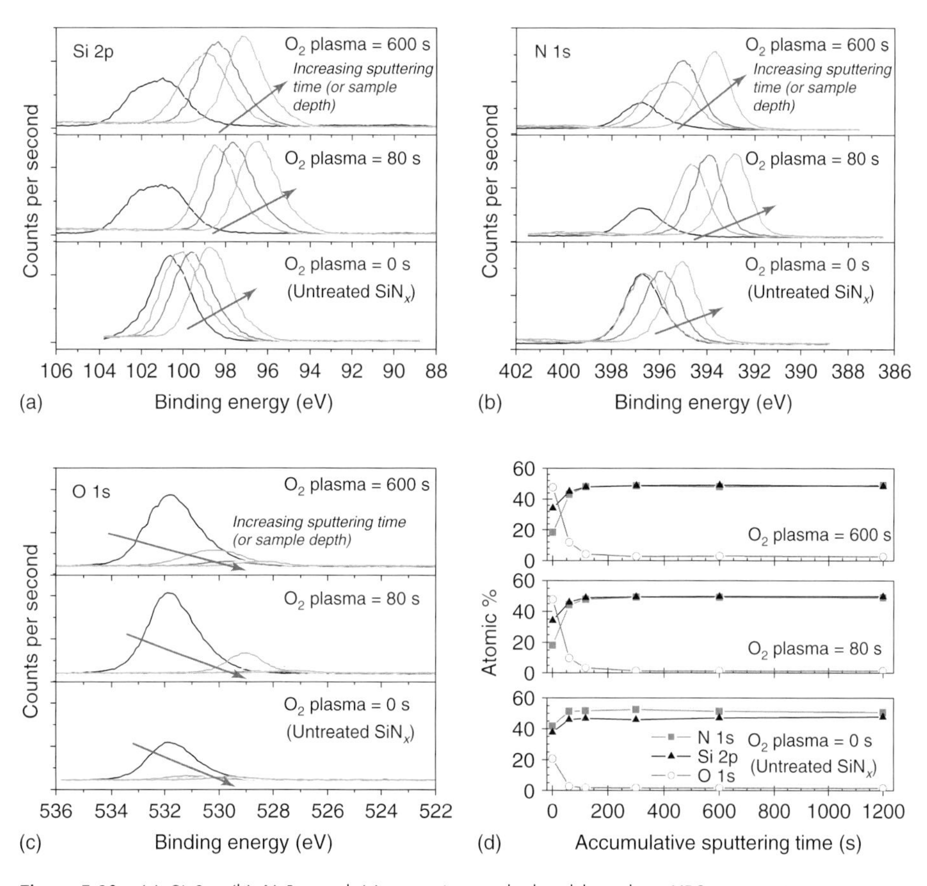

Figure 5.20 (a) Si 2p, (b) N 1s, and (c) O 1s spectra of 150 °C SiN_x ($x = 1.28$) upon successive ion sputtering in XPS depth-profiling measurement. (d) Atomic percentage as a function of sputtering time, calculated based on XPS spectra in (a–c). Three SiN_x samples were considered: as-deposited (untreated) SiN_x, and 80 and 600 s O_2 plasma treated SiN_x.

indicated that the Si 2p and N 1s peak intensities are quite constant with changes in sputtering time (or depth), suggesting homogeneous film properties from the surface region toward the bulk. O 1s is detected at the top surface of untreated SiN_x, and might originate from the ambient/environment. This O 1s peak quickly disappeared after 60 s of sputtering, suggesting this is only a surface phenomenon.

The atomic distribution in Figure 5.20d (bottom panel) showed that the untreated SiN_x sample is consistently characterized by a dominance of Si and N entities (i.e., Si–N bonds) from the surface to the bulk. This is in distinct contrast to the O_2 plasma treated SiN_x samples, displayed in the top and middle panels of Figure 5.20d. For the 80 and 600 s O_2 plasma treated SiN_x, their top surface (i.e., at sputtering time of 0 s) is dominated by O and Si. This reaffirms our previous analysis that the surface becomes oxidized after O_2 plasma exposure. In addition, their Si 2p spectrum shows features of Si-O bonds. Interestingly, this oxide-like characteristic penetrates deeper into the SiN_x film as O_2 plasma exposure duration increases. The O 1s spectra in Figure 5.20c revealed deeper penetration of oxygen into the bulk (as indicated by the intensity of the O 1s peak with increasing sputtering time) for the 600 s sample when compared to the 80 s sample. Therefore, the Si–O or oxide-like sub-region extends deeper into the film with increasing O_2 plasma exposure. Considering the Si 2p and N 1s spectra in Figure 5.20a,b, the 600 s sample showed a more gradual transition of these peak intensities with sputtering time compared to the 80 s sample; the 80 s sample displayed a more brisk transition to the steady-state peak intensity with sputtering time. This behavior corroborates the conclusion that a thicker oxide-like sub-region is present in SiN_x samples with longer O_2 plasma exposure.

5.4.3 Analysis and Discussion

Overall, a positive influence on OTFT performance was observed with O_2 plasma treatment. However, to maximize the improvements in OTFTs, the plasma exposure time must be optimized and well-controlled. Excessive exposure duration can cause a "turn-around effect" in the field-effect mobility and on/off current ratio. The turnaround effect in the variation of field-effect mobility with plasma exposure time was also observed by Lee and coworkers for a pentacene OTFT on a silicon dioxide gate dielectric [8]. Additionally, they observed a decrease in the standard deviation of device parameters with oxygen plasma exposure time, resulting in enhanced uniformity of device parameters across the wafer.

Lee and coworkers proposed that the oxide charge is responsible for the observed variation in device parameters. Upon O_2 plasma treatment, the oxide charges in the SiO_2 gate dielectric initially decrease due to UV-enhanced reduction and then increase due to the high energy particle bombardment [8]. The as-grown silicon oxide usually contains positive charges of the order of 10^{10}–10^{12} cm^{-2}, depending on the growth process. The bridging oxygen vacancy defect $O_3{\equiv}Si{\bullet\bullet}Si{\equiv}O_3$ named the E′ center is known to be responsible for the positive charges. It has been reported that the oxide charge varies depending on the plasma condition and exposure time [27]. Under the soft plasma conditions used in Ref. [27], the oxide charges decreased for 3–4 min after plasma treatment began, where the time depended on the plasma condition, and then increased again, resulting in the "turn-around" effect. A plausible origin for the decrease in oxide charge is the UV-enhanced reduction in positive charge (annihilation of E′ centers or generation of negative compensating charges by vacuum UV photons of the plasma). It appears that the change in oxide charges induced by the UV plasma component was the dominating factor up to the turn-around point. As plasma treatment proceeded, the relative influence of the UV component decreased and the effect of bombardment by high-energy particles became dominant, resulting in an increase in oxide charge. The relative contribution of these two components depended on the specific plasma conditions [2]. Therefore, the plasma treatment up to the turn-around time was found to be effective for performance enhancement in pentacene OTFTs.

The XPS results indicated an oxide-like or oxynitride-like SiN_x surface upon O_2 plasma exposure. Moreover, depth analysis by XPS revealed the presence of Si–O bonds deeper in the SiN_x as O_2 plasma exposure duration increased. Therefore, oxygen species penetrate into the SiN_x films with longer exposure duration, creating an oxide-like subregion in the SiN_x bulk, near the interface. Owing to the presence of Si-O characteristics in our SiN_x dielectric, plasma/UV-induced changes in oxide charge species, proposed by Lee *et al.* in Ref. [8], can be a relevant mechanism contributing to the observed changes in OTFT behavior as a function of O_2 plasma exposure duration.

It has been reported that plasma treatment can activate and clean surfaces, thus enabling stronger bonds with subsequent layers and reducing impurity-induced degradation in the interface's electrical properties [23]. This is reflected in our contact angle measurements of the O_2 plasma treated SiN_x surfaces, which showed that, even with short exposure time, the surfaces were activated to generate a hydrophilic surface. This implies the effectiveness of O_2 plasma treatment for surface activation is related to substituting carbon surface contaminations with hydroxy groups and chemically active dangling bonds.

The contact angle of O_2 plasma treated SiN_x displayed a weak dependence on the exposure duration, where the contact angle was smallest for 80 s of exposure (Figure 5.15). The observed contact angle trend cannot provide sufficient proof to explain the observed TFT data as a function of exposure duration. However, contact angle is sensitive to the physical state of the surface (e.g., roughness). Thus, surface roughness measurements might be able to shed light on the observed contact angle and TFT mobility trend. Figure 5.16 shows that 80 s of exposure gives minimum surface roughness for O_2 plasma/OTS treated. This matches precisely with the point of maximum mobility in Figure 5.14. With longer plasma exposure duration, the surface roughness of O_2 plasma/OTS treated samples was observed to increase. The rougher surface accounts for the reduced mobility at longer exposure duration. Because the surface roughness plays an important role in the quality of the overlying polymer semiconductor layers and the efficiency of charge transport, increased surface roughness typically degrades the quality of OTFTs.

Oxygen radicals during plasma treatment may preferentially remove weak Si–Si bonds and break Si-O bonds at the surface, leading to a rougher surface. Choi *et al.* studied the effects of O_2 plasma treatment on Si and glass substrates, and reported that the activated oxygen reacts with surface impurities, and reaction products were desorbed in the form of volatile materials as follows [23]:

$$C_xH_yO_z + (O_2^*, O) \rightarrow CO_2 + H_2O \tag{5.1}$$

where (O_2^*, O) is chemically activated oxygen. These oxygen radicals preferentially modify Si–O bonds at the top of the SiO_2 surface, and cause the generation of chemically active dangling bonds and hydroxy groups at the surface [23]. These chemical changes in bonding structure agree with the hydrophilic surface properties observed for an O_2 plasma-activated SiN_x surface.

Table 5.8 summarizes the effect of O_2 plasma exposure conditions on the electrical properties of the OTFT and the surface properties of the SiN_x dielectric. Altogether, our results suggested that the chemical states (e.g., surface energy, composition), physical states (e.g., surface roughness), and electrical states of the dielectric surface collectively influence the resulting device performance. In some instances, they operate coherently/consistently to enhance device performance. However, on other occasions, these mechanisms may compete to impose the opposite impact on device properties.

In conclusion, O_2 plasma treatment is an effective way to enhance the performance of OTFTs on a SiN_x gate dielectric if the plasma exposure time and exposure conditions are appropriately optimized and well-controlled. The effect of O_2 plasma treatment on a SiN_x dielectric is to introduce an oxide-like interface, rendering a dielectric surface more favorable for subsequent binding to alkylsilane SAMs or organic semiconductor films.

Table 5.8 Summary of the impact of O_2 plasma exposure duration and RIE power on surface properties of SiN_x and on OTFT device mobility.

O_2 plasma exposure parameter	Contact angle		Surface roughness		Effective mobility
	O_2 plasma	O_2 plasma + OTS	O_2 plasma	O_2 plasma + OTS	
Duration: with increasing exposure time					
For $t < t^*$	Small ↓	Small ↑	Small ↑	↓↓	↑↑↑
For $t > t^*$	Small ↑	Small ↑	Small ↓	↑↑	↓↓
RIE power					
As power increases	Unchanged	Unchanged	–	↑	↑

t^* denotes the "turnaround point," defined as the exposure duration that generated maximum effective mobility.

5.5 Summary and Contributions

The influence of dielectric surface treatment with OTS SAM and O_2 plasma exposure was analyzed by electrical measurements of PQT-12 OTFTs with a SiN_x gate dielectric. The interface properties were characterized to establish a correlation with the TFT characteristics. A key contribution of this investigation is that surface treatment using a combination of O_2 plasma and OTS SAM led to about a 30-fold increase in the effective field-effect mobility, with saturation mobility up to $0.22\,cm^2\,V^{-1}\,s^{-1}$. Our results showed that the SiN_x surface must be pre-treated with O_2 plasma prior to OTS SAM formation in order to generate a smooth and effective OTS layer. The adhesion of OTS on bare SiN_x is more sporadic, as indicated by a higher surface roughness. In contrast, there is a large reduction in surface roughness when OTS is formed on O_2 plasma treated SiN_x. O_2 plasma exposure provides a hydroxylated surface to facilitate attachment of OTS molecules on SiN_x.

Another important observation is the turnaround effect in O_2 plasma treatment. It is recognized that the O_2 plasma exposure time must be carefully controlled to ensure that an optimal quantity of hydroxy groups is created on the surface for maximizing the OTFT's mobility. The surface properties (studied by means of contact angle and surface roughness measurements) of SiN_x varied with the O_2 plasma exposure time. There is a turnaround effect with O_2 plasma treatment; for this particular investigation the turnaround point occurs at an exposure duration of 80 s. With 80 s of O_2 plasma exposure and OTS SAM treatment, the OTFT demonstrated the highest mobility, and the SiN_x gate dielectric exhibited the smallest surface roughness and largest contact angle.

This investigation also demonstrated that a high mobility ($\sim 0.1\,cm^2\,V^{-1}\,s^{-1}$) PQT-12 OTFT on SiN_x can be achieved even in the absence of OTS SAM. The SiN_x surface was only subjected to O_2 plasma treatment. This is a striking observation (or break through) as SAM is generally considered a critical element for OTFTs. The ability to achieve high mobility in the absence of SAM can potentially simplify the OTFT fabrication process. Table 5.9 provides a summary of the key experimental observations from this chapter.

An extension of this work to generate further device improvements or advance understanding of device behavior may include:

- Expanding the investigation to additional O_2 plasma exposure parameters, such as power, pressure, gas flow;
- Exploring fluorination as an interface treatment technique. Possible methods to functionalize the SiN_x gate dielectric surface with fluorine include using fluorinated SAMs, fluorine plasma treatment, or deposition of a fluorinated-SiN_x gate dielectric.
- Long term stress measurements to study the effect of different surface treatments on the stability of OTFTs.

Table 5.9 Summary of key observations from dielectric interface engineering experiments.

	Description	Key observations
Part A	Comparison of TFTs with and without dielectric surface treatment	μ_{FE} is enhanced in the presence of O_2 plasma and OTS (compared to untreated devices) Combination of O_2 plasma and OTS yields the highest μ_{FE} (~30× improvement over untreated device) Single O_2 plasma treatment led to ~18× increase in μ_{FE} compared to untreated device
Part B	Effect of O_2 plasma exposure conditions	O_2 plasma exposure of 80 s gives the highest μ_{FE} For longer exposure ($t > 80$ s), μ_{FE} dropped and eventually plateaued Higher plasma exposure RIE power gives higher μ_{FE}. O_2 plasma generated by RIE mode produced TFTs with higher μ_{FE} than ICP mode

References

1. Park, Y.D., Lim, J.A., Lee, H.S., and Cho, K. (2007) Interface engineering in organic transistors. *Mater. Today*, **10** (3), 46–54.
2. Lim, S.C., Kim, S.H., Lee, J.E., Kim, M.K., Kim, D.J., and Zyung, T. (2005) Surface-treatment effects on organic thin-film transistors. *Synth. Met.*, **148** (1), 75.
3. Hiroshiba, N., Kumashiro, R., Komatsu, N., Suto, Y., Ishii, H., Takaishi, S., Yamashita, M., Tsukagoshi, K., and Tanigaki, K. (2007) Surface modifications using thiol self-assembled monolayers on Au electrodes in organic field effect transistors. *Mater. Res. Soc. Symp. Proc.*, **965**, paper 0965-0S08-03.
4. Grecu, S., Roggenbuck, M., Opitz, A., and Brütting, W. (2006) Differences of interface and bulk transport properties in polymer field-effect devices. *Org. Electron.*, **7** (5), 276.
5. Tanase, C., Meijer, E.J., Blom, P.W.M., and de Leeuw, D.M. (2003) Local charge carrier mobility in disordered organic field-effect transistors. *Org. Electron.*, **4** (1), 33.
6. Jackson, T.N., Lin, Y.Y., Gundlach, D.J., and Klauk, H. (1998) Organic thin-film transistors for organic light-emitting flat-panel display backplanes. *IEEE J. Sel. Top. Quantum Electron.*, **4** (1), 100.
7. Kosbar, L.L., Dimitrakopoulos, C.D., and Mascaro, D.J. (2001) The effect of surface preparation on the structure and electrical transport in an organic semiconductor. *Mater. Res. Soc. Symp. Proc.*, **665**, paper C10.6.1.
8. Lee, M. and Song, C. (2003) Oxygen plasma effects on performance of pentacene thin film transistor. *Jpn. J. Appl. Phys.*, **42**, 4218.
9. Wasserman, S.R., Tao, Y.-T., and Whitesides, G.M. (1989) Structure and reactivity of alkylsiloxane monolayers formed by reaction of alkyltrichlorosilanes on silicon substrates. *Langmuir*, **5**, 1074.
10. Sung, M.M., Kluth, G.J., and Maboudian, R. (1999) Formation of alkylsiloxane self-assembled monolayers

on Si_3N_4. *J. Vac. Sci. Technol. A*, **17** (2), 540.

11. Song, C., Koo, B., Lee, S., and Kim, D. (2002) Characteristics of pentacene organic thin film transistors with gate insulator processed by organic molecules. *Jpn. J. Appl. Phys.*, **41**, 2730.
12. Kim, D.H., Park, Y.D., Jang, Y., Yang, H., Kim, Y.H., Han, J.I., Moon, D.G., Park, S., Chang, T., Chang, C., Joo, M., Ryu, C.Y., and Cho, K. (2005) Enhancement of field-effect mobility due to surface-mediated molecular ordering in regioregular polythiophene thin film transistors. *Adv. Funct. Mater.*, **15** (1), 77.
13. Veres, J., Ogier, S., Lloyd, G., and de Leeuw, D. (2004) Gate insulators in organic field-effect transistors. *Chem. Mater.*, **16**, 4543.
14. Ihm, K., Kim, B., Kang, T.H., Kim, K.J., Joo, M.H., Kim, T.H., and Yoon, S.S. (2006) Molecular orientation dependence of hole-injection barrier in pentacene thin film on the Au surface in organic thin film transistor. *Appl. Phys. Lett.*, **89**, 033504.
15. Lu, D., Wu, Y., Guo, J., Lu, G., Wang, Y., and Shen, J. (2003) Surface treatment of indium tin oxide by oxygen-plasma for organic light-emitting diodes. *Mater. Sci. Eng.*, **B 97**, 141–144.
16. Stephen, M.R., William, D.W., and Jerome, J.H. (1990) *Handbook of Plasma Processing Technology: Fundamentals, Etching, Deposition, and Surface Interactions*, Noyes Publications.
17. Trion Technology, Inc. (2006) Trion Phantom II RIE/ICP System Manual.
18. Tai, K.L. and Vratny, F. (1985) Method for fabricating devices with DC bias-controlled reactive ion etching. US Patent 4,496,448. *http://www.freepatentsonline.com/4496448.html.*
19. Trion Technology. (2004) ICP Source – A Brief Summary, Trion Technical Papers and Resources. *http://www.triontech.com/* (accessed 2008).
20. Chabinyc, M.L., Lujan, R., Endicott, F., Toney, M.F., McCulloch, I., and Heeney, M. (2007) Effects of the surface roughness of plastic-compatible inorganic dielectrics on polymeric thin film transistors. *Appl. Phys. Lett.*, **90**, 233508.
21. Kim, W.K. and Lee, J.L. (2007) In situ analysis of hole injection barrier of O_2 plasma-treated Au with pentacene using photoemission spectroscopy. *Electrochem. Solid State Lett.*, **10** (3), H104–H106.
22. Wang, A., Kymissis, I., Bulovic, V., and Akinwande, A.I. (2004) Process control of threshold voltage in organic FETs. 2004 IEEE International Electron Devices Meeting (IEDM) Technical Digest, pp. 381–384.
23. Choi, S.W., Choi, W.B., Lee, Y.H., Ju, B.K., and Kim, B.H. (2001) Effect of oxygen plasma treatment on anodic bonding. *J. Korean Phys. Soc.*, **38** (3), 207.
24. Kobayashi, S., Nishikawa, T., Takenobu, T., Mori, S., Shimoda, T., Mitani, T., Shimotani, H., Yoshimoto, N., Ogawa, S., and Iwasa, Y. (2004) Control of carrier density by self-assembled monolayers in organic field-effect transistors. *Nat. Mater.*, **3**, 317.
25. Calhoun, M.F., Sanchez, J., Olaya, D., Gershenson, M.E., and Podzorov, V. (2008) Electronic functionalization of the surface of organic semiconductors with self-assembled monolayers. *Nat. Mater.*, **7**, 84–89.
26. Nakahara, M., Ozawa, K., and Sanada, Y. (1994) Change in the chemical structures of carbon black and active carbon caused by CF_4 plasma irradiation. *J. Mater. Sci.*, **29**, 1646.
27. Paskaleva, A. and Atanassova, E. (2000) Bulk oxide charge and slow states in Si–SiO_2 structures generated by RIE-mode plasma. *Microelectron. Reliab.*, **40**, 2033.

6
Contact Interface Engineering

Interface control is indispensable for realization of high performance organic thin film transistors (OTFTs). The two most critical interfaces in OTFTs are the dielectric/semiconductor interface and the contact/semiconductor interface. The former interface was examined in Chapters 4 and 5. This chapter focuses on the latter. The quality of the contact/semiconductor interface is important because charge injection into the OTFT channel occurs precisely at this interface. For efficient charge injection, ohmic contact is desirable. In a transistor, ohmic behavior is characterized by a linear relation between I_D and V_{DS} at small V_{DS} biases. To achieve ohmic contacts, one of the fundamental requirements is energetic matching/alignment of the Fermi level of the source/drain metal contacts to the highest occupied molecular orbital (HOMO) or lowest unoccupied molecular orbital (LUMO) levels of the p-type or n-type organic semiconductor layer, respectively. In the case of p-type organic semiconductors, good matching is possible using contacts with high work function (e.g., gold). The Fermi level of the high work function metal electrodes lies in the proximity of the HOMO level of the p-type organic semiconductor, thus rendering a small energy barrier at the interface to facilitate efficient hole injection. This energy matching approach provides a simple guideline for contact material selection. However, in practice, physical interaction (e.g., interface dipole) between materials is more complex and may induce an energy barrier at the contact/semiconductor interface that limits charge injection [1]. A more in-depth discussion on charge injection is presented in Section 6.1.1. To improve contact properties, interface modification using self-assembled monolayers (SAMs) of charge transfer complexes can potentially lead to larger drain currents and better current saturation than non-modified contacts [2]. Use of alkanethiol SAMs for metal contact treatment is reviewed in Section 6.1.2.

The SAM contact treatment method was evaluated on poly(3,3‴-dialkyl-quarter thiophene) (PQT-12) OTFTs. The experiment is divided into two parts. Part 1 investigates the impact of Au contact modification by 1-octanethiol ($CH_3(CH_2)_7SH$) SAMs on the device performance of PQT-12 OTFTs; the results are presented in Section 6.3. Part 2 examines the influence of the execution sequence of the surface treatment steps on the device behavior; the outcomes are discussed in Section 6.4. Interface characteristics (e.g., wettability, surface roughness, chemical composition) are analyzed, where correlation with the

Organic Thin Film Transistor Integration: A Hybrid Approach, First Edition. Flora M. Li, Arokia Nathan, Yiliang Wu, and Beng S. Ong.

electrical properties of the OTFTs is made to gain a better understanding on device behavior.

6.1 Background

6.1.1 Charge Injection

The electrical performance of OTFTs is not only limited by the intrinsic carrier mobility of the organic semiconductor, but also by the efficiency of injecting and extracting charge carriers at the source and drain contacts. The contact resistance (R_C) provides a measure of the charge injection efficiency in an OTFT. A basic parameter that determines charge injection in a device is the injection barrier at the contact/semiconductor interface, as illustrated in Figure 2.18. This injection barrier is given by the energy difference between the Fermi level of the contact and the conduction or valence band of the semiconductor. Low energy barriers ($\varphi_B = IP_S - \Phi_M$) necessitate matching the work function (Φ_M) of the contact metal with the ionization potential (IP_S) of the semiconductor; this is one of the fundamental criteria for high charge injection efficiency. Organic devices are often prone to large contact resistance owing to the relatively large band-gap (~2 eV) and high ionization potential (~5 eV) of most organic materials; as result, they form Schottky barriers with various metals [3].

Figure 6.1 shows a simplified energy level diagram of the contact/semiconductor interface. When a metal and a p-type semiconductor are brought into contact, their Fermi levels (E_F) line up in equilibrium. For a low work function metal

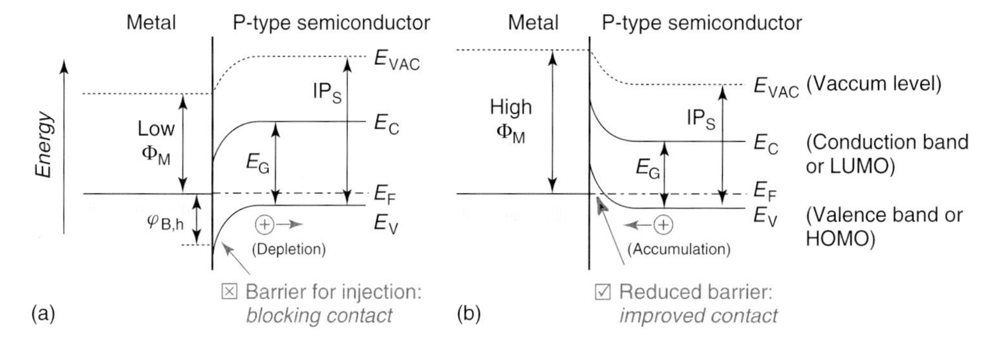

Figure 6.1 Energy level diagrams for metal and p-type semiconductor/contact interfaces (assuming negligible interface states). (a) Small work function metal leads to interface barrier and poor contact. (b) Large work function metal leads to reduced interface barrier and improved contact.

Table 6.1 Source contact resistance R_S of bottom-contact P3HT and F8T2 OTFTs made with different metal contacts ($T = 300$ K, $V_{GS} = -20$ V).

Organic semiconductor	S/D contact	Φ_M (eV)	φ_b (eV)	R_S (kΩ cm)
P3HT ($IP_S = 5.0$ eV)	Au	5.2	0.0	<5
P3HT	Ag	4.89 ± 0.10	0.1	≈15
P3HT	Cr–Au	–	–	22
P3HT	Cu	4.72 ± 0.10	0.3	320
P3HT	Cr	4.68 ± 0.10	0.3	5 400
F8T2 ($IP_S = 5.5$ eV)	Au	5.2	0.3	13 000
F8T2	Cr–Au	–	–	>70 000
P3HT	Al	4.05 ± 0.10	1.0	∞

IP_S is the ionization potential of the semiconductor, Φ_M is the measured work function of the metal, and φ_B is the estimated Schottky barrier height at the metal/organic interface (adapted from [4]).

(Figure 6.1a), the valence band of a p-type semiconductor bends downward, which creates a barrier for hole conduction and results in a high contact resistance. For a high work function metal (Figure 6.1b), the valence band of a p-type semiconductor bends upward, which reduces the barrier for hole conduction and results in an improved contact. Therefore, to alleviate contact resistance in p-type OTFTs, high work function metals should be selected for source/drain contacts. Burgi *et al.* extracted the contact resistance of poly(3-hexylthiophene) (P3HT) and poly(9,9-dioctyl-fluorene-co-bithiophene) (F8T2) OTFTs made with different metal contacts [4], and observed that metals with higher work function generate lower contact resistance for bottom-contact P3HT OTFT and F8T2 OTFT. The data also revealed that organic semiconductors with higher ionization potential (e.g., F8T2) are more susceptible to higher contact resistance [4]. These results are summarized in Table 6.1.

It is generally believed that the alignment of molecular energy levels with respect to the Fermi-level of contacts is of the utmost importance for charge carrier injection and device efficiency. Charge carrier injection barriers are often estimated by assuming vacuum level alignment across interfaces, as explained above. However, recent research studies have cast doubts on the accuracy of this simple model due to the identification/detection of interfacial dipoles across many interfaces [5, 6]. Consequently, charge injection barriers in the presence of interfacial dipoles can differ by more than 1 eV from the values estimated from the vacuum level alignment approach. Possible origins of interfacial dipoles include charge transfer, formation of chemical bonds, or a "push-back" of electrons into the metal bulk after molecule adsorption [7]. Therefore, the contact/semiconductor interface mechanisms are more complex than expected. Detailed analytical techniques such as ultraviolet photoelectron spectroscopy (UPS) can be used to study these interfacial mechanisms to allow proper analysis and modeling of OTFT behavior.

In addition to the effects of the energy level alignment and the interfacial dipole at the contact/semiconductor interface, a number of other factors can influence the charge injection efficiency and contact resistance in an OTFT, including the device configuration [8], device geometry (e.g., channel length) [4], processing technique [9], and doping level of the semiconductor layer [10]. In terms of the structural dependence of the contact resistance, smaller contact resistance was reported in top-contact OTFTs owing to the relatively larger contact areas at the contact/semiconductor interface compared to bottom-contact structures [8]; more descriptions are presented in Section 2.3.1. Process-induced damage, associated with metal deposition on top of the organic semiconductor layer to form top-contact OTFTs, can also influence contact resistance. Metals are typically deposited by evaporation or by sputtering. The evaporation of metals onto organic materials can lead to in-diffusion of the metal, changes in the morphology of the organic material, and possible disruption of chemical bonds in the polymer chains [9]. These effects can lead to large interfacial resistances in OTFTs, which can subsequently increase contact resistance and undermine device performance. Processing conditions must be optimized to inhibit chemical, physical, and morphological changes induced by metal deposition on the organic semiconductor material. Contacts formed by solution-processable organic conductors can circumvent the aforementioned process-induced contact effects. However, more extensive development is necessary to improve the quality and patterning techniques for contacts based on soluble organic conductors.

The contact resistance (R_C) in OTFTs is typically in the range 10 kΩ cm to 10 MΩ cm, and is larger than those in inorganic transistors (e.g., for amorphous silicon thin film transistors (TFTs), R_C is usually in the kΩ cm range) [4]. Unlike inorganic field effect transistors (FETs) based on silicon, the source and drain contacts in OTFTs are not easily optimized by conventional processes, such as selective semiconductor doping or metal alloying. For example, the absence of a highly doped source/drain contact regions in OTFTs compared to silicon metal-oxide-semiconductor field-effect transistors (MOSFETs), as shown in Figure 3.1, is one of the primary causes of larger contact resistance in OTFTs. If highly doped contact regions or layers can be implemented in the OTFT structure, significant enhancement in charge injection efficiency and reduction in contact resistance can be expected.

More research effort is needed to minimize contact resistance in OTFTs; large contact resistance can severely limit device performance, perhaps to an extent where the speed of organic integrated circuits may not be limited by the intrinsic carrier mobility of the organic semiconductor, but by the contact resistance of the OTFTs. Large contact effects can limit the current-carrying ability of the device, lead to underestimations of important device parameters (e.g., mobility), and restrict the utility of OTFTs in active-matrix backplanes due to the longer charging time ($\tau \sim RC$) of an individual pixel. Therefore, the nature of the interface between an organic semiconductor and a source/drain contact is critical to the performance of organic electronics.

To promote charge injection and address contact resistance issues, SAM treatment of the semiconductor/contact interface is investigated; this concept is reviewed in Section 6.1.2. Alternatively, developing techniques to enable selective doping of the source/drain contact regions of OTFTs (similar to inorganic transistors), or incorporating a charge injection layer at the contacts (similar to organic light-emitting diodes (OLEDs)) can offer additional routes to enhance the contact properties of OTFTs.

6.1.2
Alkanethiol SAM on Metals

Thiol (also referred to as *mercaptan*) is a compound that contains the functional group composed of a sulfur atom and a hydrogen atom (–SH). Alkanethiol refers to the group of compounds where the thiol group is a substituent on an alkane. SAMs of thiols on gold and other metals have emerged as one of the most important classes of surface coatings, and are used in a variety of applications [11]. Alkanethiols are widely used to prepare highly ordered monolayers whose wetting properties can be controlled by changing the chemical nature of the terminal groups. Surfaces modified by alkanethiols can be made hydrophilic by the introduction of polar groups (e.g., –OH, –COOH, $-CONH_2$), or hydrophobic by non-polar groups (e.g., $-CH_3$, $-OCH_2CF_2CF_3$, $-O(CH_2)_mCH_3$) [1]. By using more complex functionalities, SAMs can be made (bio)chemically reactive, adhesive, or biologically inert. For example, long chain alkanethiols can produce a highly packed and ordered surface, rendering a membrane-like microenvironment that is suitable for immobilizing biological molecules in biosensing applications [11]. The development of nanoelectronic devices also exploits alkanethiol SAMs, where SAMs are used to position and pattern molecular components selectively on surfaces. One example is soft lithography with SAMs for the creation of small (nano-sized) two- and three-dimensional chemical patterns on material surfaces [11]. Another application of thiol SAMs is for protection of a metal against corrosion. Thick barriers of thiols can block electron transfer and hinder the transport of water, oxygen, and aggressive ions to the metal surface. The barrier properties of SAMs are largely determined by the length and organization of the alkyl chains. Metal surfaces covered by long-chain alkanethiols have demonstrated enhanced corrosion resistance [11]. These examples exemplify the versatility of alkanethiol SAMs. Alkanethiol SAMs can provide a basis for many scientific and technological applications, including nano-fabrication, biological recognition, molecular electronics, analytical, and sensory applications; moreover, they can be used as molecular lubricants, protective coatings, or templates for crystal nucleation and growth [10].

SAMs of alkanethiols are often used to control the interactions of metal surfaces because sulfur compounds (from thiol) typically have a strong affinity to transition metal surfaces [11]. Other attractive attributes of alkanethiol SAMs include stability, ease of preparation/modification, and a high degree of order. In the liquid phase method of forming SAMs, the molecules are dissolved in an organic solvent and

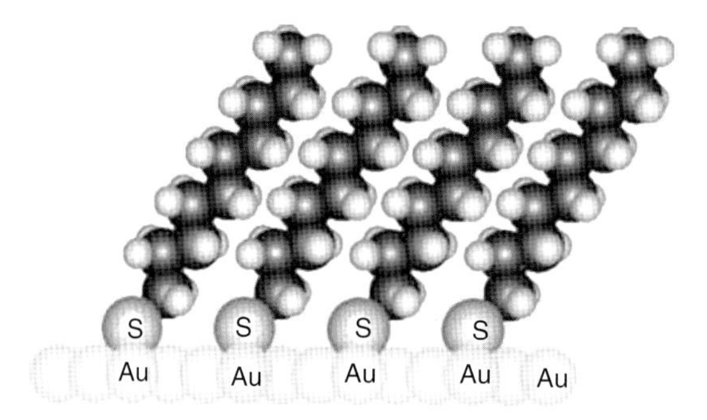

Figure 6.2 Pictorial illustration of an alkanethiol SAM on an Au surface. (Adapted from [10].)

they then adsorb spontaneously to the substrate material. When a SAM is formed by immersing a gold surface in a solution of an alkyl thiol, the –SH group of the thiol molecule experiences deprotonation (i.e., a hydrogen atom is abstracted from the thiol molecule) at the surface and the molecule binds to the surface through the S atom, forming a strong covalent Au–S bond. Figure 6.2 shows a pictorial illustration of highly ordered alkanethiol SAM on an Au surface. Stability, packing, and ordering of the SAMs are determined by two interacting factors: the interfacial linkage between sulfur and gold (i.e., surface interactions), and the lateral van der Waals attractions between hydrocarbon chains (i.e., interchain interactions). Typically, the longer the hydrocarbon chains of the deposited thiol, the greater the stability of the resulting monolayer [10].

In OTFTs, alkanethiol SAMs have been applied as a surface modifier for the Au source/drain contacts. Some researchers have reported improved molecular ordering, grain size, and adhesion of the organic semiconductor on modified Au surfaces. Thus, with proper material combination, thiol SAMs can improve charge injection at the metal contacts and enhance device performance [12]. However, there are cases where little or no influence/difference was observed with thiol SAM treatment [13, 14]. Therefore, the effects of thiol SAMs are not clear at the moment. More extensive studies are needed to clarify the real influences and establish an effective methodology for SAM treatment of source/drain contacts for practical OTFT applications.

6.2 Experimental Details

Bottom-gate bottom-contact PQT-12 OTFTs with SiN_x gate dielectric and Au source/drain contacts (with a thin Cr adhesion layer) were used for this contact interface engineering experiment. The device fabrication scheme is outlined in Figure 3.8. The basic criteria for material selection of source/drain contacts include chemical stability and energy compatibility with the organic semiconductors to

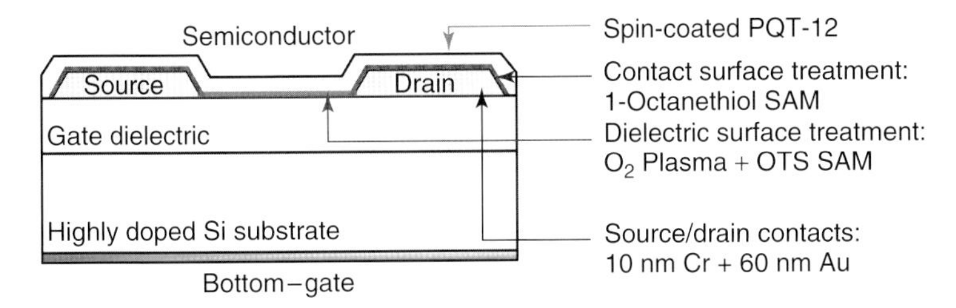

Figure 6.3 Schematic cross-section of the OTFT structure considered in this contact surface treatment study.

form ohmic contacts for efficient charge injection. Au was selected for this experiment because of its high conductivity, excellent operational stability, and, particularly, its large work function. The work function of Au ($\Phi_M \approx 5.1$ eV) is energetically similar to the ionization potential of PQT-12 ($IP_S \approx 5.2–5.3$ eV), thus hole injection from the Au source/drain contacts to the PQT-12 semiconductor is theoretically feasible. For selected devices with contact surface treatment by a thiol SAM, the Au contacts were functionalized by immersing the substrates in a 0.01 M 1-octanethiol ($CH_3(CH_2)_7SH$) solution in toluene for 30 min at room temperature. A cross-sectional diagram of the device structure is shown in Figure 6.3.

The purpose of this study was to evaluate the impact of octanethiol SAM contact surface treatment on PQT-12 OTFT characteristics, particularly the effect on device contact resistance. Two experiments were conducted, as summarized in Table 6.2. The first experiment (Section 6.3) evaluates the effects of incorporating thiol SAM treatment on four samples with different dielectric surface treatment conditions; this is an extension of the investigation reported in Section 5.3. The second experiment (Section 6.4) assesses the effect of changing the execution sequence of surface treatment steps on the device performance. The OTFT electrical characteristics and the interface characterization data are analyzed to establish correlations between the electrical, chemical, and physical characteristics of the devices; this can contribute to a better understanding of the mechanisms

Table 6.2 Contact/semiconductor interface engineering experiments.

	Description	**Detail**
Part A (Section 6.3)	Effect of addition of octanethiol SAM treatment	Compare device characteristics after adding thiol treatment, for various samples with varying dielectric surface conditions
Part B (Section 6.4)	Effect of sequence/order of SAM surface treatment	Evaluate how the execution order of SAM treatments using 1-octanethiol and OTS affect transistor performance

responsible for the changes in device characteristics due to octanethiol SAM contact surface treatment. The TFT parameters were extracted following the procedure outlined in Section 2.2. An overview of the various analytical techniques used for surface characterization is provided in Section 4.1.2.

6.3
Impact of Contact Surface Treatment by Thiol SAM

The first experiment evaluates the impact of adding a 1-octanethiol SAM contact surface treatment step to PQT-12 OTFT fabrication under four dielectric surface modification conditions:

1) No dielectric surface treatment: "none"
2) O_2 plasma exposure for 60 s
3) Octyltrichlorosilane self-assembled monolayer (OTS SAM)
4) O_2 plasma exposure for 60 s followed by OTS SAM.

The dielectric surface treatment procedures are discussed in Chapter 5. For the scenarios considered here, the thiol SAM treatment was performed after completion of the dielectric surface treatment(s), and before PQT-12 semiconductor deposition. Electrical characterization of the resulting OTFTs is reported in Section 6.3.1, and the interface characteristics are analyzed in Section 6.3.2.

6.3.1
Electrical Characterization

Figure 6.4 displays the effect of thiol SAM treatment on the mobility and on/off current ratio of a PQT-12 OTFT on a 150 °C SiN_x ($x = 1.60$) gate dielectric. For devices without dielectric surface treatment (labeled "none"), the incorporation of thiol produced an increase in mobility by $\sim$0.01 $cm^2\,V^{-1}\,s^{-1}$ (141% improvement). In contrast, for devices with dielectric pretreatment, there was a small drop in mobility by 0.01–0.014 $cm^2\,V^{-1}\,s^{-1}$ ($\sim$8–26% reduction) after thiol treatment. The presence of thiol appears to improve the on/off current ratio for most samples, as shown in Figure 6.4b. The consistency of these results was verified by repeating the study on another set of PQT-12 OTFTs with a 300 °C SiN_x ($x = 1.08$) gate dielectric. Similar trends in mobility are observed in Figure 6.5, where the presence of thiol led to an increase in mobility for devices with untreated dielectric (i.e., "none"), but a reduction in mobility for devices with pre-treated dielectrics.

A number of interesting device behaviors can be identified from an analysis of the mobility data in Figures 6.4 and 6.5:

- For devices *without* dielectric surface treatment (labeled "none"), an improvement in mobility is seen after thiol treatment in both Figures 6.4 and 6.5. It is believed that thiol treatment is favorable for previously-untreated devices (i.e., bare SiN_x and bare Au surfaces). Possible explanations for this are presented below.

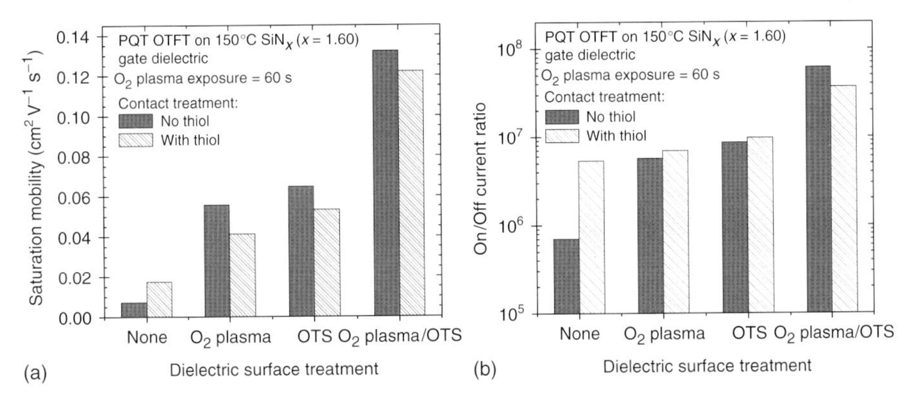

Figure 6.4 (a) Effective mobility and (b) on/off current ratio of PQT-12 OTFT with 150 °C SiN_x ($x = 1.60$) gate dielectrics in the absence ("no thiol") or presence ("with thiol") of 1-octanethiol SAM modification of Au contact surfaces. Four scenarios of gate dielectric surface treatments were considered. Measurements collected in the saturation region.

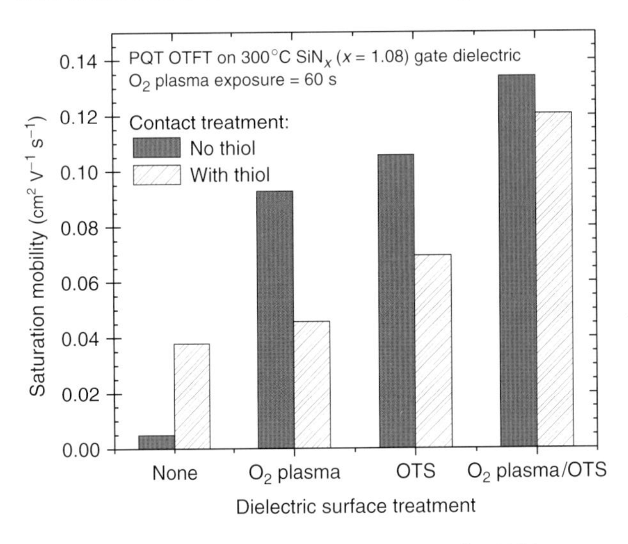

Figure 6.5 Effective mobility of PQT-12 OTFT with 300 °C SiN_x ($x = 1.08$) gate dielectrics in the absence ("no thiol") or presence ("with thiol") of 1-octanethiol SAM modification of Au contact surfaces. Four scenarios of gate dielectric surface treatments were considered. Measurements were collected in the saturation region of OTFT operation.

- **Channel related**: Improvement in the extracted mobility value may stem from (i) an actual enhancement in charge transport or (ii) the mobility's dependence on contact resistance due to an artifact/limitation of the TFT extraction model. In the former, thiol treatment may be (unexpectedly) modifying the bare SiN_x surface in a manner that enhances the dielectric/semiconductor interface and charge transport properties. However, there is a conventional belief by chemists that thiol only chemisorbs on Au surfaces, and there should be no interaction with SiN_x. This notion is challenged by the contact angle data in Figure 6.8, which revealed that thiol treatment indeed altered the SiN_x surface wettability in our experiment. More discussion on this will follow in Section 6.3.2.1. In the latter scenario, when TFT becomes less contact limited, the extraction mobility may increase due to over-simplified modeling of OTFT characteristics.

- **Contact related**: Thiol SAM treatment on bare Au contacts is effective for improving contact injection properties. Hence, the effect of contact-limited mobility is reduced, leading to noticeable device improvement compared to untreated devices.

For devices *with* dielectric pretreatment, a decrease in mobility was observed after thiol treatment in both Figures 6.4 and 6.5. Thus, thiol treatment is not as favorable for devices pre-treated with O_2 plasma and OTS SAM. Possible causes of this behavior are speculated below.

- **Channel related**: The reduction in mobility suggests interaction of the thiol with the pretreated SiN_x dielectric surface in a way that disturbs the device performance. It is speculated that traces of thiol molecules aggregated on the pretreated SiN_x surface, hindering the quality of the semiconductor channel and the dielectric/semiconductor interface. Contact angle measurements revealed that thiol treatment indeed imposed a small change in the surface wettability of an OTS-treated or O_2-plasma-treated SiN_x surface, as shown in Figure 6.8. An attempt was made to study the surface composition of thiol-treated surfaces via X-ray photoelectron spectroscopy (XPS). However, the sulfur concentration in the thiol monolayer was too minuscule to permit reliable detection and analysis using XPS.
- **Contact related**: The intention of thiol treatment is to improve charge injection at the contact/semiconductor interface, which should reduce contact resistance and lead to an overall improvement in device performance (e.g., higher current conduction). This thiol-induced improvement is not obvious for devices with a pretreated dielectric. One possibility is that the O_2 plasma and OTS dielectric treatments modified the Au contacts, thus hindering the formation of high-quality thiol SAM on Au. It is generally believed that OTS does not self-assemble on Au. This is confirmed by contact angle measurements, where negligible changes in contact angle were observed after subjecting Au to OTS treatment. On the other hand, recent studies showed evidence that O_2 plasma can modify the work function and interface dipole of Au surfaces. Kim *et al.* observed a reduction in the hole injection barrier at the interface by O_2 plasma, leading to an increase in linear mobility for pentacene OTFTs [15]. These results confirmed the (unintentional) changes in Au contact properties during dielectric surface treatment by O_2 plasma in our PQT-12 OTFT devices. However, there are currently few or no reports on the effect of thiol on O_2 plasma treated Au.

Contact resistance provides a good measure for analyzing the contact/semiconductor interface properties in OTFTs. The contact resistance was

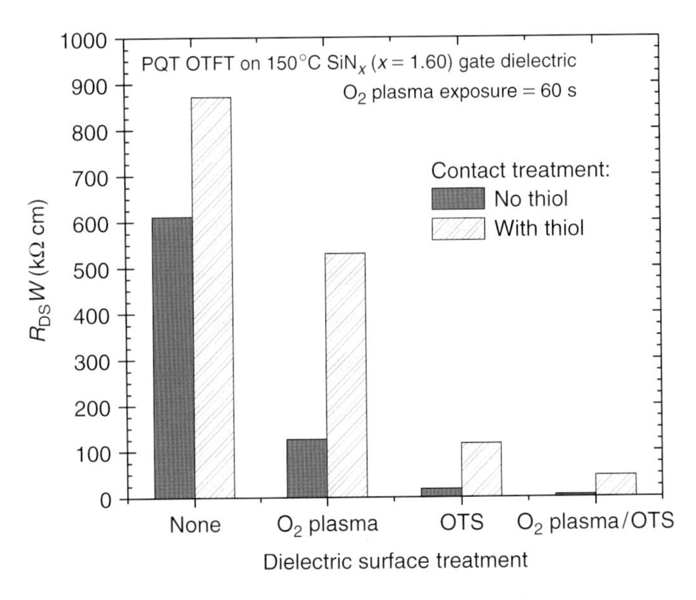

Figure 6.6 Contact resistance of PQT-12 OTFT on 150 °C SiN_x ($x = 1.60$) gate dielectrics in the absence ("no thiol") or presence ("with thiol") of 1-octanethiol SAM modification of the Au contact surfaces. Four scenarios of gate dielectric surface treatments were considered. Measurements were collected in the linear region of OTFT operation.

calculated following the extraction method published in Ref. [16] and outlined in Section 2.2.2. Figure 6.6 plots the constant (linear) component of the contact resistance in PQT-12 OTFTs with various surface treatment conditions. Devices treated with thiol exhibited higher contact resistance than devices without thiol. A consistent trend between mobility (Figure 6.4) and contact resistance (Figure 6.6) is observed and is summarized in Figure 6.7. Here, thiol-treated devices have lower mobility and higher contact resistance than their no-thiol counterparts. Interestingly, these observations are contrary to the initial expectation that thiol SAM can improve charge injection and reduce contact resistance [12].

The accuracy of the contact resistance extraction requires further examination. In most of the existing TFT models, contact resistance is often modeled with a mobility-dependent component. As a result, higher mobility devices might have a smaller extracted contact resistance. Undeniably, the extracted mobility and contact resistance data in Figure 6.6 are susceptible to error introduced by mobility-dependent contact resistance. However, the percentage change in contact resistance differs considerably from the percentage change in mobility for each pair of devices, as tabulated in Table 6.3. Therefore, we can conclude that the observed trend in contact resistance is justifiable.

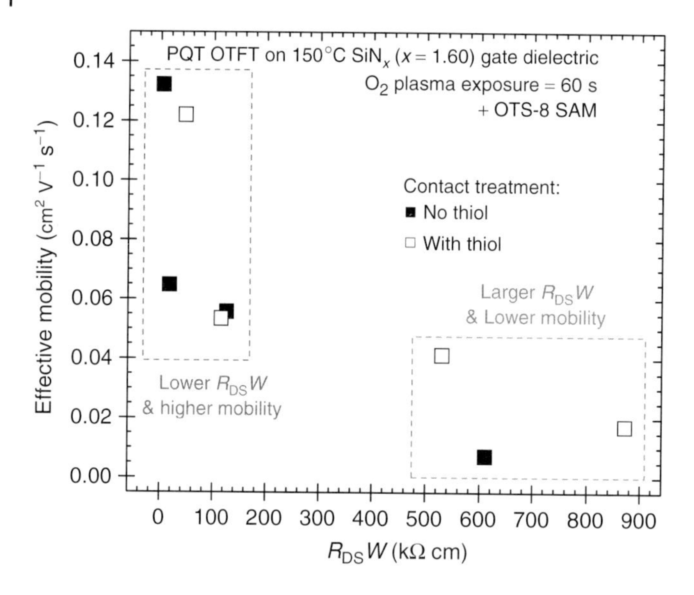

Figure 6.7 Saturation mobility versus contact resistance of PQT-12 OTFT on 150 °C SiN_x ($x = 1.60$) gate dielectrics.

Table 6.3 Percentage change in mobility and contact resistance for PQT-12 OTFT on 150°C SiN_x ($x = 1.60$), based on data in Figures 6.5 and 6.6, respectively. Percentage change is calculated as "(Thiol − noThiol)/noThiol."

Dielectric surface treatment	Change in mobility (%)	Change in $R_{DS}W$ (%)
None	141.1	42.7
O_2 plasma	−25.8	545.1
OTS	−20.8	320.5
O_2 plasma/OTS	−7.6	905.2

Our observation that thiol-treated devices have higher contact resistance is contradictory to the general belief that SAM would reduce contact resistance. One possible reason for this deviation is that the thiol SAM might be imperfect due to pre-processing conditions. As such, instead of enhancing contact injection, it might introduce a barrier that hinders charge injection [17], and thus leads to higher contact resistance. More detailed experimentation is required to confirm these hypotheses. In addition, employment of alternate TFT contact resistance models should be employed to confirm the results reported here [15, 18–21]. In the next section, the interface properties of the thiol-treated surfaces are examined to seek insight into the observed OTFT behavior.

6.3.2
Interface Characterization

6.3.2.1 Contact Angle

Contact angle measurements were done on various SiN_x samples with or without 1-octanethiol SAM treatment in order to study the impact of thiol on the dielectric surface properties and correlate the results with the observed device characteristics. It is generally conceived that alkanethiol SAM modifies Au surfaces only, and there should be minimal or no interaction with SiN_x [1]. However, contrary to conventional intuition, Figure 6.8 shows a consistent reduction in the contact angle on various SiN_x surfaces after 1-octanethiol SAM treatment. This reduction in contact angle for thiol-treated devices correlates well with the decrease in mobility (Figures 6.4 and 6.5), since a dielectric surface with a higher contact angle (or hydrophobicity) is typically linked to higher device mobility [2, 7, 13].

However, the interaction between SiN_x and thiol is unclear. Does thiol actually chemisorb to form a SAM to provide a "real" surface modification? Or does the change in contact angle merely reflect the presence of traces of thiol residues on the SiN_x surface? At present, scientific reports/evidence on thiol–SiN_x bonding cannot be found; most literature reports thiol–Au bonding only (since the sulfur molecules in thiol have a strong affinity for Au). More extensive experiments are needed to clarify the interaction between thiol and SiN_x.

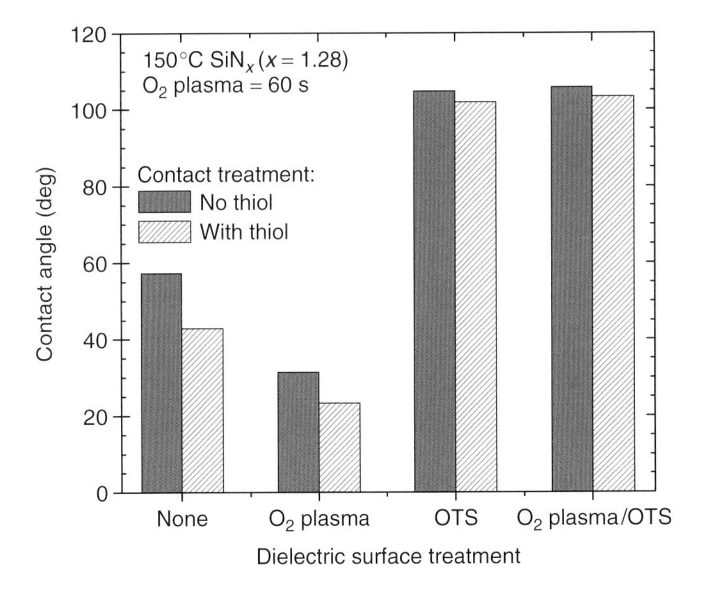

Figure 6.8 Water contact angle of 150 °C SiN_x (x = 1.28) surface after various types of dielectric surface treatment, and in the absence or presence of 1-octanethiol exposure.

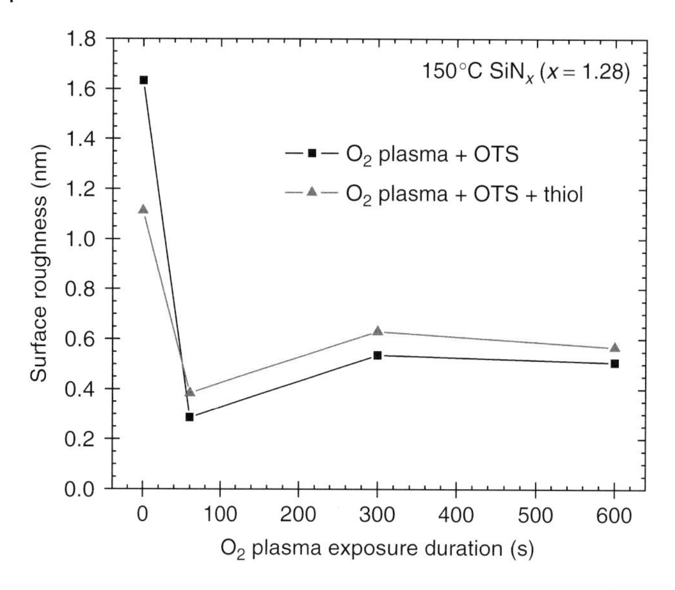

Figure 6.9 Mean surface roughness measured by atomic force microscopy (AFM) for SiN_x surfaces after O_2 plasma/OTS SAM surface treatment, in the absence or presence of 1-octanethiol SAM treatment.

6.3.2.2 **Surface Roughness**

The effect of 1-octanethiol SAM treatment on the surface roughness of SiN_x is illustrated in Figure 6.9. Comparing the data for "O_2 plasma + OTS" and "O_2 plasma + OTS + thiol", thiol treatment led to an increase in surface roughness for samples with O_2 plasma and OTS treatments. This trend was observed consistently for samples subjected to different O_2 plasma exposure durations. Although experimental error must be considered, the consistency observed in each set of samples presents a convincing argument that the thiol treatment step is indeed modifying the SiN_x surface physically or chemically, as noted by changes in surface roughness and contact angle, respectively.

6.3.2.3 **Chemical Composition**

Figure 6.10 compares the elemental distribution on the SiN_x surface without and with the thiol treatment step. A small decrease in Si 2p and C 1s intensities is observed after thiol treatment, which may be related to the low Si and C content in thiol. However, the percentage change is small (~1–6% after addition of thiol), thus the data must be interpreted with caution taking experimental errors into consideration. The intensity of the sulfur (S 2p) peak from the thiol compound was too low to be accurately detected by XPS. Thus, we cannot draw any conclusion from these XPS data.

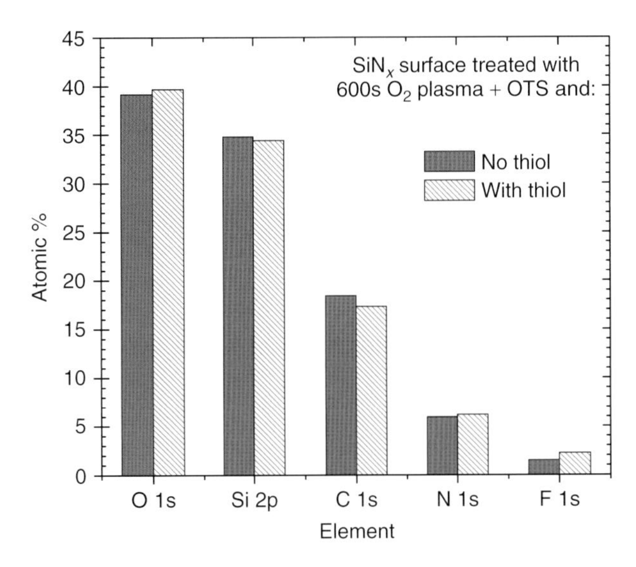

Figure 6.10 Atomic percentage detected by XPS on SiN_x surface after O_2 plasma/OTS SAM surface treatment, in the absence or presence of 1-octanethiol treatment. (Experimental error = $\pm 5\%$).

6.3.3 Analysis

The key observations upon incorporating 1-octanethiol SAM contact surface treatment in the fabrication of PQT-12 OTFTs are:

- Reduced field-effect mobility and increased contact resistance in the OTFT
- Reduced contact angle and increased surface roughness of the SiN_x gate dielectric surface.

It has been reported that OTFT mobility decreases when the dielectric surface becomes less hydrophobic (i.e., smaller contact angle) and when the surface roughness increases [22]. Therefore, the interface characterization data display good correlation to justify the OTFT electrical characteristics.

The most striking observation in this experiment is that the incorporation of thiol treatment actually degrades device performance, which is contrary to expectation. Therefore, the device performance was compromised by the addition of a thiol SAM for the PQT-12 OTFT devices. It has been reported that the application of thiol SAMs on Au contacts can improve OTFT performance [11, 12, 23]. This improvement was attributed to an enhancement in semiconductor film morphology near the contact/semiconductor interface or the effect of the interfacial dipole induced by the SAM. However, such improvement was not observed in the present experiment, particularly for the case where devices were subjected to dielectric surface treatment prior to thiol SAM contact surface treatment. The increase in contact resistance after thiol treatment suggests the formation of a contact barrier layer at the Au/PQT interface. The reduction in mobility after thiol treatment for the pretreated devices indicates that the combination of "O_2 plasma + thiol" and "OTS + thiol" unfavorably modifies device interfaces in a manner that limits charge injection or charge transport. The precise mechanisms responsible for the reduced performance in thiol-treated devices are currently unclear. A number of possible explanations for the observed device behavior are presented below.

- **Modification of Au from O_2 plasma exposure**: Previous reports on successful device improvement by thiol treatment are assumed to be on a bare Au surface. However, for the fabrication of our bottom-gate bottom-contact devices, the Au surface is subjected to O_2 plasma and OTS modification during the dielectric treatment steps. These pretreatment steps may influence how the thiol self-assembles onto the Au surface. The effect of O_2 plasma exposure on Au was reported by Kim *et al.* who observed that O_2 plasma treated Au contacts led to enhanced hole injection and increased mobility for pentacene OTFTs when compared to bare Au contacts [15]. The band structure at the Au/pentacene interface was studied quantitatively, and they found that the work function of Au increased from 4.65 to 5.28 eV as the Au surface was treated with O_2 plasma. The corresponding interface dipoles were -0.30 eV for bare Au and -0.71 eV for O_2-Au. Accordingly, the hole injection barrier at the Au/pentacene interface was reduced from 0.45 to 0.15 eV by the O_2 plasma before the deposition of pentacene, leading to an increase in linear field-effect mobility for pentacene OTFTs [15]. This report confirmed that the Au contacts in the PQT-12 OTFTs were indeed modified by O_2 plasma exposure, and, more importantly, O_2 plasma treatment should enhance the contact properties of our TFTs. This agrees with the dramatic reduction in contact resistance shown in Figure 6.6 for the O_2 plasma treated sample (no thiol) when compared to the untreated device. Therefore, O_2-Au contacts display superiority over bare-Au contacts in PQT-12 OTFTs.
- **Ineffective formation of thiol SAM on O_2 plasma treated Au**: Thiol on O_2-Au may behave differently from thiol on bare-Au. Reports on device enhancement by thiol SAMs were based on bare Au surfaces [11, 12, 23]. On the other hand, the effect of thiol SAMs on O_2-Au surfaces for OTFTs has not been published. Lahio *et al.* studied the influence of initial oxygen (contamination from the atmosphere) on the formation of thiol layers [11]. They observed that the adsorption of thiol removes oxygen from the surface (i.e., the oxygen concentration is reduced after thiol treatment), and that the presence of initial oxygen on Au reduces the amount of thiol on the surface. These results provide evidence that the presence of oxygen indeed affects the adsorption of thiol on Au surfaces. More extensive studies are needed to examine this behavior or the formation of thiol SAMs on an O_2-Au surface.
- **Thiol-induced contact barrier**: The extracted contact resistance values in Figure 6.6 displayed a consistent increase in contact resistance after thiol treatment, regardless of the type of dielectric surface treatment. It is hypothesized that the thiol

SAM introduced a barrier layer that hinders charge injection at the contacts, thus resulting in increased contact resistance. This notion of a thiol-induced contact injection barrier is substantiated by a recent study on pentacene OTFTs with Au electrodes modified by 1-hexadecanethiol ($C_{16}H_{33}SH$) [17], which revealed the presence of tunneling barriers due to the insulating $C_{16}H_{33}SH$ at the interfaces between the Au contacts and pentacene thin film [17].

- **Unoptimized processing conditions**: Device improvements may be possible if thiol SAM treatment recipes are reoptimized for devices subjected to O_2 plasma or OTS pretreatments. The thiol processing conditions used in our experiment were designed for a bare Au surface. However, since our device fabrication scheme involves other pretreatment steps that might inadvertently alter the Au surface prior to thiol treatment, special considerations are required. Hiroshiba *et al.* found that OTFT characteristics are very sensitive to the preparation condition of the thiol SAM modified surfaces (e.g., chemical formulation and substituent groups of the thiol compound, concentration and preparation duration, etc.) [12]. Longer treatment duration may form a "perfect" but thicker thiol layer, which may create a barrier for carrier injection in OTFTs. In contrast, imperfect modifications with a short thiol preparation time delivered improved TFT action and improved charge injection compared to untreated devices! In our case, the thiol treatment conditions may not be optimal for the devices under study, and thus may create an additional barrier for carrier injection. More in-depth studies to understand the interfacial kinetics at the thiol/O_2/Au or thiol/OTS/O_2/Au junctions are needed to develop a suitable contact treatment recipe.

Based on the above observations, it is concluded that 1-octanethiol SAM treatment should be omitted for bottom-gate bottom-contact PQT-12 OTFTs fabricated using the processing conditions reported here. More thorough studies must be conducted to clarify the underlying mechanisms responsible for the reduced performance upon incorporation of the thiol treatment step. Nonetheless, it is believed that with proper treatment conditions, thiol SAMs can potentially enhance contact properties and improve OTFT device performance.

6.4 Impact of Execution Sequence of Surface Treatment

This section examines the question: "does the order or sequence in which surface treatments are executed affect device performance?" In this experiment, two sets of devices were prepared by varying the execution order of 1-octanethiol and OTS SAM treatments, as summarized in Table 6.4. The sample labeled "Thiol/OTS" was prepared by performing thiol treatment on the prepatterned substrate first, followed by OTS treatment. The sample labeled "OTS/Thiol" was first treated with OTS SAM, followed by a thiol treatment step. Plasma enhanced chemical vapor deposited (PECVD) SiN_x gate dielectrics were considered in this experiment. All surfaces were pretreated with O_2 plasma (60 s) prior to OTS and 1-octanethiol SAM treatments.

Table 6.4 Experiment on execution sequence of OTS and thiol surface treatments.

Sample name	Treatment sequence	Key observations
OTS/Thiol (i.e., OTS first, Thiol on top)	OTS first Thiol second PQT-12 last	Higher μ_{FE} and I_{ON}/I_{OFF} Larger $\theta_{contact}$ Smaller surface roughness
Thiol/OTS (i.e., Thiol first, OTS on top)	Thiol first OTS second PQT-12 last	Smaller μ_{FE} and I_{ON}/I_{OFF} Smaller $\theta_{contact}$ Larger surface roughness

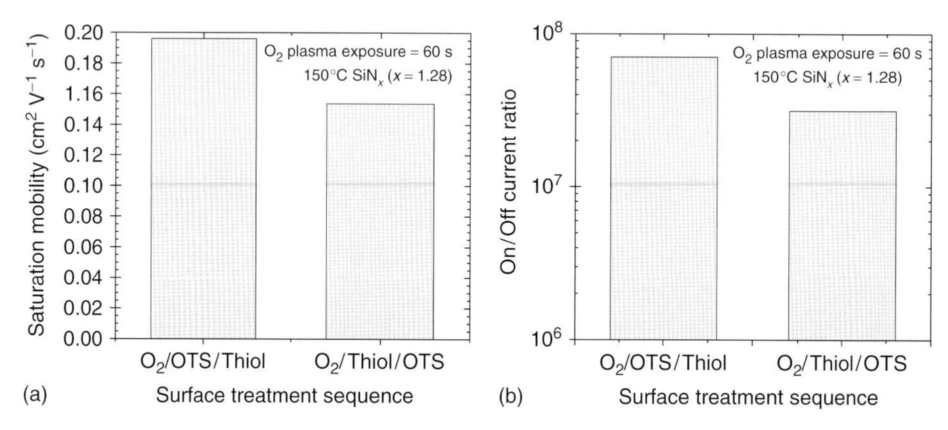

Figure 6.11 Comparison of (a) effective mobility and (b) on/off current ratio of PQT-12 OTFTs subjected to different execution sequences of OTS and Thiol SAM treatment.

6.4.1 Electrical Characterization

The average effective field-effect mobility and on/off current ratio in the saturation region of OTFTs with various geometries were measured and the results are displayed in Figure 6.11. The results showed that OTS/Thiol devices delivered a higher mobility (by 27% or 1.3×) and on/off current ratio (by 124% or 2.3×) than Thiol/OTS devices. One speculation for the superior characteristics with OTS/Thiol devices is that OTS SAM is deposited directly on the SiN_x surface in this configuration. As such, a better quality OTS SAM can be formed to effectively improve the dielectric/semiconductor interface (assuming thiol does not adhere to the OTS surface). In the Thiol/OTS configuration, residues of thiol might remain on the SiN_x surface, which might hinder the formation of a highly ordered OTS SAM. Consequently, this can interfere with the formation of a well-ordered PQT-12 semiconductor layer in the channel, thus reducing device performance.

6.4.2 Interface Characterization

To account for the observed device behavior, interfacial studies were conducted to examine the surface chemistry of OTS/Thiol and Thiol/OTS samples.

6.4.2.1 Contact Angle

Results from the contact angle measurements are shown in Figure 6.12. The OTS/Thiol surface displays a larger contact angle (θ_{contact}) than the Thiol/OTS surface. This agrees with the observed mobility trend, where a higher contact angle is linked to higher mobility. However, the difference in contact angle ($\Delta\theta_{\text{contact}}$) between the two samples is very small (less than 1°); thus, these data must be analyzed with caution taking experimental error into consideration.

6.4.2.2 Surface Roughness

The OTS/Thiol sample displays lower surface roughness than Thiol/OTS, as shown in Figure 6.13. This is concurrent with the OTFT data, where higher mobility in the OTS/Thiol sample is linked to lower surface roughness. The higher surface roughness of the Thiol/OTS sample may be linked to possible thiol residues on the SiN_x surface that obstruct the formation of a high quality OTS SAM. As a result, the Thiol/OTS sample has a higher surface roughness and lower contact angle; these characteristics correlate well with the lower mobility observed in the corresponding OTFT.

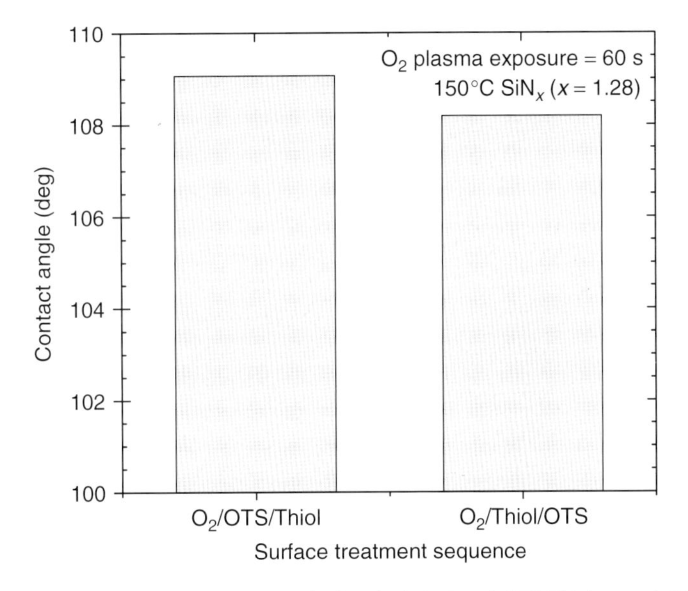

Figure 6.12 Water contact angle for Thiol/OTS and OTS/Thiol treated SiN_x surfaces.

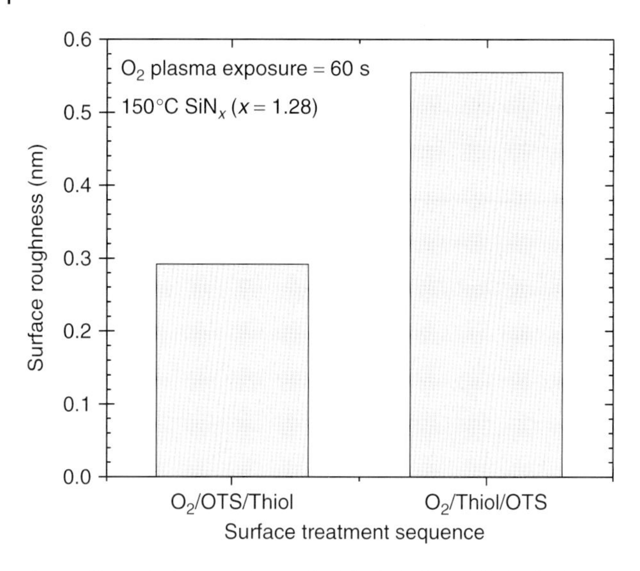

Figure 6.13 Surface roughness for Thiol/OTS and OTS/Thiol treated SiN_x surfaces.

6.4.2.3 **Chemical Composition**

XPS was used to analyze the chemical composition of the OTS/Thiol and Thiol/OTS surfaces. Figure 6.14 shows that the atomic compositions for the two samples are very similar. Thiol/OTS has a larger C 1s concentration, characteristic of the top OTS layer. However, it was difficult to collect reliable data from XPS to quantitatively

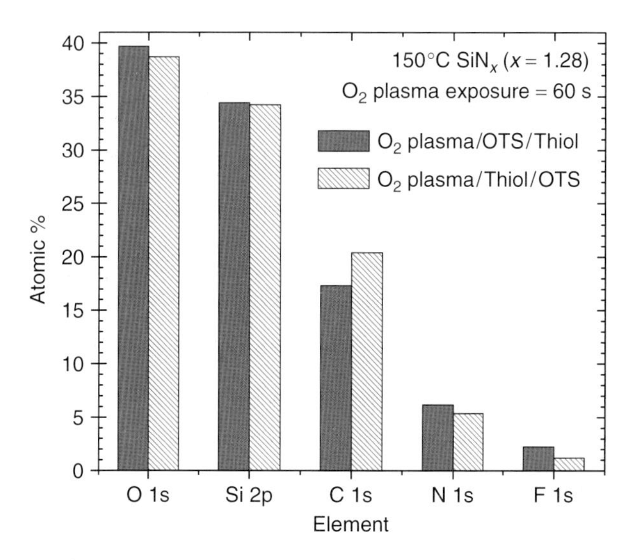

Figure 6.14 Atomic percentage detected by XPS on the surface of SiN_x for different execution sequences of OTS and 1-Octanethiol SAM treatments.

compare the characteristics of the thiol SAM on the treated surfaces. The atomic concentration of sulfur from the Thiol SAM was too minuscule to generate a strong XPS signal. Thus, no conclusions can be drawn from the XPS measurements.

Overall, this experiment demonstrates that the execution order or placement of a thiol layer affects the surface energy of the dielectric surface. The key observations are summarized in Table 6.4. For the SiN_x gate dielectric and OTFT process sequence used in this experiment, it is preferable to perform the thiol treatment after the OTS SAM treatment (i.e., OTS/Thiol is better than Thiol/OTS). However, combining the observations from Sections 6.3 and 6.4, it is best to omit the thiol treatment step for our PQT-12 OTFTs on SiN_x gate dielectric to deliver better device performance.

6.5 Summary and Contributions

The influence of Au contact surface treatment with 1-octanethiol SAM was analyzed by electrical and interface measurements of PQT-12 OTFTs. For devices on "bare/untreated SiN_x", thiol SAM surface treatment led to increased field-effect mobility by up to 140%. However, for devices with dielectric surface pre-treatment, mobility decreases of up to 25% were found after the addition of thiol. It is interesting to observe that the percentage reduction in mobility, or sensitivity, is smaller for OTS-treated devices than for O_2 plasma-treated devices. Perhaps devices with an OTS-treated surface are less sensitive to thiol treatment; since OTS is already hydrophobic, there is a higher chance that thiol does not adhere to or interact with OTS. On the other hand, devices with only O_2 plasma are more sensitive to the addition of thiol; this can be attributed to modification of the bare Au surface by oxygen plasma prior to thiol treatment. While a correlation is observed between field-effect mobility and contact angle, XPS measurements were unable to provide useful insight because the atomic concentration of sulfur from the thiol monolayer was below the detection limit.

The key observations from this contact interface engineering investigation are summarized in Table 6.5. Combining these results, the recommendation is to exclude the 1-octanethiol treatment step to achieve better device performance for PQT-12 OTFTs on SiN_x gate dielectrics. The 1-octanethiol SAM treatment conditions applied in this work might not be optimal for substrates pre-treated with O_2 plasma and OTS SAM. It is recommended that future work considers optimizing the processing parameters for thiol treatment, and/or exploring other thiol compounds that can provide a better fit to the material system being investigated.

Although we do not have a precise understanding of the mechanisms responsible for the device trends observed here, this experiment allows us to gain new knowledge on how different surface treatment conditions affect device performance and to gain new insight into ways to generate device improvements. Indeed, the experimental results and analyses presented here can provide some useful

Table 6.5 Summary of key observations from contact interface engineering experiments.

	Description	Key observations
Part A	Effect of thiol SAM for contact surface treatment	Thiol improves μ_{FE} for devices without any other pretreatments (compared to untreated devices) Addition of thiol decreases mobility and increases contact resistance for devices with prior dielectric surface treatments Recommendation to omit thiol treatment step for devices considered here. Further optimization of the contact surface treatment process (e.g., processing parameters) or investigation of a more fitting thiol compound should be considered
Part B	Effect of execution sequence of surface treatments	OTS/Thiol configuration (i.e., OTS first and thiol last) gives higher mobility, higher contact angle and smaller surface roughness than Thiol/OTS configuration (i.e., thiol first, OTS last)

guidelines/insights for further optimization of interface modification recipes for OTFTs, with the potential to generate more intriguing device improvements.

References

1. Watkins, N.J., Le, Q.T., Zorba, S., Yan, L., Gao, Y., Nelson, S.F., Cuo, C.S., and Jackson, T.N. (2001) Photoemission characterization of interfaces between Au and pentacene. *Proc. SPIE*, **4466**, 1.
2. Kim, S.H., Lee, J.H., Yang, Y.S., and Zyung, T. (2003) Device fabrications of organic thin-film transistors. *Mol. Cryst. Liq. Cryst.*, **405**, 137.
3. Chen, Y., Shih, I., and Xiao, S. (2004) Effects of $FeCl_3$ doping on polymer-based thin film transistors. *J. Appl. Phys.*, **96** (1), 454.
4. Burgi, L., Richards, T.J., Friend, R.H., and Sirringhaus, H. (2003) Close look at charge carrier injection in polymer field-effect transistors. *J. Appl. Phys.*, **94** (9), 6129.
5. Ishii, H., Sugiyama, K., Ito, E., and Seki, K. (1999) Energy level alignment and interfacial electronic structures at organic/metal and organic/organic interfaces. *Adv. Mater.*, **11**, 605.
6. Crispin, X., Geskin, V., Crispin, A., Cornil, J., Lazzaroni, R., Salaneck, W.R., and Brédas, J.L. (2002) Characterization of the interface dipole at organic/metal interfaces. *J. Am. Chem. Soc.*, **124**, 8131.
7. Koch, N., Elschner, A., Johnson, R.L., and Rabe, J.P. (2005) Energy level alignment at interfaces with pentacene: metals versus conducting polymers. *Appl. Surf. Sci.*, **244**, 593.
8. Halik, M., Klauk, H., Zschieschang, U., Schmid, G., Radlik, W., Ponomarenko, S., Kirchmeyer, S., and Weber, W. (2003) High-mobility organic thin-film transistors based on α,α'-didecyloligothiophenes. *J. Appl. Phys.*, **93** (5), 2977.

9. Lee, T.W., Zaumseil, J., Bao, Z., Hsu, J.W.P., and Rogers, J.A. (2004) Organic light-emitting diodes formed by soft contact lamination. *Proc. Natl. Aacd. Sci.*, **101** (2), 429.
10. Rep, D.B.A., Morpurgo, A.F., and Klapwijk, T.M. (2003) Doping-dependent charge injection in regioregular poly(3-hexylthiophene). *Org. Electron.*, **4**, 201.
11. Laiho, T. and Leiro, J.A. (2006) Influence of initial oxygen on the formation of thiol layers. *Appl. Surf. Sci.*, **252**, 6304.
12. Hiroshiba, N., Kumashiro, R., Komatsu, N., Suto, Y., Ishii, H., Takaishi, S., Yamashita, M., Tsukagoshi, K., and Tanigaki, K. (2007) Surface modifications using thiol self-assembled monolayers on Au electrodes in organic field effect transistors. *Mater. Res. Soc. Symp. Proc.*, **965**, paper 0965-0S08-03.
13. Wang, J., Gundlach, D.J., Benesi, A.J., and Jackson, T.N. (1999) High mobility polymer thin film transistors based on copolymers of thiophene and 3-hexylthiophene. 41st Electronic Materials Conference Digest, p. 16.
14. Kim, S.H., Lee, J.H., Lim, S.C., Yang, Y.S., and Zyung, T. (2004) Improved contact properties for organic thin-film transistors using self-assembled monolayer. *Jpn. J. Appl. Phys.*, **43**, L60.
15. Street, R.A. and Salleo, A. (2002) Contact effects in polymer transistors. *Appl. Phys. Lett.*, **81**, 2887.
16. Servati, P. (2004) Amorphous silicon TFTs for mechanically flexible electronics. PhD thesis. University of Waterloo, Canada.
17. Kawasaki, N., Ohta, Y., Kubozono, Y., and Fujiwara, A. (2007) Hole-injection barrier in pentacene field-effect transistor with Au electrodes modified by $C_{16}H_{33}SH$. *Appl. Phys. Lett.*, **91**, 123518.
18. Ahnood, A., Servati, P., Esmaeili-Rad, M.R., Li, F.M., Sazonov, A., and Nathan, A. (2008) Extraction of non-ohmic contact resistance in nanocrystalline silicon thin film transistors. *Spring MRS Meeting Abstract A16.6. San Francisco, March 24–28, 2008.*
19. Gundlach, D.J., Zhou, L., Nichols, J.A., Jackson, T.N., Necliudov, P.V., and Shur, M.S. (2006) An experimental study of contact effects in organic thin film transistors. *J. Appl. Phys.*, **100**, 024509.
20. Natali, D., Fumagalli, L., and Sampietro, M. (2007) Modeling of organic thin film transistors: effect of contact resistances. *J. Appl. Phys.*, **101**, 014501.
21. Chabinyc, M.L. (2008) Characterization of semiconducting polymers for thin film transistors. *J. Vac. Sci. Technol.*, **B 26**, 445.
22. Kline, R.J., DeLongchamp, D.M., Fischer, D.A., Lin, E.K., Heeney, M., McCulloch, I., and Toney, M.F. (2007) Significant dependence of morphology and charge carrier mobility on substrate surface chemistry in high performance polythiophene semiconductor films. *Appl. Phys. Lett.*, **90**, 062117/1.
23. Kawasaki, M., Imazeki, S., Ando, M., and M., Ohe (2005) Bottom contact organic thin-film transistors with thiol-based SAM treatment. *Proceedings – Electrochemical Society, v PV 2004–2015, Thin Film Transistor Technologies VII – Proceedings of the International Symposium*, p. 249.

Further Reading

Kim, W.K. and Lee, J.L. (2007) In situ analysis of hole injection barrier of O_2 plasma-treated Au with pentacene using photoemission spectroscopy. *Electrochem. Solid State Lett.*, **10** (3), H104.

Kosbar, L.L., Dimitrakopoulos, C.D., and Mascaro, D.J. (2001) The effect of surface preparation on the structure and electrical transport in an organic semiconductor. *Mater. Res. Soc. Symp. Proc.*, **665**, paper C10.6.1.

Lim, S.C., Kim, S.H., Lee, J.E., Kim, M.K., Kim, D.J., and Zyung, T. (2005) Surface-treatment effects on organic thin-film transistors. *Synth. Met.*, **148** (1), 75.

Song, C., Koo, B., Lee, S., and Kim, D. (2002) Characteristics of pentacene organic thin film transistors with gate insulator processed by organic molecules. *Jpn. J. Appl. Phys.*, **41**, 2730.

Veres, J., Ogier, S., Lloyd, G., and de Leeuw, D. (2004) Gate insulators in organic field-effect transistors. *Chem. Mater.*, **16**, 4543.

7 OTFT Circuits and Systems

As discussed in Chapter 1, organic thin film transistors (OTFTs) are aimed for applications requiring large-area coverage, structural flexibility, low-temperature processing, and low cost. Applications envisioned for OTFTs include large-area flexible displays, electronic paper (e-paper), disposable electronics, low-cost and low-end electronic devices such as radio frequency identification (RFID) tags and smart cards [1]. While inorganic semiconductors already serve in many of these systems, all of these applications can benefit from the potential cost reduction and simpler integration offered by organic semiconductor technology. To implement practical organic electronic devices, integration of OTFTs into circuits to perform switching, logic, or amplification functions is compulsory. Some of the desirable functional specifications imposed on OTFTs for circuit applications are reviewed in Section 7.1. This chapter discusses the development of OTFT circuits, with emphasis on two main application fields: active matrix display backplanes and RFID tags; these technologies are reviewed in Sections 7.2.1 and 7.2.2, respectively. A number of OTFT circuits were fabricated, incorporating the technical knowledge acquired from our investigations on gate dielectric, interface treatment, and integration strategies; the experimental outcomes are presented in Section 7.3.

7.1 OTFT Requirements for Circuit Applications

Desirable qualities/characteristics of an OTFT include: a high field-effect mobility, high on/off drain current ratio, low leakage current, minimal threshold voltage shift, and sharp subthreshold slope [2]. These general requirements are discussed in Section 2.2.3. Depending on the specific circuit application, the demands imposed on certain thin film transistor (TFT) parameters are more stringent. Table 7.1 reviews application-specific demands for TFT. For active-matrix liquid crystal displays (AMLCDs), TFT behaves as a switch, and thus, minimal leakage is the most critical requirement. On the other hand, the current drive and on/off current ratio are of great importance in active-matrix organic light-emitting diode (AMOLED) displays, since the TFT must provide sufficient current output to drive the organic light-emitting diode (OLED) for light emission. For RFID tags, speed is

Organic Thin Film Transistor Integration: A Hybrid Approach, First Edition. Flora M. Li, Arokia Nathan, Yiliang Wu, and Beng S. Ong.

Table 7.1 A concise comparison of OTFT characteristics demanded by various applications.

OTFT parameter	AMLCD (TFT as switch)	AMOLED (TFT as active driver)	RFID tags, smart cards
Operating speed	Not as critical	Not as critical	High$^{\diamond}$
Leakage (I_{leak})	Low$^{\diamond}$	Low$^{\diamond}$	Low
Current drive (I_{ON})	Not as critical	Large$^{\diamond}$	Large
Threshold voltage (V_T)	Low	Low	Low
On/off ratio (I_{ON}/I_{OFF})	Large	Large	Large
Stability	High	High	High

The most critical parameter(s) in each application is identified by the symbol $^{\diamond}$.

the most critical parameter. In most cases, the threshold voltage (V_T) should be as low as possible for large dynamic range (in displays) and for low voltage operation. Good device stability is demanded in all applications. These application-specific requirements can be addressed via choice of material, device structure, and/or the associated fabrication method, as analyzed in this section. In particular, the discussion focuses on the speed, current-driving capability, leakage current, and stability requirements for OTFTs.

7.1.1
Speed

In RFID and memory devices, the operating speed of the integrated circuit (IC) is a critical parameter. OTFT circuits must operate sufficiently fast to deliver the data rates demanded by these applications. The maximum speed at which a transistor circuit can operate is limited by the time it takes for the charge carriers to transit from the source contact through the channel to the drain contact [3]. This is known as the *transit time* (τ), which is also described as the time it takes for the accumulation layer to be emptied of charges through the drain after V_G has been switched off. Transit time is often expressed using the maximum switching frequency(f_{max}):

$$\frac{1}{\tau} = f_{max} \approx \frac{\mu \cdot V_{DS}}{L^2} \tag{7.1}$$

Note that the value of f_{max} calculated with the above formula provides an upper bound. In practice, f_{max} is lower due to parasitic capacitances and other non-ideal effects.

For an OTFT based on a soluble polymer semiconductor, assuming $\mu_{FE} = 10^{-1}$ $cm^2\,V^{-1}\,s^{-1}$, $V_{DS} = 10\,V$, and $L = 10\,\mu m$, then $f_{max} \approx 1$ MHz. Other scenarios are summarized in Table 7.2. Therefore, the upper speed limit for these OTFTs is of the order of 1 MHz due to limitations on transistor mobility, gate length, and patterning methods. (For comparison, inorganic electronics work with gigahertz

Table 7.2 Calculation of maximum intrinsic switching frequency for OTFTs.

Scenario	μ_{FE} ($cm^2\ V^{-1}\ s^{-1}$)	V_{DS} (V)	L (μm)	Calculated f_{max}
1 (e.g., poly(3-hexylthiophene) (P3HT), shadow mask)	0.01	10	100	1 kHz
2 (e.g., P3HT)	0.01	10	10	100 kHz
3 (e.g., PQT-12)	0.1	10	10	1 MHz
4 (e.g., pentacene)	1	10	10	10 MHz

frequencies. Thus, it is clear that organic electronics should be aimed for low-cost and low-end products rather than high-performance electronics.) From an application perspective, the refresh rate of reflective displays is around 40–85 Hz, thus OTFT technology should be able to fulfill these requirements. In the case of RFID tags, the logic circuit component operates around 100 kHz, which is deliverable by existing OTFTs. However, the rectification stage of RFID tags is expected to operate at 13.56 MHz, thus faster OTFTs are needed.

As indicated by Equation 2.4, high μ_{FE} and small L are required to improve speed (i.e., increase f_{max}). μ_{FE} is largely determined by the quality and microstructure of the organic semiconductor material. A higher degree of molecular ordering results in a higher μ. Researchers have been actively working on enhancing mobility via a variety of approaches, including the synthesis/development of new materials, improving interfaces and material systems, and so on [4]. Since $f_{max} \propto 1/L^2$, reducing L should have a large impact on speed improvement. L is usually dictated by the fabrication technique or processing technology. In a conventional lateral OTFT structure, it is difficult to define L less than 5 μm as it becomes challenging to delineate the source and drain contacts without extensive overlap with the gate (ideally, the overlap should be minimized to ensure low parasitic capacitance) [3]. Researchers are actively exploring innovative processing techniques to overcome these resolution limitations. A vertical OTFT structure, as shown in Figure 7.1, can provide an alternative means of achieving L in the submicron scale. In a vertical OTFT, L is defined by the dielectric film thickness, which can be readily controlled by precise timing of the deposition process. Therefore, a vertical OTFT structure can facilitate further improvement in switching speed. A preliminary demonstration of a vertical OTFT can be found in Refs. [5, 6].

It should be noted that Equation 2.4 represents an upper limit for f_{max}. Parasitic capacitances (e.g., overlap capacitance from the overlap of the source/drain electrodes with the gate electrode) may be present in the transistor, which charge/discharge when V_{GS} is switched on/off, thus limiting the switching frequency of an OTFT. In order to maximize speed, it is important to minimize these parasitic capacitances,

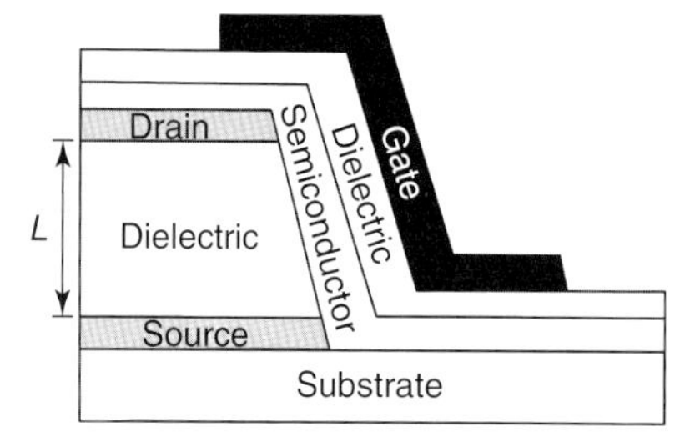

Figure 7.1 Cross sectional diagram of a vertical OTFT structure.

via proper device design, an optimized patterning procedure, and suitable material choices. In addition to achieving maximal speed, other desired OTFT characteristics for RFID applications include: small overlap capacitance, good off-current behavior (important since RFID tags are power-constrained devices), and good stability (important for analog circuit operation).

7.1.2 Leakage

In active matrix display backplanes (e.g., AMLCD, AMOLED), the desirable OTFT characteristics include low off-current (i.e., leakage), low gate leakage current, sufficient drain current, small gate overlap capacitance, and good device uniformity [2]. Minimum leakage current is particularly important in order to retain charge in the pixel during the "off" state of the transistor. To minimize leakage current, it is important to control the quality of the gate dielectric, and minimize the interface state density and interfacial stress. Also, isolation between transistors on the same substrate is needed by patterning the semiconductor layer, in order to reduce parasitic leakage and cross-talk.

7.1.3 Current Drive Capacity

In AMOLED displays, the output current drive of the OTFT is also critical. Since OLEDs are current-driven devices, the OTFT must have sufficient output current to provide the necessary brightness for the OLED element. To get a large output drain current, the transconductance (g_m) of the transistor must be considered. g_m represents the change in I_D for a given change in V_G. For large g_m (and thus large output current), μ_{FE} and the W/L ratio of the transistor should be large, as dictated by the following relationship:

$$g_m = \left(\frac{\partial I_D}{\partial V_G}\right)_{V_D = \text{const}} = \frac{WC_i}{L}\mu_{FE} V_D \tag{7.2}$$

Alternatively, since $C_i = \varepsilon_0 \varepsilon_r A/d$, choosing a high-$k$ dielectric or using a thinner gate dielectric can increase C_i, thus promoting larger g_m.

7.1.4
Stability

Most OTFT applications demand good device stability. This necessitates maintaining good stability of the organic material, even under environmental stress or operational stress. Most organic materials are sensitive to environmental parameters, including air, humidity, moisture, oxygen, light, and so on. Tailoring the chemical structure of the organic material during synthesis is a possible route to improving the intrinsic environmental stability. Encapsulation methods provide an extrinsic technique to address stability issues. Operational instability in OTFTs, such as threshold voltage shift (ΔV_T), should also be taken into account. ΔV_T is often observed in OTFTs, and becomes significant after prolonged gate bias. ΔV_T is typically caused by the creation of metastable defects in the band-gap, and charge trapping in the gate and passivation dielectric layers. The extent of charge trapping is determined by the quality of the dielectric layer and the dielectric/semiconductor interface state density. The interface treatment techniques, examined in Chapters 5 and 6, provide possible routes to enhancing the interface quality, and perhaps improving device stability.

7.2
Applications

7.2.1
Displays

One of the most promising applications of OTFTs is their use as an on-pixel switching element in active-matrix displays, similar to the functions that are currently fulfilled by a-Si:H TFTs in AMLCD and AMOLED display applications. Active-matrix backplane electronics based on OTFTs are attractive because of the large-area capability and low-cost advantage of organic technology, as well as the opportunity to realize a new generation of flexible, lightweight displays, and e-paper [7]. The incorporation of OTFTs in active-matrix displays has recently been demonstrated with a polymer-dispersed liquid crystal or an electrophoretic material as a display element [8, 9]. Development of OTFT-driven AMOLED displays is also in progress, and integrated OTFT-OLED smart pixels have been reported [7, 10, 11]. OTFT-OLED integration is appealing because it suggests the potential to manufacture inexpensive and flexible display modules with completely functional organic materials.

Table 7.3 summarizes some of the display prototypes demonstrated using OTFTs; most of the existing prototypes were built using pentacene OTFTs. The e-paper and liquid crystal display (LCD) were made with an OTFT matrix array and the AMOLED with dot patterns [12]. Since the OTFT serves as an active element, the AMOLED is very sensitive to non-uniformity in the OTFT performance across the array, which can lead to non-uniformity in the display's brightness. The non-uniformity often

Table 7.3 Reported display prototypes using OTFTs [12].

Application	Semi conductor	Specification	Author organization	References
E-paper	Polyfluorene-based polymer (inkjet printing)	60 × 80 pixel on poly(ethylene terephthalate) (PET)	Plastic Logic (UK) and E-ink (USA)	[13]
E-paper	Pentacene (solution-process)	QVGA on poly (ethylene naphthalate) (PEN)	Philips (Netherlands)	[8]
LCD	Pentacene	1.4 in 80 × 80 RGB on glass	Hitachi (Japan)	[14]
LCD	Pentacene	64 × 128 on plastic	ERSO/ITRI (Taiwan)	[15]
LCD	Pentacene	12 in full color XGA on glass	Samsung Elec. (Korea)	–
OLED	Pentacene	8 × 8 pixels on glass	Pioneer (Japan)	[16]
OLED	Pentacene	4 × 4 pixels on PC	NHK (Japan)	[17]

arises from the grain size distribution of polycrystalline organic semiconductors. On the other hand, OTFTs are more accommodating for LCD or e-paper because they act mainly as switches and the key performance requirement is a high on/off current ratio.

In this chapter, we demonstrate OTFT pixel circuits for active matrix display backplanes based on the hybrid OTFT fabrication schemes. The circuit designs were inspired by existing pixel circuits built using a-Si:H TFTs [9]. Some of the pixel circuit architectures and the results for OTFT-based pixel circuits are discussed in Section 7.3.5.

7.2.2 RFID Tags

This section provides a review of RFID tags, along with design considerations, although no experimental results are presented using the hybrid OTFT structures discussed here. In recent years, there has been significant interest in the development of RFID tags for their versatile detecting and tracking capabilities. RFID technology provides an automatic way to collect product, place, time, or transaction data quickly and easily without human intervention or error. This technology is expected to dramatically improve automation, inventory control, distribution, shipment, tracking, and purchasing operations, provided it is cheap enough to be widely deployed [18]. Depending on the intended application, the RFID system will operate at different frequency bands, as listed in Table 7.4. In general, a higher system frequency allows for a longer read range, but with the trade-off of higher cost. For lower-end applications, RFID tags are viewed as a promising alternative

Table 7.4 Frequency bands and applications of RFID systems [18].

Frequency band	RFID system characteristics	Example applications
Low: 100–500 kHz	Short read range, inexpensive	Access control, animal identification, inventory control
Intermediate: 10–15 MHz (13.56 MHz)	Medium read range	Access control, smart cards
High: 850–950 MHz, 2.4–5.0 GHz	Long read range, high reading speed, line of sight required, expensive	Railroad car monitoring, toll collection systems

to today's barcode technology. RFID tags can provide more comprehensive data collection/storage and can be read from a distance; in contrast, a barcode provides limited identification information and requires in-line detection [18].

A typical RFID system consists of a "reader" that uses an antenna to transmit radio energy to interrogate a "tag" or "transponder". In its simplest form, an RFID tag is composed of an antenna attached to an IC. The IC carries information that identifies an item to which the tag is attached. Antenna coils are used to magnetically couple the RF energy from the reader into the tag. This energy, emitted by the reader, is used to power the tag and provide bi-directional communications between the reader and the tag. After extracting the data stored in the tag's IC, the retrieved information is directed back to the reader, from where it can be fed to a computer for processing [18]. The set-up of a simple RFID system is shown in Figure 7.2a. Figure 7.2b describes the major functional blocks inside an RFID tag.

Organic RFID tags are still in their early stage of development, with some initial progress reported in Ref. [19–22]. They operate in the low to medium frequency ranges (see Table 7.4). For these lower-end applications, the logic circuitry of an RFID tag (i.e., "Control Logic" component in Figure 7.2b) usually operates in the vicinity of 100 kHz [19]. These data rates are expected to be deliverable by OTFTs. However, the RFID tag has a front end (i.e., "RF Interface" in Figure 7.2b) that must handle rectification and operate at the frequency of the incoming RF signal (e.g., 13.56 MHz). The rectification stage at the interface needs to convert the absorbed RF energy into DC power to run to the entire RFID tag. Designing organic circuits to operate at this high frequency is a major challenge; this will require clever and innovative designs to overcome the speed limitations of OTFTs.

Figure 7.3 illustrates a simple approach to implementing the rectification stage of RFID tags which uses diodes in a half-wave rectifying configuration. This configuration is commonly used in silicon-based RFID tags. However, organic semiconductor based diodes often fall short of the required performance necessary to establish sufficient rectification at high frequencies (>100 kHz) and high voltages. In many cases, the peak input voltage in a tag can reach well over 50 V

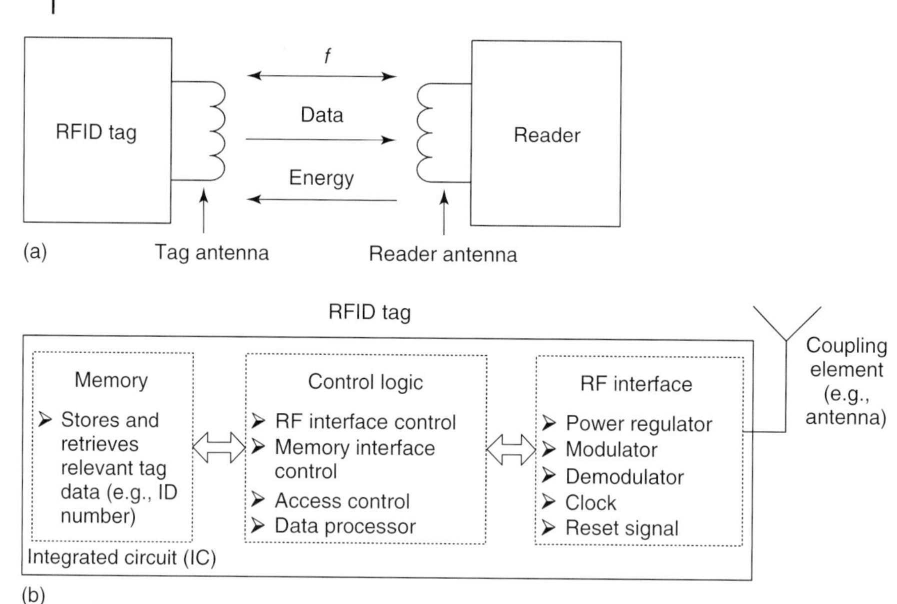

Figure 7.2 (a) Typical set-up of a RFID system. (b) Key modules/components of a RFID tag [18, 19].

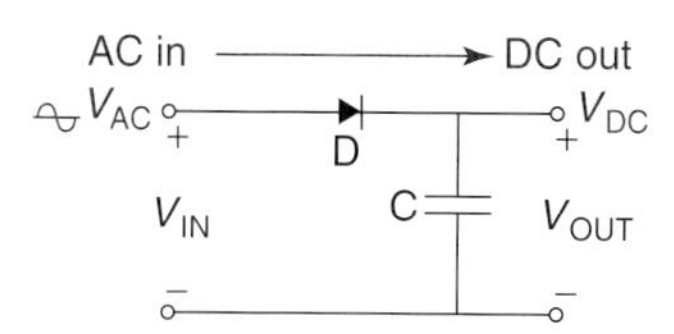

Figure 7.3 A simple diode-based half-wave rectification stage for RFID tag applications [20].

[20]. Thus, the rectification stage has to provide sufficient breakdown resistance at these levels, while maintaining a sufficient output voltage to operate the tag's logic circuitry. In order to provide sufficient forward bias current to run the circuit, the diode needs to be relatively large to compensate for the low vertical mobilities seen in most organic diodes. Rectification stages based on organic diodes have been reported by PolyIC [22].

To address the limitations of organic-based diodes, the company 3M introduced an AC powering scheme for the tag that eliminated the rectification stage [20]. This concept is illustrated in Figure 7.4, where the 1-bit OTFT RFID tag uses AC power from a tuned antenna to directly power the logic circuitry, without an intermediate rectification stage. The RF interface of the tag consists of an LC resonant tank, which absorbs RF energy from the reader antenna to generate the AC power required for the tag circuitry. A basic property of the LC tank is its ability to store energy in the form of AC power that oscillates/resonates at a particular frequency,

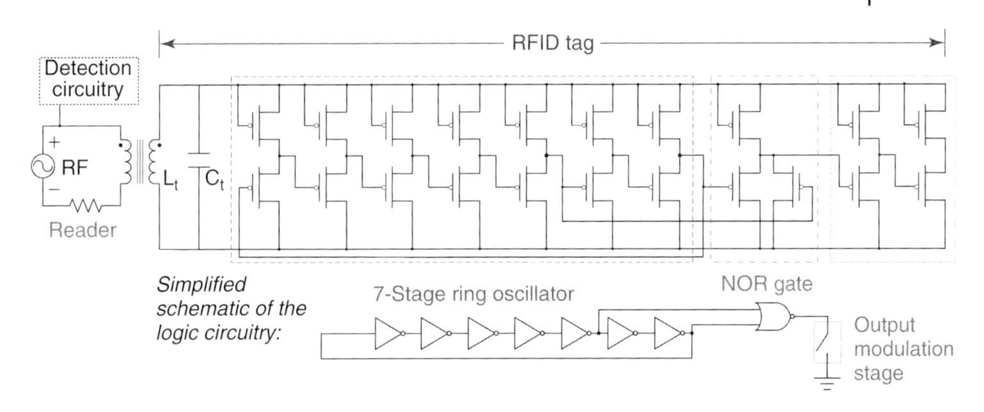

Figure 7.4 Circuit schematic of a 1-bit OTFT-based RFID tag. Power is coupled into the tag circuit using inductive coupling between the reader and the tag. L_t and C_t form a resonant tank that powers the tag circuitry directly, without a separate rectification stage. The lower figure shows a scaled-down schematic of the logic circuitry [20].

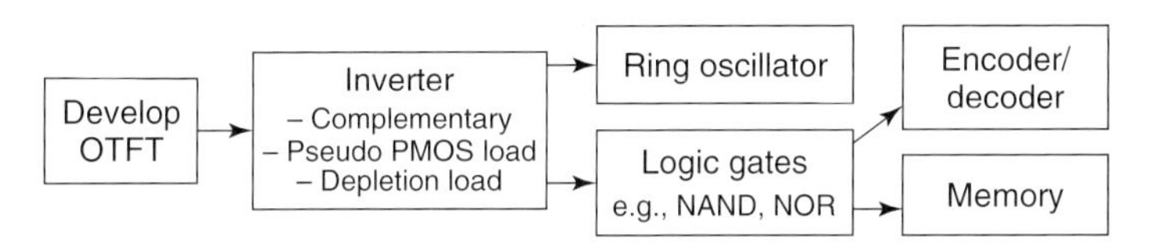

Figure 7.5 Development strategy for OTFT-based RFID tag circuitry.

$f = 1/\left(2\pi\sqrt{LC}\right)$. A ring oscillator and NOR gate, powered directly from the LC tank, generates a pulse signal whenever the two input nodes of the NOR gate are at logic level '0'; this occurs once per cycle. This pulse signal is then buffered and sent to a large inverter at the output modulation stage. This output stage serves to amplitude modulate the RF signal absorbed by the tag's antenna according to the data stored in the tag. The modulated signal is subsequently detected by the reader, and relevant information can be retrieved. This 1-bit OTFT RFID tag design has been realized using pentacene OTFTs patterned with polymeric shadow masks and powered by near-field coupling at a RF of 125 kHz [20]. An 8-bit RFID transponder circuitry, based on an extension of this design, has also been reported [20]. Higher operating frequencies can be achieved with an RFID tag using this AC powering scheme, when compared to an RFID tag that uses organic diodes in a rectification stage configuration.

Figure 7.5 outlines the development of OTFT-based RFID circuitry, in which the key circuit components (as identified in Figure 7.2b) are included. Integration of these components, with an appropriate RF interface and antenna, constitutes the next step to realizing organic-based RFID tags.

7.3
Circuit Demonstration

The integration of OTFTs for circuit applications necessitates a well-developed and reliable fabrication process that can consistently produce OTFTs with good performance, careful selection of material systems for optimal device performance, a systematic approach for device characterization, and a comprehensive model that can predict device behavior. Synergy between these research efforts (with results reported in the preceding chapters) enables the fabrication of the simple OTFT circuits presented in this section. Preliminary experimental results are presented for inverters, current mirrors, ring oscillators, and various display pixel circuits to demonstrate the feasibility of our integration approach. Detailed circuit characterization, analysis, and optimization are needed to enable the design and fabrication of more complex and application-specific circuits.

7.3.1
Fabrication Schemes

The OTFT circuits reported here were implemented with fully-patterned bottom-gate bottom-contact PQT-12 OTFTs. These circuits were fabricated using three processing schemes:

- **Scheme 1**: four-mask photolithography scheme with direct patterning of the organic semiconductor layer, where photoresist was deposited directly on the organic semiconductor for photolithographic patterning (see Figure 3.9).
- **Scheme 2**: four-mask photolithography scheme with indirect patterning of the organic semiconductor layer, where the organic semiconductor was passivated with a parylene buffer layer prior to photolithographic patterning (see Figure 3.9).
- **Scheme 3**: hybrid photolithography–inkjet printing scheme, where the organic semiconductor layer was deposited/patterned by inkjet printing (see Figure 3.14).

Detailed discussion of these fabrication schemes is presented in Chapter 3. Each method has its own strengths and limitations. Device performance may vary depending on the fabrication scheme used, which can lead to slight variations in the circuit characteristics. Figure 7.6 is a photograph of an array of OTFT circuits fabricated using the photolithography scheme with indirect patterning using parylene on a 3 in glass wafer.

Please note the main objective of the following results is to demonstrate functional OTFT circuits, evaluate the initial performance, and study the feasibility of the fabrication approaches. The intent is to compile and integrate the knowledge/experience gained throughout this doctoral research on fabrication techniques, dielectric material optimization, interface modification, and device characterization to produce organic circuits for practical applications. The results reported here are the first demonstrations, to date, of PQT-12 OTFT circuits. As such, circuit performance has not been optimized, and there is definitely room for improvement.

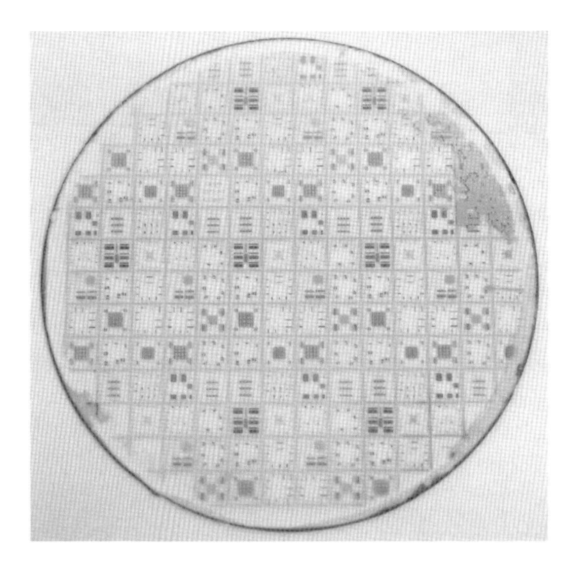

Figure 7.6 Photograph of a variety of OTFT circuits on a 3 in glass wafer, fabricated using the integration schemes presented here. The photograph illustrates circuits fabricated using a photolithography approach with parylene passivation (scheme 2).

7.3.2
Inverters

The simple inverter circuit is a key building block for many digital electronic circuits and for analog amplifying circuits. An inverter circuit outputs a voltage representing the opposite logic-level or polarity as its input. Figure 7.7a displays a circuit schematic and top-view photograph of a p-type silicon metal oxide semiconductor (PMOS) inverter with saturated load, built using a p-channel PQT-12 OTFT; its cross-sectional diagram is shown in Figure 7.7c to illustrate the interconnections. An alternative inverter design with a depletion-mode transistor is displayed in Figure 7.7b. In both designs, T1 serves as the driver (or input transistor) and T2 acts as the load. In the ideal case, T1 is a transistor with infinite off resistance (for $|V_{GS}| < |V_T|$) and a finite on-resistance (for $|V_{GS}| > |V_T|$). When V_{IN} is at a voltage that turns T1 on (e.g., $V_{IN} = -40\,V = V_{SS}$), V_{OUT} is pulled up toward V_{DD} (i.e., V_{IN} = low, V_{OUT} = high). When V_{IN} is at a voltage that turns T1 off (e.g., $V_{IN} = 40\,V = V_{DD}$), V_{OUT} is pulled down toward V_{SS} (i.e., V_{IN} = high, V_{OUT} = low). Thus, the inverter circuit operates in two stable states.

The choice of load affects the inverter characteristics (e.g., gain, voltage level, noise margin) [23, 24]. For the circuit in Figure 7.7a with a saturated load, $V_{GS\_Load} = V_{DS\_Load}$; thus, the load transistor is always in saturation ($|V_{GS} - V_T| < |V_{DS}|$) and gives a relatively constant/steady drain current. For the circuit in Figure 7.7b with depletion load, $V_{GS\_Load} = 0$; the load transistor is always "on" since the p-type OTFT

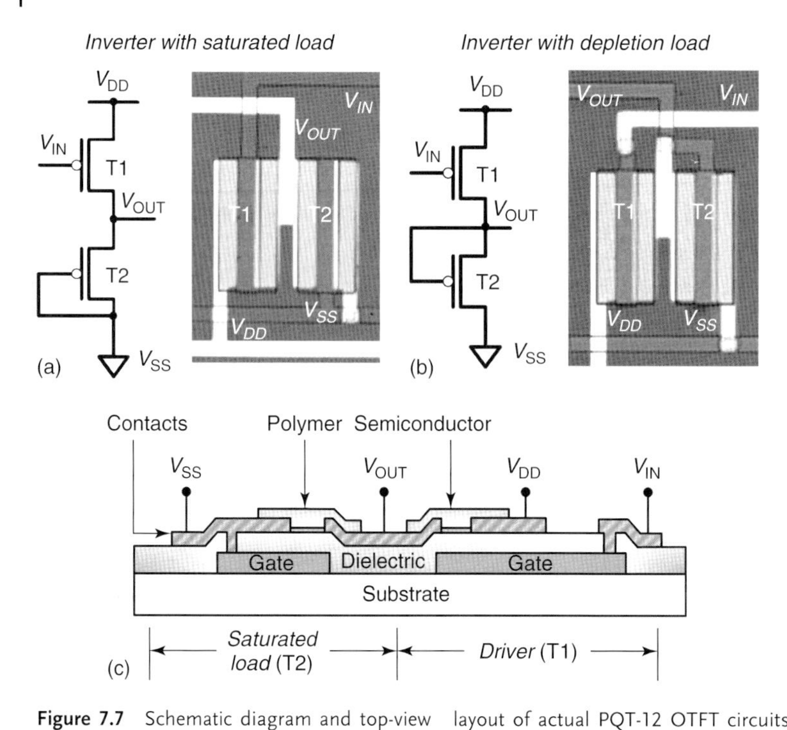

Figure 7.7 Schematic diagram and top-view photograph of OTFT inverter circuits with (a) saturation load and (b) depletion load. (c) Cross-section diagram of an inverter with saturated load. The photographs in (a) and (b) correspond to the top-view layout of actual PQT-12 OTFT circuits ($W/L = 200\,\mu m/25\,\mu m$ for both T1 and T2) fabricated by a photolithography approach with parylene buffer/passivation layer (scheme 2).

typically has a positive turn-on voltage. Since $V_{GS\_Load} = 0$ and V_{DS_Load} changes approximately between "$V_{DD} - V_{SS}$" and V_{SS}, the load transistor is subjected mainly to drain stress, which should experience a much weaker bias-induced V_T shift (compared to gate-induced bias stress) [25]. Reduction of bias-induced stress effects is crucial for the stability and lifetime of OTFT circuits. Please refer to Refs. [23, 24] for detailed circuit analysis and comparison of the different inverter designs.

Figures 7.8 and 7.9 show the transfer voltage characteristics of a PQT-12 inverter with saturated load and with depletion load, respectively. These devices were fabricated using the photolithography scheme with photoresist passivation. Inversion operation is evident from these curves, and the inverter has sufficiently large gain. However, due to the positive switch-on voltage of PQT-12 OTFTs, the input and output levels do not match. Threshold voltage shift, leakage current, and the non-ideal subthreshold behavior of the OTFTs also contributed to deviation from the ideal inverter characteristics. Hysteresis was observed in the inverter characteristics, possibly due to mobile charges in the gate dielectric and material instability.

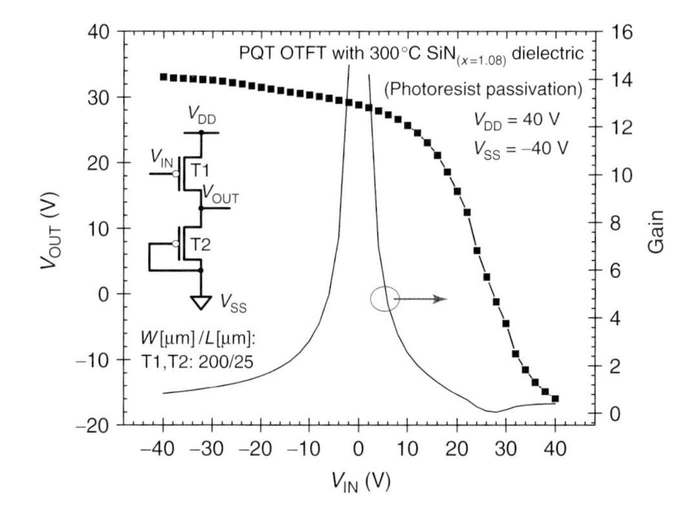

Figure 7.8 Voltage-transfer characteristics of a p-type inverter circuit with saturation load, fabricated using photolithographically-defined PQT-12 OTFT with photoresist as passivation (scheme 1).

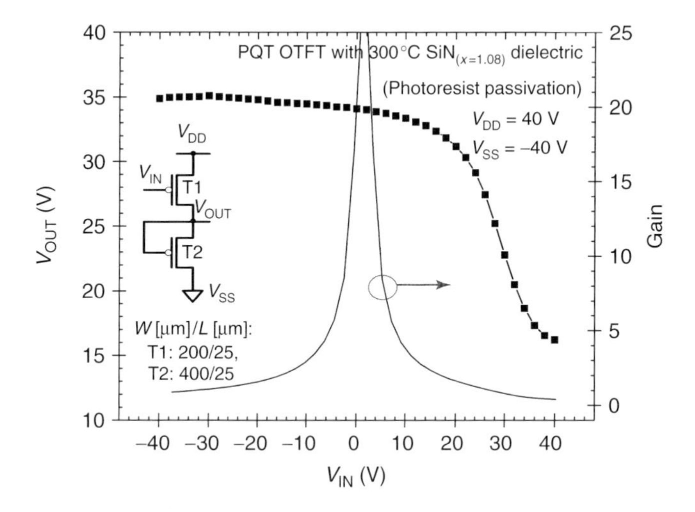

Figure 7.9 Voltage-transfer characteristics of a p-type inverter circuit with depletion load, fabricated using photolithographically-defined PQT-12 OTFT with photoresist as passivation (scheme 1).

Figures 7.10 and 7.11 present the voltage transfer characteristics of the PQT-12 OTFT inverter with saturated load, fabricated using photolithography with a parylene buffer/passivation layer and by an inkjet-printed PQT-12 layer, respectively. These results demonstrate the feasibility of fabricating functional OTFT inverter circuits using the various processing approaches developed in this research. The

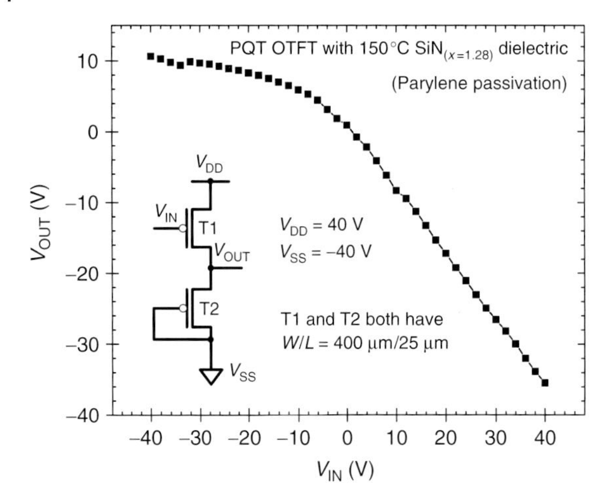

Figure 7.10 Voltage-transfer characteristics of a p-type inverter circuit with saturation load, fabricated using the photolithography approach with parylene passivation (scheme 2).

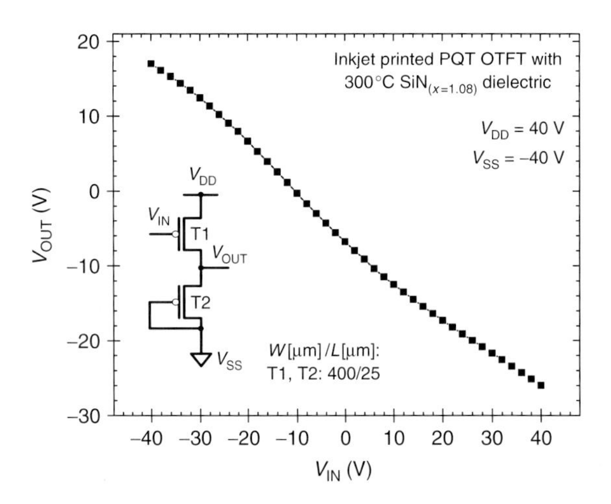

Figure 7.11 Voltage-transfer characteristics of a p-type inverter circuit with saturation load, fabricated with inkjet printed PQT-12 layer (scheme 3).

non-idealities observed in these electrical characteristics are due to lack of process optimization in these preliminary devices.

In summary, functional PQT-12 OTFT inverters are demonstrated here. To the author's knowledge, this is one of the first reports of a PQT-12 inverter circuit in the scientific literature. These initial demonstrations provide a basic idea of the OTFT inverter characteristics. Device parameters can be extracted from these early experiments and can be used to generate device models for circuit analysis and simulation purposes. With the ability to predict device and circuit behavior, proper circuit design techniques can be used (e.g., transistor sizing, layout) to create circuits with desirable inverter characteristics. Alternate circuit topography can be considered to address the shortcomings of OTFTs. For instance, to allow circuits to operate over a wide voltage range, regardless of switch-on voltage and hysteresis, inverters with active load, and integrated level shifting can be considered [26]. A complementary metal oxide semiconductor (CMOS)-type inverter, implemented with complementary p-type and n-type OTFTs, is expected to lead to improved device performance. Ambipolar OTFTs are a promising choice for implementing complementary logic circuits [16].

7.3.3 Current Mirrors

Conceptually, an ideal current mirror is simply an ideal current amplifier. Figure 7.12 illustrates the simplest configuration of a metal oxide semiconductor (MOS) current mirror. In the ideal case,

$$\frac{I_{OUT}}{I_{IN}} = \frac{(W/L)_2}{(W/L)_1} \tag{7.3}$$

where $(W/L)_1$ and $(W/L)_2$ are the channel width to channel length ratios for T1 and T2, respectively. The ideal case assumes T1 and T2 are well matched (i.e., $k = \mu_{FE} C_i$ is identical) and $V_{DS,1} = V_{DS,2}$.

Figures 7.13 and 7.14 display the transfer characteristics of PQT-12 OTFT current mirror circuits. These preliminary devices demonstrated "current mirror" or "current amplification" effects. The circuit in Figure 7.13 was constructed with two transistors of identical geometries, thus, theoretically, $I_{OUT}/I_{IN} = 1$. The experimental value for I_{OUT}/I_{IN} is 0.927, which is close to the theoretical value. The circuit in Figure 7.14 employed transistors with $(W/L)_2/(W/L)_1 = 2$; thus,

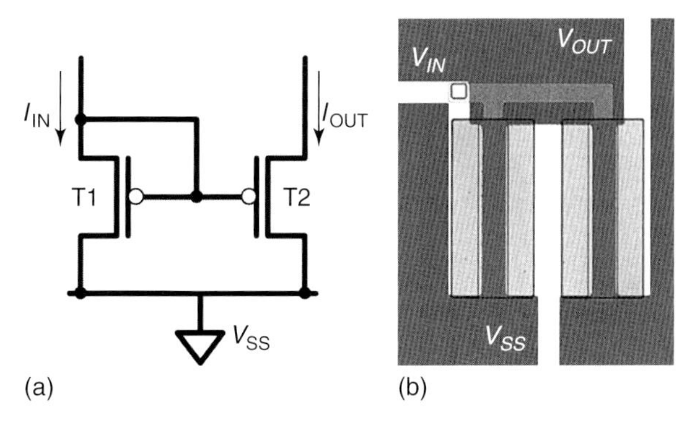

Figure 7.12 Current mirror: (a) circuit schematic and (b) micrograph showing top view of an actual circuit fabricated using hybrid photolithograph–inkjet method (scheme 3). Photo shows $W/L = 200/25$ for both T1 and T2.

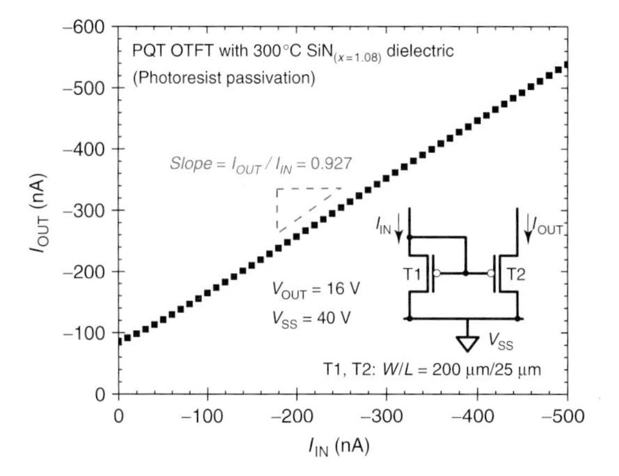

Figure 7.13 Transfer characteristic of current mirror circuits built using PQT-12 OTFTs on SiN_x gate dielectric with $(W/L)_2/(W/L)_1 = 1$. The circuit was fabricated using the photolithography approach with photoresist passivation (scheme 1).

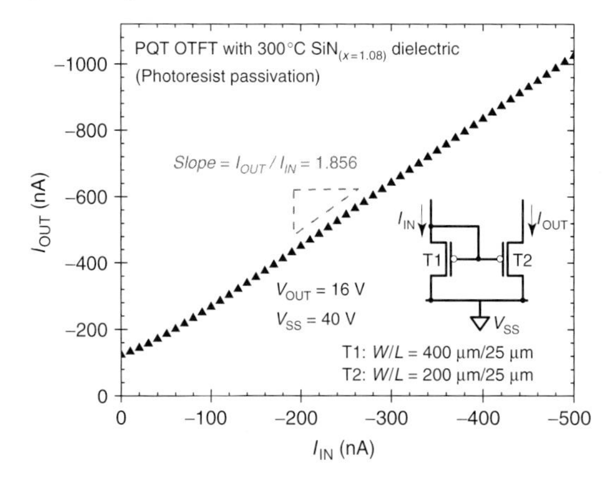

Figure 7.14 Transfer characteristic of current mirror circuits built using PQT-12 OTFTs on SiN_x gate dielectric with $(W/L)_2/(W/L)_1 = 2$. The circuit was fabricated using the photolithography approach with photoresist passivation (scheme 1).

theoretically, $I_{OUT}/I_{IN} = 2$. This is well-matched with the experimental value of 1.856. Therefore, proper current amplification function was achieved with these OTFT current mirror circuits. The deviation of the measured data from ideal transfer characteristics may be due to imperfections in the OTFT devices (e.g., large V_T, leakage current, poor subthreshold slope, parasitic leakage, parasitic capacitances), and process-induced mismatches between devices (e.g., material non-uniformity, offset in mask alignment during photolithography exposure). Further process and design optimization should deliver improved device performance.

7.3.4
Ring Oscillators

A ring oscillator is composed of an odd number of inverter stages, whose output oscillates between two voltage levels, representing high (V_{DD}, logic 1) and low (V_{SS}, logic 0). The inverters are connected in a chain, and the output of the last inverter is fed back into the first, as shown in Figure 7.15. Theoretically, a ring oscillator only requires power to operate; above a certain threshold voltage, oscillations begin spontaneously. The frequency of the oscillation depends on the gate delay of the inverter. To increase the frequency of oscillation, one may increase the applied voltage, use a smaller ring oscillator, adjust the W/L of the transistors, or employ techniques to reduce the gate delay of the inverter circuit. A ring oscillator circuit is commonly used to test the device delay for a given process. It can also be used to provide clocking functionality, a key element in RFID tags.

Figure 7.16 illustrates the ring oscillator circuits fabricated using PQT-12 OTFTs. Two types of inverter circuits are shown for the implementation of a five-stage ring

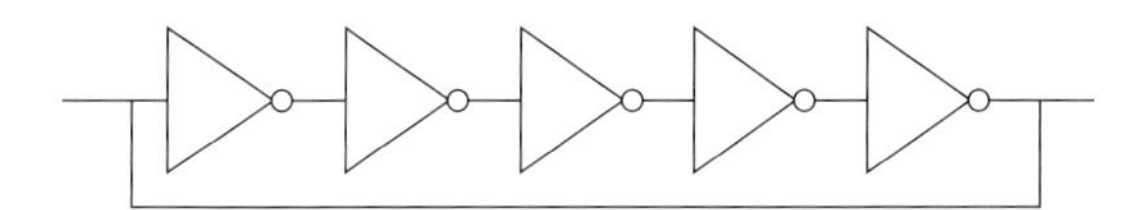

Figure 7.15 Five-stage ring oscillator symbol representation.

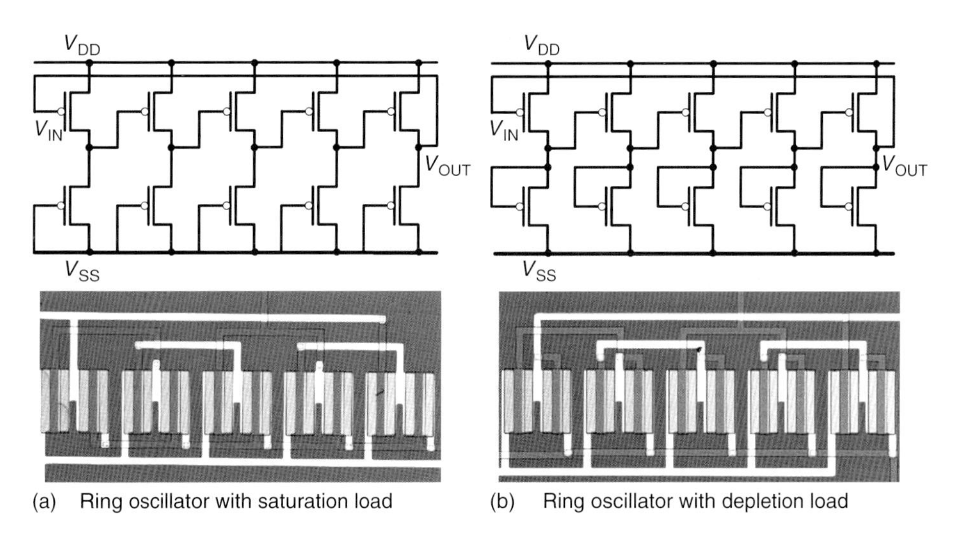

Figure 7.16 (a) Five-stage ring oscillator with saturation load. (b) Five-stage ring oscillator with depletion load. Circuit schematic and micrograph of the top-view of the fabricated structure (using scheme 2) are also shown ($W/L = 200\ \mu m/25\ \mu m$ for the transistors shown).

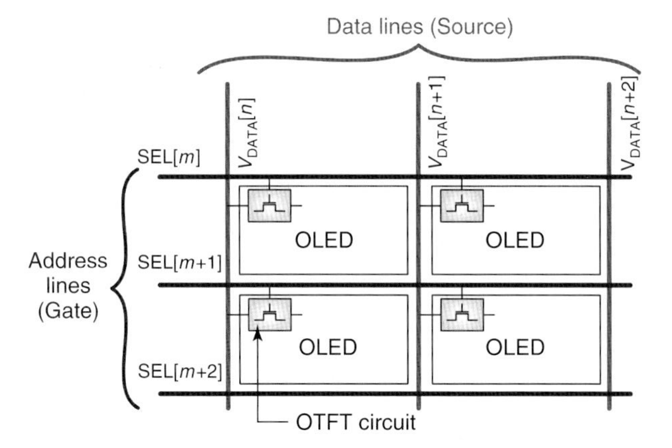

Figure 7.17 Active-matrix addressed backplane.

oscillator: (i) an inverter with saturation load, and (ii) an inverter with depletion load. For high accuracy measurements, a higher sensitivity and lower capacitance probe is needed to avoid capacitive loading effects that might restrict the accuracy of the measurements. An output buffer stage can be implemented to improve the sensitivity of the circuit/measurement set-up.

7.3.5 Display Pixel Circuits

Active matrix addressing, which is needed for high information content formats, involves a layer of backplane electronics based on TFTs to provide the bias voltage and drive current needed in each OLED pixel [9]. Figure 7.17 illustrates the concept of an AMOLED display backplane. An OTFT backplane is a feasible candidate for small-area displays such as those needed in pagers, cell phones, and other mobile devices, where performance and speed requirements are less stringent. The lower mobility associated with OTFTs can be compensated by scaling up the drive transistor in the pixel to provide the needed drive current, without necessarily compromising the aperture ratio. A number of AMOLED pixel circuits, originally designed for a-Si:H TFT backplanes, are adapted to demonstrate the feasibility of OTFT pixel circuits in this research.

7.3.5.1 Conventional 2-TFT Pixel Circuit

The simplest pixel driver circuit is the 2-TFT voltage-programmed circuit shown in Figure 7.18. T2 serves as the switching transistor and T1 serves as the drive transistor. During the programming cycle, SEL is at a voltage that turns T2 on, and the voltage of V_{DATA} is transferred to C_S and charges up the gate node of T1

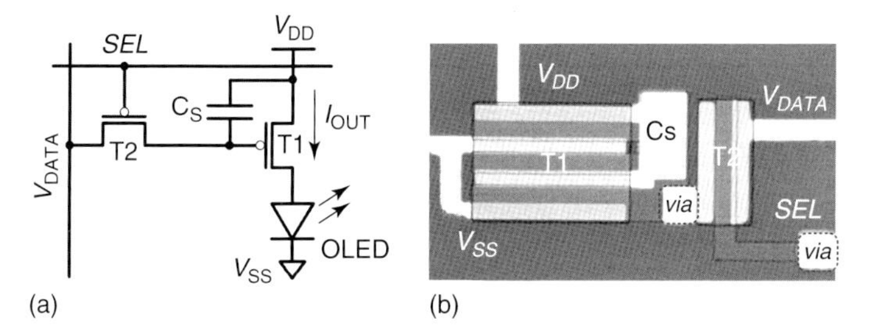

Figure 7.18 Conventional 2-TFT AMOLED pixel circuit: (a) schematic diagram, (b) micrograph of the top-view of the fabricated structure (using scheme 2).

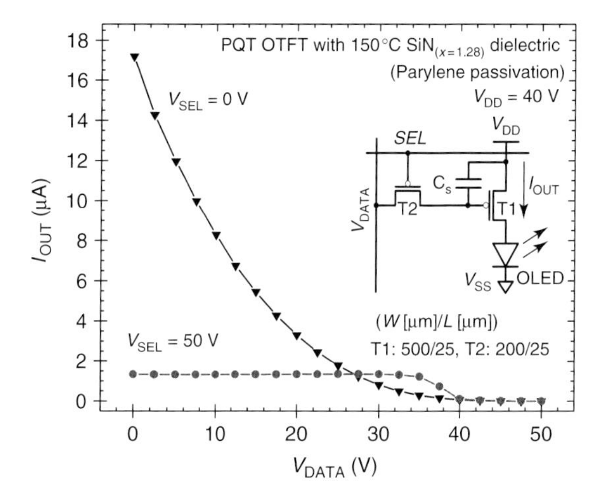

Figure 7.19 Current–voltage characteristic of a conventional 2-T pixel circuit for AMOLED, built using PQT-12 OTFTs on SiN_x gate dielectric. Circuit was fabricated using the photolithography approach with parylene passivation (scheme 2).

to a certain programming voltage (V_p). During the driving cycle, SEL is low and a current related to the programming voltage passes through the OLED.

Figure 7.19 demonstrates the static operation of a conventional 2-T pixel circuit constructed with PQT-12 OTFTs, fabricated using a photolithographically defined polymer semiconductor layer with a parylene buffer/passivation layer. When $V_{SEL} = 0$ V, T2 turns on and V_{DATA} is passed to C_S; this charges up C_S and turns on T1. As a result, T1 generates a current I_{OUT} to drive the OLED and the pixel is "on". On the other hand, when $V_{SEL} = 50$ V, T2 turns off; T1 should be off theoretically, and hence a smaller I_{OUT} was measured. In this case, the pixel is considered "off". However, I_{OUT} during the off-state (i.e., $V_{SEL} = 50$ V) may originate from a number

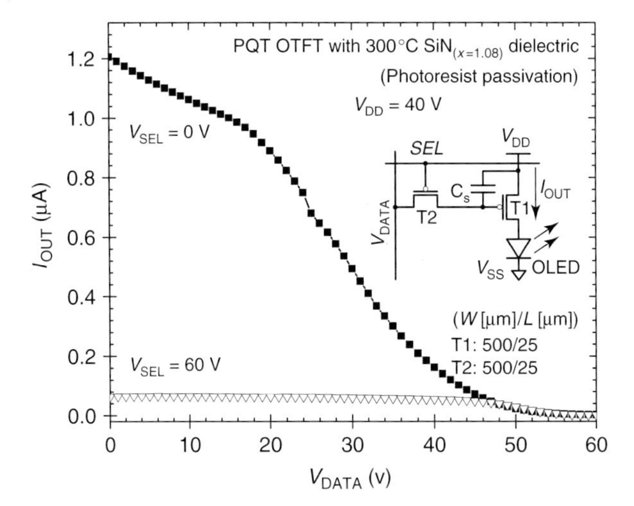

Figure 7.20 Current–voltage characteristic of a conventional 2-T pixel circuit for AMOLED, built using PQT-12 OTFTs on SiN_x gate dielectric. Circuit was fabricated using the photolithography approach with photoresist passivation (scheme 1).

of sources, including leakage current through T2 which undesirably charges up C_S, leakage current of T1, and residual charges in C_S from the previous programming cycle.

The same pixel circuit was fabricated using another photolithography scheme, whereby the PQT-12 semiconductor layer was patterned directly by photoresist. The output characteristic of this circuit is displayed in Figure 7.20. The operation observed here is similar to that described above: when $V_{SEL} = 0\,V$, the pixel is "on" and a sizeable I_{OUT} conducts through OLED; when $V_{SEL} = 50\,V$, the pixel is "off" and a significantly smaller I_{OUT} is detected. Comparatively, the photoresist-passivated OTFT circuit (in Figure 7.20) has a lower current drive than the parylene-passivated OTFT circuit (in Figure 7.19). As discussed in Chapter 3, devices fabricated using the photoresist-passivated approach are more susceptible to process-induced degradation, and hence, reduced device performance. AC or timing measurements are in progress to provide a more in-depth evaluation of the circuit performance.

Key advantages of this conventional 2-T voltage-programmed pixel circuit include its simplicity, the ability to accommodate high aperture ratio displays (due to low transistor count per pixel), and low power. However, this simple design does not compensate for the V_T shift that is inherent in OTFTs. ΔV_T is most pronounced in the drive TFT (i.e., T1) of an OLED pixel due to its continuous "on" state operation. The bias stress causes $|V_T|$ to increase over time, and leads to a reduction in the drive current. These changes translate to a gradual decrease in the brightness of the OLED with time, which undesirably degrades the quality of the display. Therefore, a circuit that can compensate for ΔV_T is required to maintain the performance

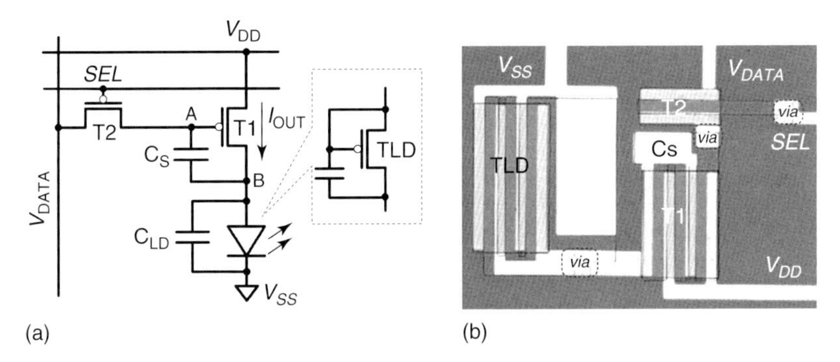

Figure 7.21 Compensating 2-TFT AMOLED pixel circuit: (a) schematic diagram, (b) micrograph of the top-view of the fabricated structure (using scheme 2). A diode-connected TFT (TLD) and a capacitor (C_{LD}) are used to emulate the OLED, Also, since the TLD is under stress, its threshold voltage increases, which resembles the OLED voltage shift.

and improve the lifetime of the OTFT-driven OLED display. Pixel circuits with a built-in compensation function are presented next.

7.3.5.2 **Compensating 2-TFT Pixel Circuit**

Although the conventional 2-TFT AMOLED voltage-programed pixel circuit can provide high resolution and high yield, this 2-TFT pixel circuit is prone to image retention over time due to V_T shift in the OTFTs. An alternative stable driving scheme that can compensate for V_T shift in the 2-TFT pixel circuit is considered here, derived from Ref. [27]. This compensating driving scheme not only preserves the simplicity of the 2-TFT pixel, it also demonstrates high uniformity and improved stability. Figure 7.21a presents a schematic of the compensating 2-TFT pixel circuit. A diode-connected TFT (TLD) and a capacitor (C_{LD}) are used to emulate the OLED. Since TLD is subject to bias stress, its threshold voltage may shift, resembling the OLED voltage shift in practical circuits [27]. PQT-12 OTFTs were used to implement this compensating 2-TFT pixel circuit design; a top-view microphotograph of a fabricated circuit is depicted in Figure 7.21b.

Figure 7.22 presents the static output characteristics of the compensating 2-T pixel circuit constructed with PQT-12 OTFTs, fabricated using a photolithographically defined polymer layer with a parylene buffer/passivation layer. When $V_{SEL} = -10\,V$, T2 turns on and V_{DATA} is passed to C_S; this charges up C_S and turns on T1. As a result, T1 generates a current I_{OUT} to drive the OLED and the pixel is "on". On the other hand, when $V_{SEL} = 50\,V$, T2 turns off; hence, a small I_{OUT} was measured. In this case, the pixel is considered "off". I_{OUT} during the off-state (i.e., $V_{SEL} = 50\,V$) can be due to a number of factors: a leakage current through T2 which undesirably charges up C_S, leakage current of T1, residual charges in C_S from a previous programming cycle. Overall, the plot shows proper operation of the circuit.

To demonstrate the true benefits of this compensating pixel circuit, timing and stability measurements are required. The sophisticated driving scheme of this pixel

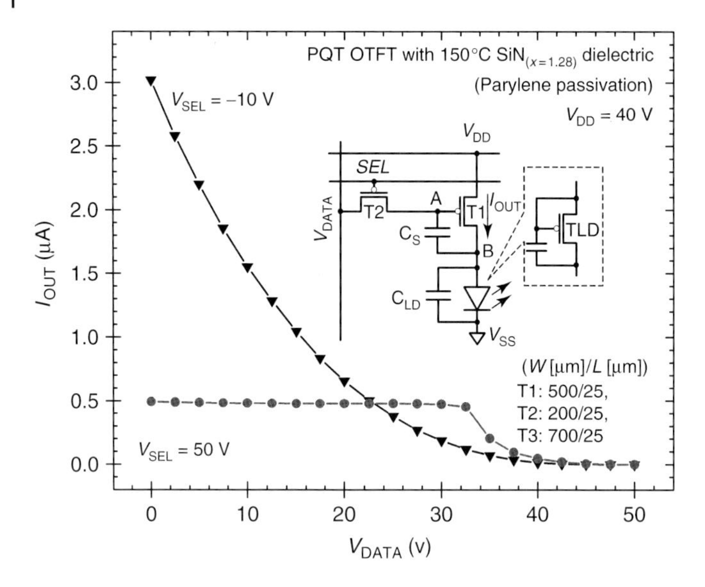

Figure 7.22 Input–output transfer characteristic of a compensating 2-T pixel circuit for AMOLED, built using PQT-12 OTFTs on SiN_x gate dielectric. The circuit was fabricated using the photolithography approach with parylene passivation (scheme 2).

is reported in Ref. [27]. One major challenge is related to the wire bonding and packaging of OTFT circuit dies in a dual in-line package (DIP) for circuit testing using a test board with properly-designed driving circuits. The standard wire bonding recipes caused degradation in OTFT performance, possibly due to stress and/or handling. Efforts to tailor a bonding recipe for OTFT circuits are currently underway. More discussion on back-end processing for organic electronics is available in Section 7.4.2.

A similar compensating circuit was implemented using a-Si TFT technology. Experimental results indicated that this compensating 2-TFT pixel circuit and its corresponding driving scheme delivered improved stability compared to the conventional 2-T pixel circuit [27]:

- After 15 days of operation, OLED current degradation in the new driving scheme was less than 11%, compared to over 50% degradation for the conventional driving scheme.
- After a 70% change in the temperature, the current remained approximately constant in the new driving scheme, compared to up to 300% increase in current in the conventional driving scheme. The current in the new driving scheme is less sensitive to temperature variations due to an internal feedback mechanism.

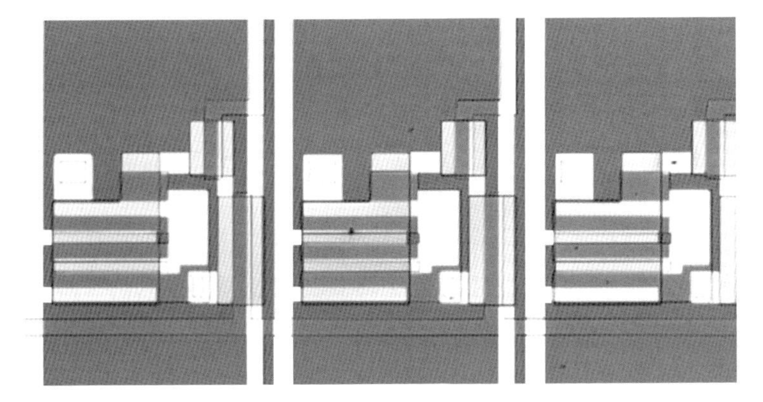

Figure 7.23 Photograph showing a section of a 4 × 4 array of AMOLED PQT-12 OTFT pixel circuits, based on a compensating 2-TFT circuit design (using fabrication scheme 2).

Therefore, the compensating 2-TFT pixel circuit offers improved stability against a bias stress-induced V_T shift in TFT and voltage shift of the OLED. Its simple 2-TFT configuration is favorable for displays demanding a high aperture ratio. The main disadvantage is its complex driving scheme, and thus it is not as suitable for large area displays. Timing and stability measurements are in progress to provide a more in-depth evaluation of the circuit performance.

4 × 4 arrays of PQT-12 OTFT pixel circuits have been fabricated using conventional 2-TFT pixel circuit design and the compensating 2-TFT pixel circuit design. Figure 7.23 displays a photograph from a section of such a 4 × 4 array. Electrical measurements are in progress.

7.3.5.3 **4-TFT Current Mirror Pixel Circuit**

To overcome the problem of OLED current degradation due to a bias-induced V_T-shift in the conventional 2-TFT pixel circuit, a current-programmed pixel circuit based on the current mirror circuit family can be used [28]. Figure 7.24a illustrates a current mirror-based V_T-shift compensated 4-TFT OLED pixel circuit. T1 and T2 are drive transistors that constitute a current mirror pair, and T3 and T4 are switch transistors. When SEL becomes high (i.e., a pixel is selected by gate drivers), T3 and T4 conduct. Initially, I_{DATA} flows through T4 and charges up the storage capacitor C_S until T2 starts to conduct. The gate voltage of T2 keeps increasing until all of I_{DATA} passes through T3 and T2. This current is then mirrored to pass through T1 since the gates of T1 and T2 are connected. Thus, the OLED gets the desired current [9]. In this manner, the drive TFT does not have to remain "on" continuously, thus reducing the effects of bias-induced V_T-shift. Furthermore, since the V_{GS} of both drive TFTs is the same, the threshold voltages of T1 and T2 will shift equally and the output current will not be affected. Therefore, the OLED current in this circuit is independent of threshold voltage or mobility variation in the drive TFT provided it stays in the saturation region of operation. A 4-TFT current mirror pixel circuit

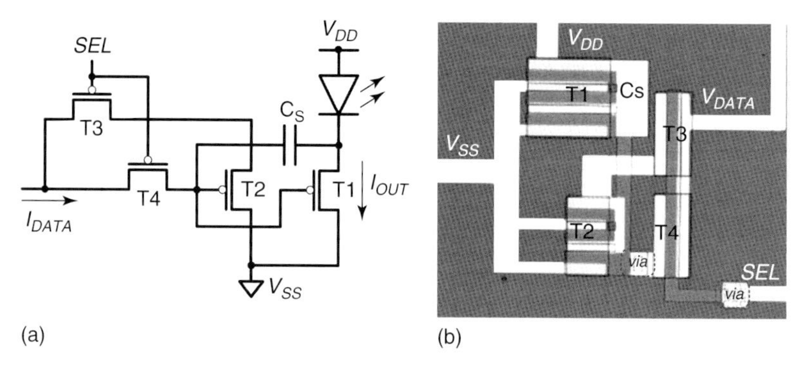

Figure 7.24 4-TFT current mirror AMOLED pixel circuit: (a) schematic diagram, (b) photomicrograph of the top-view of the fabricated structure (using scheme 2).

was fabricated using PQT-12 OTFTs. A photomicrograph displaying the top-view of the circuit is given in Figure 7.24b.

In summary, this current-programmed current mirror-based V_T-shift compensated 4-TFT OLED pixel circuit presents advantages in terms of OLED current stability, insensitivity to V_T and V_{OLED}, as well as insensitivity to temperature variation. The major shortcoming is that current programming is slow, thus this design is not suitable for medium and large area displays. A number of novel TFT pixel circuit designs and pixel addressing schemes have recently been proposed for a-Si TFT backplane technology to deliver enhanced performance [9, 27–31]. Since a-Si and organic semiconductors share a number of similar traits (in which both are disordered semiconductors), one should adapt these innovative pixel circuit designs for OTFT technology. This will provide a means to gain a better understanding of the strengths/weaknesses of OTFT pixel circuits, which will assist the development of reliable and high-performance OTFT pixel circuits for integration with AMOLED or other display backplanes.

7.4
Summary, Contributions, and Outlook

Organic circuits constructed using fully-patterned PQT-12 OTFTs with plasma enhanced chemical vapor deposited (PECVD) gate dielectrics have been demonstrated. The results on organic circuits presented here are one of the first/early demonstrations of PQT-12 organic circuits in the scientific literature. Although further process optimization is necessary to improve circuit performance, the preliminary results here signify a promising outlook for the integration of OTFTs into practical electronic applications in the near future. These outcomes also confirmed the practicality of the photolithography and inkjet printing fabrication approaches for organic circuit implementation. As a next step, in addition to improving material, device, and circuit performance, two novel recommendations to advance OTFT technology are proposed below.

7.4.1 Active-Matrix Backplane Integration

Following successful demonstration and evaluation of various OTFT pixel circuit designs, subsequent research efforts can be devoted to the optimization of OTFT pixel circuits and the development of OTFT pixel arrays for realization of display backplanes. The application of pixel circuits based on dual-gate OTFTs for vertically-integrated display backplanes is an interesting concept to explore, as depicted in Figure 7.25. In AMOLED displays, the data are stored in the pixel using a TFT circuit, which connects the OLED display device to the address, data, and power supply lines. In such applications, the *aperture ratio*, which is defined as the ratio of the active light-emitting area to the total pixel area, is a critical performance parameter. A higher aperture ratio is desirable as it improves the lifetime of the OLED, enhances the device efficiency, and ensures display quality [32]. One approach for maintaining a high aperture ratio, even in the case of higher on-pixel integration density, is to vertically stack the active layers of the OLED on the backplane electronics, as illustrated in Figure 7.25. However, the presence of a continuous back electrode can induce a parasitic channel in TFTs, giving rise to a high leakage current. A dual-gate TFT structure can effectively reduce this leakage component, because the voltage on the top-gate can be chosen to minimize the parasitic conduction induced in the TFTs [32]. Therefore, the use of dual-gate OTFTs for vertical integration of display applications can potentially improve the aperture ratio of the device, and permit shielding of parasitic effects in the vertically integrated backplane electronics.

Likewise, this vertical integration concept can lead to an intriguing implementation of sensor arrays for a variety of applications. Discrete OTFT-based gas sensors

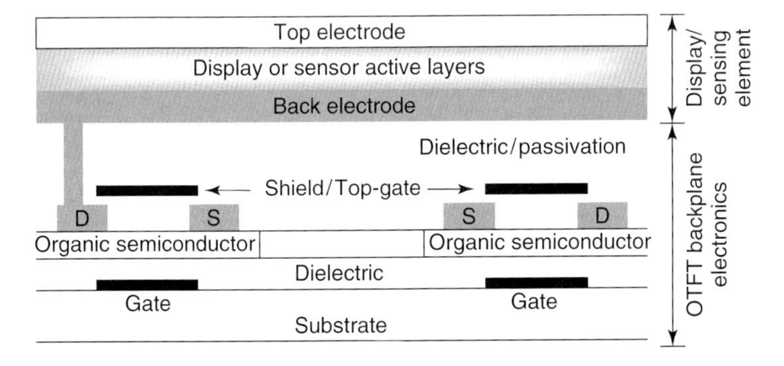

Figure 7.25 Architecture of a vertically integrated AMOLED display, featuring pixels with dual-gate OTFTs.

[33], light sensors, biosensors [34], temperature and respiration sensors for health monitoring [35], and humidity sensors [36], have been reported. Array implementation of these OTFT sensors using vertically stacked structures is expected to lead to enhanced sensitivity and/or aperture ratio. More interestingly, the concept of active-matrix arrays can be extended to nanocomposite TFT devices. Nanocomposite TFTs based on mixtures of polymer semiconductors, carbon nanotubes, Si nanowires, and ZnO nanowires have been fabricated [37, 38]. The unique properties of nanocomposite materials will open up new and exciting opportunities for diversified application areas.

7.4.2 Back-End Process Integration: Bonding and Packaging

Most of the current/existing research in organic electronics has been directed toward front-end processing, which includes the fabrication of transistors and circuits. To deliver a total manufacturing package for the commercialization and deployment of organic electronics products, solutions for OTFT-compatible back-end process integration are needed. Back-end processes include wafer dicing, die attaching/bonding, wire bonding, and encapsulation. In today's microelectronics industry, IC packaging is the final stage of semiconductor device fabrication, followed by IC testing. Packaged ICs are mounted on a printed circuit board (PCB), where on-board or peripheral driving circuits are designed to deliver the necessary electronic device/system functionality.

In this research, while attempting to wire-bond and package OTFT circuits in DIP for extensive electrical testing (see Figure 7.26), degradation in OTFT

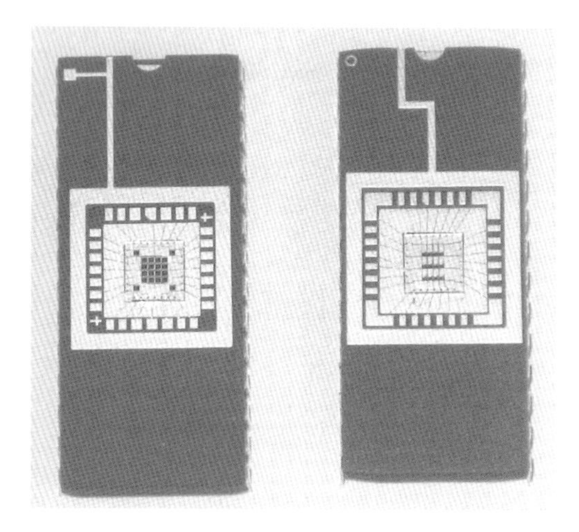

Figure 7.26 Photograph of packaged and wire-bonded OTFT circuits in a dual inline package (DIP).

performance was observed. Standard back-end processing recipes, from silicon ICs packaging, were applied. Experiments revealed that organic thin film materials are sensitive to these standard back-end processing conditions, including mechanical stress, pressure, sonication, heating, solvents, and handling. Efforts to tailor the wire bonding recipe for OTFT circuits are currently underway; initial process optimization showed improved device characteristics (i.e., less degradation). More research effort is needed to develop OTFT-specific back-end integration processes, where the sensitivities/requirements of organic electronics materials are properly addressed.

References

1. Cantatore, E. (2000)Organic materials: a new chance for electronics?*Proceedings of the SAFE/IEEE Workshop*, November 27.
2. Nathan, A. (2001) *Large Area Thin Film Electronics*, University of Waterloo, Waterloo, Ontario, Canada.
3. Clemens, W., Fix, W., Ficker, J., Knobloch, A., and Ullmann, A. (2004) From polymer transistors toward printed electronics. *J. Mater. Res.*, **19**(7), 1963.
4. Facchetti, A. (2007) Organic semiconductors for organic transistors. *Mater. Today*, **10**(3), 37.
5. Stutzmann, N., Friend, R.H., and Sirringhaus, H. (2003) Self-aligned, vertical-channel, polymer field-effect transistors. *Science*, **299**, 1881.
6. Chen, Y. and Shih, I. (2007) Fabrication of vertical channel top contact organic thin film transistors. *Org. Electron.*, **8**, 655.
7. Gelinck, G.H. *et al.* (2004) Flexible active-matrix displays and shift registers based on solution-processed organic transistors. *Nat. Mater.*, **3**, 106.
8. Van Veenendaal, E., Schrijnemakers, L., Van Mil, M., Van Lieshout, P., Touwslager, F., Gelinck, G., and Huitema, E. (2004) Rollable displays and integrated drivers based on organic electronics. *Proceedings of the 11th International Display Workshops (IDW '04)*, p. 371.
9. Nathan, A., Kumar, A., Sakariya, K., Servati, P., Karim, K.S., Striakhilev, D., and Sazonov, A. (2004) Amorphous silicon back-plane electronics for OLED displays. *IEEE J. Sel. Top. Quantum Electron.*, **10** (1), 58.
10. Li, Z.L., Yang, S.C., Meng, H.F., Chen, Y.S., Yang, Y.Z., Liu, C.H., Horng, S.F., Hsu, C.S., Chen, L.C., Hu, J.P., and Lee, R.H. (2004) Patterning-free integration of polymer light-emitting diode and polymer transistor. *Appl. Phys. Lett.*, **84** (18), 3558.
11. Schon, J.H., Berg, S., Kloc, C., and Batlogg, B. (2000) Ambipolar pentacene field-effect transistors and inverters. *Science*, **287**, 1022.
12. Jang, J. and Han, S.H. (2005) High-performance OTFT and its application. *Curr. Appl. Phys.*, **6**, e12–e21.
13. Reynolds, K., Burns, S., Banach, M., Brown, T., Chalmers, K., Cousins, N., Creswell, L., Etchells, M., Hayton, C., Jacobs, K., Menon, A., Ramsdale, C., Reeves, W., Watts, J., von Werne, T., Mills, J., Curling, C., Sirringhaus, H., Amundson, K., and McCreary, M.D. (2004) Printing of polymer transistors for flexible active matrix displays. *Proceedings of the 11th International Display Workshops (IDW '04)*. p. 367.
14. Ando, M., Kawasaki, M., Imazeki, S., Sekiguchi, Y., Hirota, S., Sasaki, H., Uemura, S., and Kamata, T. (2005) *The Proceedings of the 1st International TFT Conference (ITC '05)*, p. 178.
15. Ho, J.C., Huang, L.Y., Hu, T.S., Hsieh, C.C., Hwang, W.K., Wang, Y.W., Lin, W.L., Cheng, H.Y., Lin, T.H., Hsiao, M.C., Wang, Y.K., Wu, P.S., and Lee, C.C. (2004) *Society for Information Display 2004 (SID '04)*, p. 1298.

16. Chuman, T., Ohta, S., Miyaguchi, S., Satoh, H., Tanabe, T., Okuda, Y., and Tsuchida, M. (2004) *Society for Information Display 2004 (SID '04)*, p. 45.
17. Inoue, Y., Fujisaki, Y., Suzuki, T., Tokito, S., Kurita, T., Mizukami, M., Hirohata, N., Tada, T., and Yagyu, S. (2004) Active-matrix OLED panel driven by organic TFTs. *Proceedings of the 11th International Display Workshops (IDW '04)*, p. 335.
18. (a) Jackson, T. (2005) Organic Semiconductors: Beyond Moore's Law. *Nature Materials* **4**, 581–582. (b) Dressen, D., (2004) Considerations for RFID Technology Selection, *Atmel Applications Journal*, pp. 35–47. *http://www.atmel.com/dyn/resources/Prod_documents/secrerf_3_04.pdf* Access Year: 2010 (c) Wyld, D.C. (2006) "RFID 101: the next big thing for management". *Management Research News*, **29** (4), pp. 154–173. *More info: http://www.emeraldinsight.com/journals.htm?articleid=1554341&show=html*
19. Rotzoll, R., Mohapatra, S., Olariu, V., Wenz, R., Grigas, M., Shchekin, O., Dimmler, K., and Dodabalapur, A. (2005) 13.56 MHz organic transistor based rectifier circuits for RFID tags. *Mater. Res. Soc. Symp. Proc.*, **871E**, paper I11.6.1.
20. (a) Baude, P.F., Ender, D.A., Haase, M.A., Kelley, T.W., Muyres, D.V., and Theiss, S.D. (2003) Pentacene-based radio-frequency identification circuitry. *Appl. Phys. Lett.*, **82** (22), 3964; (b) Baude, P.F., Ender, D.A., Kelley, T.W., Muyres, D.V., and Theiss, S.D. (2004) Pentacene-based RFID transponder circuitry. *Proceedings of the 62nd Device Research Conference (DRC) Digest*, p. 227.
21. (a) Subramanian, V., Chang, P.C., Lee, J.B., Molesa, S.E., and Volkman, S.K. (2004) Printed organic transistors for ultra-low-cost RFID applications. IEEE Polytronics 2004, pp. 67–71; (b) Redinger, D., Molesa, S., Farshchi, R., and Subramanian, V. (2004) An ink-jet deposited passive components for RFID. *IEEE Trans. Electron Devices*, **51** (12), 1978.
22. Zipperer, D., Clemens, W., Ullmann, A., Boehm, M., and Fix, W. (2005) Polymer based rectifiers and integrated circuits for printable RFID tags. 2005 Materials Research Society Spring Meeting, Symposium I: Organic Thin-Film Electronics, I11.5.
23. Neamen, D.A. (2001) *Electronic Circuit Analysis and Design*, 2nd edn, McGraw-Hill Science Engineering.
24. Jaeger R.C. and Blalock, T. (2003) *Microelectronic Circuit Design*, 2nd edn, McGraw-Hill.
25. Zilker, S.J., Detcheverry, C., Cantatore, E., and de Leeuw, D.M. (2001) Bias stress in organic thin-film transistors and logic gates. *Appl. Phys. Lett.*, **79**, 1124.
26. Klauk, H., Halik, M., Zschieschang, U., Eder, F., Schmid, G., and Dehm, C. (2003) Pentacene organic transistors and ring oscillators on glass and on flexible polymeric substrates. *Appl. Phys. Lett.*, **82**, 4175.
27. Chaji, G.R. and Nathan, A. (2006) A stable voltage-programmed pixel circuit for a-Si:H AMOLED displays. *J. Display Technol.*, **2**, 347.
28. Nathan, A., Striakhilev, D., Servati, P., Sakariya, K., Sazonov, A., Alexander, S., Tao, S., Lee, C.-H., Kumar, A., Sambandan, S., Jafarabadiashtiani, S., Vygranenko, Y., and Chan, I.W. (2004) a-Si AMOLED display backplanes on flexible substrates. *Mater. Res. Soc. Symp. Proc.*, **814**, paper I3.1.1.
29. Chaji, G.R. and Nathan, A. (2005) A novel driving scheme for high-resolution large-area a-Si:H AMOLED displays. IEEE 48th Midwest Symposium on Circuit and Systems, p. 782.
30. Chaji, G.R. and Nathan, A. (2007) Parallel addressing scheme for voltage-programmed active-matrix OLED displays. *IEEE Trans. Electron Device*, **54**, 1095.
31. Nathan, A., Chaji, G.R., and Ashtiani, S.J. (2005) Driving schemes for a-Si and LTPS AMOLED displays. *J. Display Technol.*, **1**, 267.
32. Servati, P., Prakash, S., Nathan, A., and Py, C. (2002) Amorphous silicon driver circuits for organic light-emitting diode

displays. *J. Vac. Sci. Technol. A*, **20**, 1374.

33. Chang, J.B., Liu, V., Subramanian, V., Sivula, K., Luscombe, C., Murphy, A., Liu, J., and Fréchet, J.M.J. (2006) Printable polythiophene gas sensor array for low-cost electronic noses. *J. Appl. Phys.*, **100**, 014506.
34. Farinola, G.M., Torsi, L., Naso, F., Zambonin, P.G., Valli, L., Tanese, M.C., Hassan Omar, O., Giancane, G., Babudri, F., and Palmisano, F. (2007) Chemical design, synthesis and thin film supramolecular architecture for advanced performance chemo- and bio-sensing organic field effect transistors. *Proceedings of the 2nd International Workshop on Advances in Sensors and Interface*, pp. 1–2.
35. Jung, S., Ji, T., and Varadan, V.K. (2006) Point-of-care temperature and respiration monitoring sensors for smart fabric applications. *Smart Mater. Struct.*, **15**, 1872.
36. Zhu, Z.T., Mason, J.T., Dieckmann, R., and Malliaras, G.G. (2002) Humidity sensors based on pentacene thin-film transistors. *Appl. Phys. Lett.*, **81**, 4643.
37. Beecher, P., Servati, P., Rozhin, A., Colli, A., Scardaci, V., Pisana, S., Hasan, T., Flewitt, A.J., Robertson, J., Hsieh, G.W., Li, F.M., Nathan, A., Ferrari, A.C., and Milne, W.I. (2007) Ink-jet printing of carbon nanotube thin film transistors. *J. Appl. Phys.*, **102**, 043710.
38. Hsieh, G.W., Beecher, P., Li, F.M., Servati, P., Colli, A., Fasoli, A., Chu, D., Nathan, A., Ong, B., Robertson, J., Ferrari, A.C., and Milne, W.I. (2008) Formation of composite organic thin film transistors with nanotubes and nanowires. *Physica E Low Dimens. Syst. Nanostruct.* **40**, 2406–2413.

Further Reading

Kawase, T., Sirringhaus, H., Friend, R.H. *et al.* (2003) Inkjet printing of polymer thin film transistors. *Thin Solid Films*, **438**, 279.

Rogers, J.A., Bao, Z., Meier, M., Dodabalapur, A., Schueller, O.J.A., and Whitesides, G.M. (2000) Printing, molding, and near-field photolithographic methods for patterning organic lasers, smart pixels and simple circuits. *Synth. Met.*, **115**, 5.

Servati, P., Karim, K.S., and Nathan, A. (2003) Static characteristics of a-Si:H dual-gate TFTs. *IEEE Trans. Electron Devices*, **50**, 926.

Sirringhaus, H., Tessler, N., and Friend, R.H. (1998) Integrated optoelectronic devices based on conjugated polymers. *Science*, **280**, 1741.

8
Outlook and Future Challenges

The development of functional organic materials has been an active area of research over the past two decades. Novel materials continue to evolve, bringing device performance and system functionality to new heights. Because of their processability advantages, compatibility with large area flexible substrates and their unique material properties with potentially multidimensional/versatile functionality, organic materials bring exciting opportunities for flexible, light weight, low cost, and disposable electronics. Indeed, the research community has taken significant strides in the development of new materials (e.g., the demonstration of high mobility organic semiconductors such as regioregular poly(3-hexylthiophene) (P3HT) [1] by Bell Labs, pentacene with mobilities reaching 3–5 cm^2/V-s by 3 M [2]; air-stable materials, for example, Xerox's poly(3,3'''-dialkyl-quarterthiophene) (PQT) [3], Merck's poly(2,5-bis(3-alkylthiophen-2-yl)thieno[3,2-*b*]thiophene) (PBTTT) [4], recent PQT analogs [5, 6], TIPS (6,13-bis(triisopropylethynyl)) pentacene [7], analogs in the polymer matrix [8]; n-type semiconductors that are air stable and with high mobility, for example, fluorinated compounds at Bell Lab [9], Polyera's n-type polymer [10]), fabrication processes (inkjet printing [11] and SAP printing [12] by Cambridge University for large area arrays), and systems (e.g., the organic thin film transistor (OTFT) display backplane by Philips [13], although this was patterned using photolithography).

However, for OTFTs to gain wider acceptance and usage there are several issues that need to be overcome. The first lies in the shelf-life under operation. Although OTFTs promise low-cost manufacturing and many groups have demonstrated all printed transistors, there are issues associated with the yield and reproducibility, particularly for completed circuits. Passivation is mandatory to extend the shelf-life but requirements on the passivation dielectric are even more stringent than that used in the silicon industry, due to the sensitivity of organic devices to oxygen and moisture. Here, the solution may be to adopt the thin film passivation matrix that is being considered for top-emitting organic light emitting diode displays. The other issue lies in keeping up with consumer demand for reduced energy consumption, and, hence, low power operation. In comparison with silicon-based thin film transistors (TFTs), OTFTs operate under much higher voltage, which poses challenges in terms of the power supply and energy storage requirements, particularly in mobile device applications.

Organic Thin Film Transistor Integration: A Hybrid Approach, First Edition. Flora M. Li, Arokia Nathan, Yiliang Wu, and Beng S. Ong.

In the preceding chapters, we described the advancement of OTFT research from an engineering and integration perspective, in which established materials and techniques were exploited/engineered to deliver higher device performance and to demonstrate the feasibility of practical organic circuits. In particular, the following theme areas were identified:

- **Device performance**: optimization of plasma enhanced chemical vapor deposited (PECVD) gate dielectric properties and investigation of interface engineering methodologies to enhance PQT-12 OTFT performance;
- **Device manufacture**: development of OTFT fabrication strategies to enable reliable device fabrication and circuit integration;
- **Device integration**: demonstration of the integration of PQT-12 OTFTs into functional circuits for display and other applications.

The major outcomes from these investigations coupled with future challenges are highlighted next.

8.1 Device Performance

In the process of enhancing field-effect mobility by tuning the gate dielectric composition and the interface properties, we can draw the following conclusions from investigations of bottom-gate bottom-contact PQT-12 OTFTs on PECVD gate dielectrics:

- The film composition of the silicon nitride (SiN_x) gate dielectric affects the electrical performance of PQT-12 OTFTs. Overall improvement in mobility, on/off current ratio, and gate leakage current was observed as the silicon content in the SiN_x gate dielectric was increased. Interface characterization confirmed that a silicon-rich SiN_x gate dielectric, with appropriate surface modifications, presents a more desirable dielectric/semiconductor interface (with higher contact angle, lower surface roughness, and more Si–O bonds) for OTFT fabrication than a nitrogen-rich SiN_x gate dielectric.
- OTFTs fabricated with a 150 °C PECVD SiN_x gate dielectric showed improved mobility and on/off current ratio compared to devices with a 300 °C SiN_x gate dielectric. PQT-12 OTFTs on plastic substrates with a 150 °C SiN_x gate dielectric were demonstrated.
- OTFTs with a PECVD SiO_x gate dielectric showed higher mobility ($0.34\,cm^2\,V^{-1}\,s^{-1}$) than those with PECVD SiN_x, indicating superior SiO_x/PQT interface properties. However, PECVD SiO_x was more susceptible to dielectric breakdown at lower fields and higher gate leakage current. Further material optimization is necessary to strengthen the electrical integrity of SiO_x films. PQT-12 OTFTs on a plastic substrate with a 180 °C PECVD SiO_x gate dielectric were demonstrated.
- Interface treatment of a PECVD SiN_x gate dielectric by a combination of O_2 plasma exposure and an octyltrichlorosilane self-assembled monolayer

(OTS SAM) delivered the highest field-effect mobility for PQT-12 OTFT, in comparison to untreated SiN_x, O_2-plasma-only-treated SiN_x, and OTS-only-treated SiN_x. Interface characterization showed that the combination of O_2 plasma and OTS treatments rendered a dielectric surface with a large contact angle (i.e., low surface energy), low surface roughness, and an abundance of Si–O bonds, which are desirable interface qualities for high performance OTFTs.
- Investigation of the impact of the exposure duration of O_2 plasma for dielectric surface treatment on OTFT performance revealed a "turn-round" effect, where device characteristics (i.e., mobility and on/off current ratio) peaked at an exposure duration of ~80 s. A SiN_x dielectric surface treated with 80 s of O_2 plasma exposure and OTS SAM displayed the largest contact angle, lowest surface roughness, and largest concentration of Si–O bonds, when compared to samples subjected to either shorter or longer O_2 plasma treatment. These surface attributes coincided with the peak device performance observed at 80 s of O_2 plasma exposure, where μ_{FE} as high as $0.22\,cm^2\,V^{-1}\,s^{-1}$ was achieved. In addition, device mobility increased with higher reactive ion etcher (RIE) power of the O_2 plasma exposure.
- For devices with dielectric surface pre-treatments, the addition of 1-octanethiol contact treatment led to a reduction in device mobility and an increase in contact resistance. The 1-octanethiol SAM may be creating a charge injection barrier at the contact/semiconductor interface.

Thus, for the experimental conditions considered here, we can conclude that a combination of PECVD SiN_x dielectric composition together with surface treatment produces high performance PQT-12 OTFTs. *The SiN_x is silicon-rich and deposited at 150°C, with 80 s of O_2 plasma exposure and OTS SAM as dielectric surface treatments, and with the omission of 1-octanethiol contact surface treatment.* Indeed the low-temperature SiN_x dielectric provides an economical means for first generation OTFTs and circuit integration, by enabling use of the well-established PECVD infrastructure, yet not compromising the performance and scalability over large areas. However, this does not imply that inorganic dielectrics are necessarily the long-term solution. Research in organic dielectrics is progressing and it will just be a matter of time before they meet the reliability requirements of gate dielectrics from the standpoint of breakdown characteristics, pinhole density and distribution, leakage current, roughness, bandgap, and band offsets.

8.2 Device Manufacture

The establishment of robust, reliable, and scalable OTFT integration strategies is a prerequisite for full-scale deployment of organic electronics manufacturing. One of the key technological challenges lies in the patterning of organic-based thin films. A number of OTFT fabrication schemes were discussed in the preceding chapters, including an all-photolithography processing scheme and a hybrid photolithography–inkjet printing scheme, and the key conclusions were:

- Photolithography enables precise definition/patterning of device layers with good resolution and registration. However, the adoption of photolithography for organic electronics fabrication requires judicious/meticulous planning and design of the processing sequence. Indeed photolithographic patterning of the organic semiconductor layer by incorporating a compatible buffer/passivation layer provides a viable approach to the fabrication of OTFTs. Fully-patterned bottom-gate parylene-passivated PQT-12 OTFTs displayed mobilities comparable to unpassivated/unpatterned devices, validating the robustness of this photolithography approach. A variety of photolithographically-defined fully-patterned and fully-encapsulated bottom-contact OTFTs in bottom-gate, top-gate, and dual-gate configurations were successfully fabricated, demonstrating the scalability/adaptability of the photolithography process for OTFT fabrication.
- The hybrid photolithography–inkjet printing fabrication scheme combines the high-resolution and multilayer-registration capabilities of photolithography with the direct and non-disruptive patterning of organic-based thin films by inkjet printing. OTFTs produced by this hybrid scheme with an inkjet printed organic semiconductor exhibited reduced mobility compared to OTFTs with a spin-coated organic semiconductor layer. This can be attributed to non-uniformity, inhomogeneity, and discontinuity of the inkjet printed organic semiconductor layer, although device improvements can be achieved through improved control and optimization of printing parameters and ink formulation.

While inkjet printing is envisioned to be one of the key manufacturing standards for future generations of organic electronics, the photolithography, and inkjet printing integration schemes present an interim alternative that enables immediate fabrication of highly integrated OTFT circuits and facilitates timely evaluation of organic device/circuit behavior.

8.3 Device Integration

PQT-12 OTFT circuits, including inverters, current mirrors, and various display pixel circuits, were successfully fabricated based on optimized recipes for PECVD gate dielectric and interface treatments, as derived from the systematic investigations described in the earlier chapters. Proper circuit functionality was demonstrated, which proves the feasibility of practical organic circuits using the integration approaches described here, although the performance of OTFT circuits can benefit from further process, device, and material optimization. Researchers should pay particular attention to the development of compatible backend integration processes, especially with plastic substrates, to provide a total manufacturing solution for the commercialization of organic electronics in the near future.

The future looks very promising for organic electronics. Continued growth and development in this technological field can be anticipated, fueled by the promise of new products and applications that can be derived from electrically and optically active organic and hybrid materials. In order to fully realize the benefits of organic electronics, long-term research efforts, and innovation are needed to provide functional organic materials with enhanced performance, processability, and, very importantly, stability and lifetime. As material properties and integration strategies advance and mature over the coming years, organic electronics may create new application areas requiring low cost, mechanical flexibility, and large area compatibility. There is little doubt that, with industry and academia pooling their resources, organic, or plastic electronics will become part of our lives in the not-too-distant future.

References

1. Bao, Z., Dodabalapur, A., and Lovinger, A.J. (1996) Soluble and processable regioregular poly(3-hexylthiophene) for thin film field-effect transistor applications with high mobility. *Appl. Phys. Lett.*, **69**, 4108–4110.
2. Kelly, T.W., Boardman, L.D., Dunbar, T.D., Muyres, D.V., Pellerite, M.J., and Smith, T.P. (2003) High-performance OTFTs using surface-modified alumina dielectrics. *J. Phys. Chem. B*, **107**, 5877.
3. Ong, B.S., Wu, Y., Liu, P., and Gardner, S. (2004) High-performance semiconducting polythiophenes for organic thin-film transistors. *J. Am. Chem. Soc.*, **126**, 3378.
4. McCulloch, I., Heeney, M., Bailey, C., Genevicius, K., Macdonald, I., Shkunov, M., Sparrowe, D., Tierney, S., Wagner, R., Zhang, W.M., Chabinyc, M.L., Kline, R.J., McGehee, M.D., and Toney, M.F. (2006) Liquid-crystalline semiconducting polymers with high charge-carrier mobility. *Nat. Mater.*, **5**, 328–333.
5. Pan, H., Li, Y., Wu, Y., Liu, P., Ong, B.S., Zhu, S., and Xu, G. (2007) Low-temperature, solution-processed, high-mobility polymer semiconductors for thin-film transistors. *J. Am. Chem. Soc.*, **129**, 4112.
6. Fong, H.H., Pozdin, V.A., Amassian, Malliaras, G.G., Smilgies, D.M., He, M., Gasper, S., Zhang, F., and Sorensen, M. Tetrathienoacene copolymers as high mobility, soluble organic semiconductors. *J. Am. Chem. Soc.*, **130**, 13202.
7. Payne, M.M., Parkin, S.R., Anthony, J.E., Kuo, C.C., and Jackson, T.N. (2005) Organic field-effect transistors from solution-deposited functionalized acenes with mobilities as high as 1 cm^2/V-s. *J. Am. Chem. Soc.*, **127**, 4986.
8. Hamilton, R., Smith, J., Ogier, S., Heeney, M., Anthony, J.E., McCulloch, I., Bradley, D.D.C., Veres, J., Anthopoulos, T.D. (2009) High performance polymer-small molecule blend organic transistors. *Adv. Mater.*, **21**, 1166.
9. Katz, H.E., Lovinger, A.J., Johnson, J., Kloc, C., Seigrist, T., Li, W., Lin, Y.-Y., and Dodabalapur, A. (2000) A soluble and air-stable organic semiconductor with high electron mobility. *Nature*, **404**, 478.
10. Yan, H., Chen, Z., Zheng, Y., Newman, C., Quinn, J.R., Dotz, F., Kastler, M., and Facchetti, A. (2009) A high-mobility electron-transporting polymer for printed transistors. *Nature*, **457**, 679.
11. Sirringhaus, H., Kawase, T., Friend, R.H., Shimoda, T., Inbasekaran, M., Wu, W., and Woo, E.P. (2000) High-resolution inkjet printing of all-polymer transistor circuits. *Science*, **290**, 2123.
12. Sele, C.W., Werne, T., Friend, R.H., and Sirringhaus, H. (2005) Lithography-free, self-aligned inkjet printing with sub-hundred-nanometer resolution. *Adv. Mater.*, **17**, 997.
13. Gelinck, G.H. *et al.* (2004) Flexible active-matrix displays and shift registers based on solution processed organic transistors. *Nat. Mater.*, **3**, 106–110.

Index

Organic Thin Film Transistor Integration: A Hybrid Approach, First Edition. Flora M. Li, Arokia Nathan, Yiliang Wu, and Beng S. Ong.

p